SCIENTIFIC EXPLORATION
of the
SOUTH PACIFIC

Proceedings of a Symposium held during the
Ninth General Meeting of the
Scientific Committee on Oceanic Research,
June 18–20, 1968, at the
Scripps Institution of Oceanography, La Jolla, California

WARREN S. WOOSTER, *Editor*

NATIONAL ACADEMY OF SCIENCES
WASHINGTON, D.C.   1970

Standard Book Number 309-01755-6

Available from
Printing and Publishing Office
National Academy of Sciences
2101 Constitution Avenue
Washington, D.C. 20418

Library of Congress Catalog Card Number 72-603750

Frontispiece: "Venus Fort, Erected by the Endeavour's People to Secure Themselves during the Observations of the Transit of Venus, at Otaheite," Plate IV from *A Journey to the South Seas, in His Majesty's Ship the Endeavour. Faithfully Transcribed from the Papers of the late Sydney Parkinson, Draghtsman to Sir Joseph Banks*, 2nd edition, edited by J. C. Lettsom, London, C. Dilly, 1784.

The Scientific Committee on Oceanic Research (SCOR) is a component of the International Council of Scientific Unions, charged with furthering scientific activity in all branches of oceanic research. Its present membership consists of 39 marine scientists nominated by national committees on oceanic research in 29 countries and by a number of interested ICSU organizations.

During its Ninth General Meeting, June 17–21, 1968, at the Scripps Institution of Oceanography in La Jolla, California, SCOR held a Symposium on Scientific Exploration of the South Pacific. The symposium was organized by Warren S. Wooster and consisted of the 19 scientific papers recorded in the present volume. These papers constitute a comprehensive review of many aspects of the oceanography of this vast and little-known region.

Most of the expenses of the invited speakers were met by travel grants from the U.S. National Academy of Sciences, made possible by a grant from the U.S. National Science Foundation. The Academy also arranged for the technical editing and publication of this volume. Extensive editing of some manuscripts was carried out by Mr. Albert N. Bove of the NAS staff, in consultation with Warren S. Wooster.

CONTENTS

Luis R. A. Capurro

INTRODUCTORY REMARKS

For this select audience perhaps it is obvious to say that oceanography is a planetary science. However, what I wish to stress is the fact that the main objective of this science is the study of the World Ocean.

A quick look at the research effort in the field of marine sciences during the last 15 years shows clearly that oceanographers have already expanded the scope of their research programs. They have been steadily moving from regional oceanographic cruises to long-range expeditions covering considerable expanses of the ocean. The ever-increasing number of international cooperative expeditions—to which countries and marine laboratories have committed a great deal of their oceanographic facilities and manpower in the study of areas often far away from their own shores—is a clear indication that the concept of the World Ocean has become the focus of their scientific motivation.

The Scientific Committee on Oceanic Research (SCOR) has been sensitive to the main objectives of oceanographic science; one of its main concerns has been to have an up-to-date evaluation of the knowledge of the World Ocean and to try to improve this understanding by identifying potential areas of scientific interest amenable to international cooperative attack.

I still have a clear recollection of the first SCOR meeting at the Woods Hole Oceanographic Institution in 1957, when SCOR was created. During discussions of various attractive high-priority programs for marine research that were then under way, one of the members of the Committee called attention to how little was known about the Indian Ocean and how many significant scientific problems could be attacked there.

The outcome of this suggestion was the well-known International Indian Ocean Expedition. The results of this international endeavor have, in my opinion, considerably increased our understanding of the World Ocean. It is diffi-cult to see how we could have made such progress had we not had such a cooperative effort. I am aware that there is some criticism of the concept of large cooperative expeditions, but the benefits to be derived from such international efforts dampen the effect of those criticisms.

Pursuant to our objective of identifying scientific problems that, through concerted international action, could add to our understanding of the World Ocean, it became clear that two prominent problems required proper consideration. The first one dealt with ocean variability, and SCOR sponsored the Symposium on Variability in the Ocean, which took place in Rome in 1966; the results will be available very soon. The important action that in part emanated from that symposium was the creation of two IOC Working Groups, one on Integrated Global Ocean Stations System and the other on Ocean Variability. The Scientific Committee on Oceanic Research also has a very active group working on continuous current velocity measurements.

The second problem was that the impressive oceanic area of the South Pacific was practically unknown. The Symposium on Scientific Exploration of the South Pacific was a follow-up of recognition by SCOR that work in this area was much needed. These Proceedings contain discussions on the oceanography (physical, chemical, biological, geological, and meteorological) of the South Pacific. Distinguished scientists review the present state of knowledge and provide some guide for future action.

Our main objectives in sponsoring this symposium were to review the existing knowledge, to identify outstanding scientific phenomena in the region, and to discuss how these phenomena can be investigated.

I am confident that the papers presented in this volume will contribute significantly to our knowledge of the hitherto embarrassingly unknown part of the Pacific Ocean.

I
CLIMATOLOGY OF
THE SOUTH PACIFIC

Andrei S. Monin

INSTITUTE OF OCEANOLOGY,

USSR ACADEMY OF SCIENCES, MOSCOW

WEATHER AND CLIMATE OSCILLATIONS

Atmospheric observations in the Southern Ocean area have revealed certain problems relative to the applicability of the physical laws governing weather and climate oscillations.

One of the phenomena observed is the existence of a circumpolar eastward current, practically uninterrupted by obstacles, in conjunction with large-scale waves in the atmosphere, called gyroscopic or Rossby waves, which are present in their purest nondistorted form and therefore closely resemble those observed in the sun's atmosphere (see Ward, 1964). The statistical laws indicate that such waves cross any fixed meridian, on the average, every 4 days, while the position and intensity of the entire eastward current will change approximately every 2 weeks (the so-called index cycle). Further, the climate of this region is dominated by the effects of the huge Antarctic ice cap, which is the major relic of the Pleistocene ice ages.

These phenomena are most strongly pronounced in the Southern Ocean area, but they are typical of the entire planet. In this paper they are considered as characteristics of the atmosphere of the whole earth.

SPECTRUM OF WEATHER OSCILLATIONS

The spectrum of oscillation periods of meteorological fields ranges from hundredths of a second to hundreds of thousands of years.

The region of the shortest periods is clearly seen from the spectrum of wind velocity (Figure 1). Here, we can see several oscillation periods.

1. The region of *micrometeorology*, or turbulence, with periods from hundredths of a second to minutes and with the maximum of the spectrum at the period of 1 minute.

2. The region of *mesometeorology*, with periods from minutes to hours and with the spectral minimum* at periods of about 20 minutes (here weather begins).

3. The region most important for weather, the *synoptic* region, with periods from hours to days and with the spectral maximum at periods of 4–5 days (propagation of Rossby waves). The synoptic region also contains the spectral line corresponding to the *diurnal* oscillations of meteorological fields.

The intermediate periods begin with

4. The region of *global* weather oscillations, having periods of weeks to months, which are perhaps of the greatest significance for long-term weather forecasts. We may note here (Figure 2), first, the index cycle, with the spectral maximum at the period of 12 days and, second, possible oscillations in the "ocean-cloudiness-atmosphere" system, predicted from the theoretical model by B. L. Gavrilin and A. S. Monin (1967).

5. Seasonal oscillations, that is, spectral lines corresponding to the yearly period and its harmonics.

6. The region of year-to-year weather oscillations, among which are the 26-monthly oscillations first discovered in the equatorial stratosphere and the possible oscillations in the ocean-atmosphere system, which are related to the periods of large oceanic gyres. Among these oscillations are the 3½-year auto-oscillations of the Gulf Stream, which were observed by V. V. Shuleykin before 1940, and the oscillations in the Atlantic and Pacific ocean areas that have been studied for several years by J. Bjerknes; year-to-year

---

*A possible explanation of the mesometeorological minimum was suggested by V. N. Kolesnikova and A. S. Monin (1965).

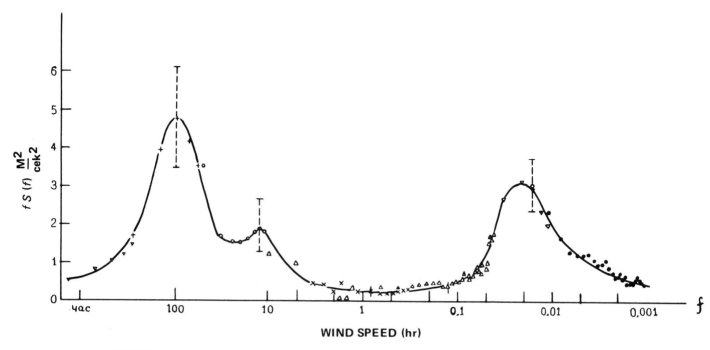

FIGURE 1 Horizontal wind-speed spectrum. (Reprinted with permission from Van der Hoven, 1957.)

oscillations related to the oceanic gyres in the northern Pacific were recently discovered by V. G. Kort.

Among the long-period oscillations are the following:

7. Intracentennial oscillations.

8. Extracentennial oscillations.

9. Pleistocene ice ages.

It is natural to assume that, beginning with the intermediate periods, the oscillations of meteorological fields are of a global character and that their spectra are usually at least approximately similar to each other. For instance, V. N. Kolesnikova and A. S. Monin (1966) found that the year-to-year oscillations of temperature and of surface radiative heat fluxes constitute 15–30 percent of their intrayear oscillations.

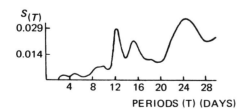

FIGURE 2 Spectrum of global oscillations. (Reprinted with permission from Monin, 1956.)

It should be noted that the hypothesis that the 11-year cycle of solar activity is manifested in the weather apparently is not corroborated by the spectral analysis of the oscillations of meteorological fields. An example is shown in Figure 3, which represents the curves of the oscillations of mean monthly temperature in Moscow and Leningrad and gives the Wolf numbers (indicating sunspot activity) for 1820–1950. One can see that the most pronounced temperature oscillations have considerably shorter periods than the Wolf numbers and that no relation exists between the phases of temperature oscillations and the Wolf numbers. The spectra of these oscillations (Figure 4), constructed by V. N. Kolesnikova and A. S. Monin (1966), fully confirm this impression; the 11-year period characteristic of Wolf numbers is not reflected in temperature spectra, which show only a diffused maximum over the range of 2–5 years. Similar results have been obtained by others—by G. W. Brier (1961), for example—for precipitation spectra.

SPECTRUM OF CLIMATE OSCILLATIONS

If the oscillations considered thus far (turbulence and weather) are regarded as short-period oscillations, *climate* can be defined as a *statistical regime of the short-period oscillations of meteorological fields.*

The inclusion of year-to-year oscillations in the concept of weather and not of climate means that climatic averaging should be performed by periods that are large relative to the periods of year-to-year oscillations, i.e., by periods of at

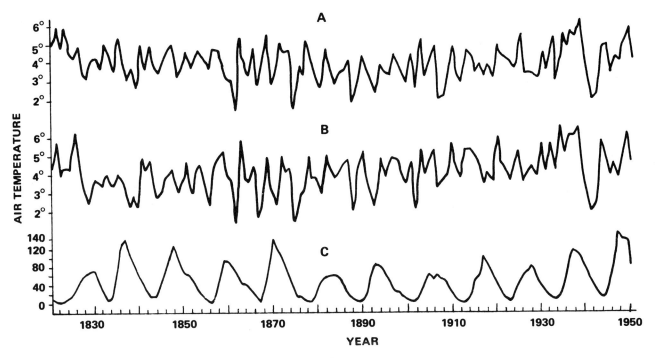

FIGURE 3   Mean yearly air temperatures in Moscow (a) and Leningrad (b), and Wolf numbers (c), 1820–1950.

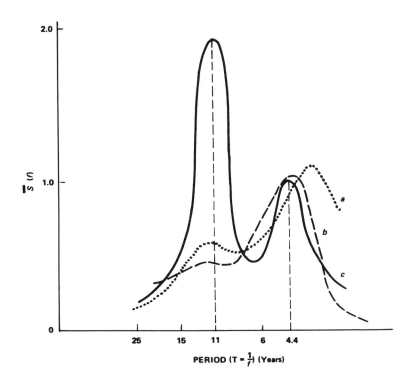

FIGURE 4   Spectra of oscillations shown in Figure 3. Moscow (a), Leningrad (b), Wolf numbers (c).

least several decades. In recent years some climatologists who think in geographic terms have tried to find a basis for determining climatic averages of some meteorological fields, as for instance, the fields of surface heat fluxes, using data for only a few years and even for separate years. But this would mean either the neglect of year-to-year oscillations, which is evidently wrong, or their inclusion in the concept of climate, which is probably unreasonable.

Climate itself is subject to long-period oscillations. The latter include, first, *intracentennial* oscillations, a vivid example of which is given by the climatic warming observed in the first half of the twentieth century; Figure 5 shows temperature curves of the last 100 years, and Figure 6 is a map showing differences in the average winter temperatures for 1920–1939 and 1900–1919. The map shows an essential difference in the climatic behavior of the continents and oceans; the sharply warmer winters of the Arctic were ac-

companied by slightly colder winters on the continents and in the tropical regions of the Atlantic and Indian oceans. An explanation of this climatic warming is a challenge to modern climatologists because detailed meteorological observations are available here. The first step can be the study of statistics of weather processes typical of the period of warming. Thus, according to Dzerdzievsky (1956), in 1930–1950 zonal circulation was observed more often than in 1900–1930, and meridional circulation was observed less often.

Extracentennial oscillations, illustrated by data on climatic oscillations in postglacial time are illustrated by Figure 7. This figure shows the oscillations in the length and height of the névé line of Norwegian and Icelandic glaciers (Ahlmann, 1953). Assuming that the last glaciation ended in about the sixty-fifth century B.C., the following features can be related to their probable time of occurrence:

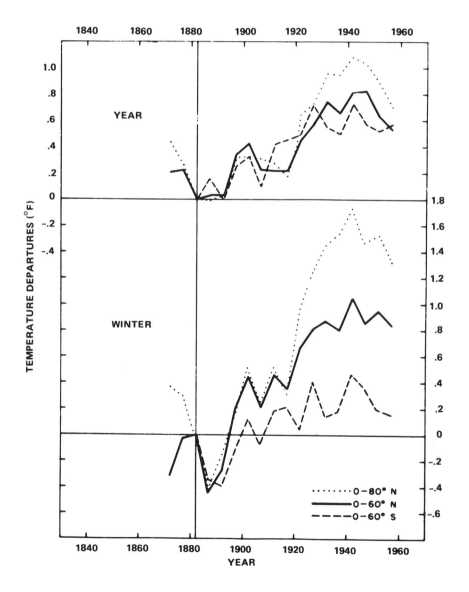

FIGURE 5 Departures of mean 5-year temperatures (°F) of last 100 years from their values in 1880–1884. (Annals of the New York Academy of Sciences, vol. 95, Art. 1, J. M. Mitchell. Copyright © The New York Academy of Sciences; 1961. Reprinted by permission.) Upper curves for the entire year, lower curves for winter only.

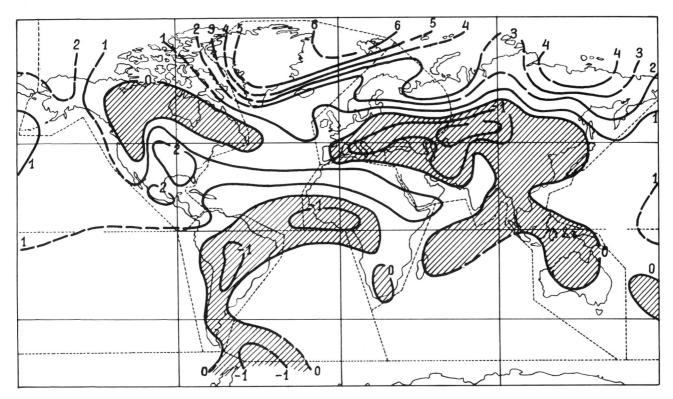

FIGURE 6 Departures of mean winter temperatures (°F) of 1920–1939 from those of 1900–1919. Regions of cooling are shaded. (Annals of the New York Academy of Sciences, vol. 95, Art. 1, J. M. Mitchell. Copyright © The New York Academy of Sciences; 1961. Reprinted by permission.) Regions of cooling are shaded.

Century	Feature
40-20 B.C.	"climatic optimum," with Arctic probably melted
10 B.C.-3 A.D.	"sub-Atlantic" period of cold climate
A.D. 4-10	warmer "Viking age," with Arctic possibly melted
A.D. 13-14	cold climate
A.D. 15-16	warm climate
A.D. 17-19	appreciably colder, "little ice age"
A.D. 20	warming

Bjerknes attributes the "little ice age" to ocean-atmosphere interaction, with feedback between the weakening of circulation in midlatitudes and the weakening of heat transfer from the ocean to the atmosphere.

It is interesting that these extracentennial oscillations do not show any noticeable periodicities. In particular, it should be noted that the assumed solar-activity cycle of 80–90 years, which is based on curves of the type shown in Figure 8 (oscillations of 23-year moving average values of Wolf numbers), seems to be unsupported. The well-known tendency to consider a three- to five-fold occurrence as "periodicity" is not supported by mathematical statistics; here, the number of occurrences—only two—is still less.

One should not try to explain the extracentennial oscillations by changes in the earth's orbit, in the distribution of land and sea, or in the heights of mountains; no such changes have been observed in postglacial time.

Finally, we consider the Pleistocene ice ages: Gunz-Nebraska (520–490 thousand years B.C.), Mindel-Kansas (430–370 B.C.), Riss-Illinois (130–100 B.C.), and Würm-Wisconsin (40–18 B.C.). Figure 9 represents the shortened chronology constructed by Emiliani (1955) from the materials on stratigraphy of marine sediment cores from the Atlantic Ocean, according to which the whole of Pleistocene occurred within 300 thousand years.

The noteworthy features of the glacial stages were their global character and the presence of fine structure (i.e., extracentennial oscillations) of glacial and, especially, interglacial stages.

Numerous supposed causes of the glacial stages have been advanced. External factors are:

1. Astronomical—changes of the earth's orbital elements, such as eccentricity, inclination of the axis, wobbling of equinoctial points (Milankovich, Zeiner)

2. Astrophysical—a decrease (Flint) or increase (Simpson, Willett) of solar radiation, or secular changes of solar activity (Huntington and Fisher, Opik, and others).

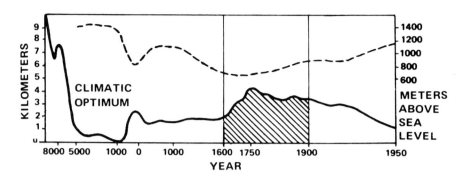

FIGURE 7 Length (solid line, in km) and height (dashed line, in m) of Norwegian and Icelandic glaciers. (After Ahlmann, 1953; reprinted with permission.)

Internal factors are:

1. Continental drift (Wegener) or polar wandering
2. Mountain elevation (Markov, Bruks)
3. Variations of carbon dioxide content of the atmosphere (Arrhenius, Chamberlain)
4. Variations of the amount of volcanic dust (Humphreis)
5. Albedo change and the role of polar ice caps (Bruks)

All this makes one marvel that during 90 percent of Post-Cambrian time there was no glaciation.

Mention should also be made of the hypothesis (by Opik, for instance) about ancient glaciations, each of 50 million years' duration separated by 200-million-year nonglacial intervals: Huronian, Algonkinian, Eocambrian, and Permo-Carboniferous glaciations, and the Pleistocene. In my opinion, on such a scale, the geologic-geochemical environment is so inhomogeneous that it is hardly possible to speak about some periodicities of climate.

Leaving the ancient glaciations aside, Figure 10 illustrates the entire spectrum of weather and climate oscillations. The supposed spectrum at periods larger than the index cycle is represented here by a dashed line.

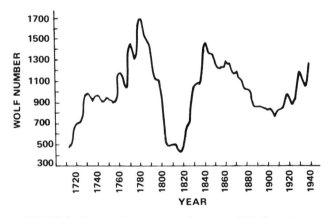

FIGURE 8 Twenty-three-year moving mean of Wolf numbers.

PLANETOLOGICAL APPROACH

The misfortune of all the above-mentioned hypotheses about the causes of the glacial stages (and thus about theories of the earth's climate) is that they are unsubstantiated and involve a groundless overestimation of the role of single factors. So far, we have no scientific answers to questions about the significance of these factors.

How would climate change with changes of the "solar constant"? Will the ever-increasing production of carbon dioxide from fuel combustion lead to its accumulation in the atmosphere and to climatic warming due to the intensification of the "hot-house effect," or will there be only an increase in the absorption of carbon dioxide by the oceans and photosynthesizing plants? Does the accumulation of volcanic dust in the atmosphere result in warmer or colder climates? What could be the effect of the Arctic and Antarctic ice-melting—restoration of ice cover (which apparently took place in the Arctic after the "climatic optimum" and the "Viking epoch"), or irreversible climatic warming?

It is absolutely clear that such questions can be answered only by a detailed mathematical calculation based on a scientifically sound physical model. Even now some categorical answers remain that are without benefit of any serious calculation—survivals of the time of "heroic geography." But now sufficient answers are produced only with the use of good computers capable of "heroic calculations."

A suitable approach for any planet may be worked out somewhat as follows:

Let its radius, the duration of a day and a year, the inclination of the equator to the ecliptic, acceleration of gravity at the surface, solar constant, albedo, and the mass and composition of the atmosphere be given. In this case, our first, and possibly most difficult, question would be: Is climate uniquely determined by these parameters or, with the given parameters are several steady climatic states possible (so to say, "potential pits" on the energy graph), between which a system wanders (for instance, glacial and interglacial stages)?

Thus, what will climate be at the preassigned "external parameters"? One of the most important parameters (Ta-

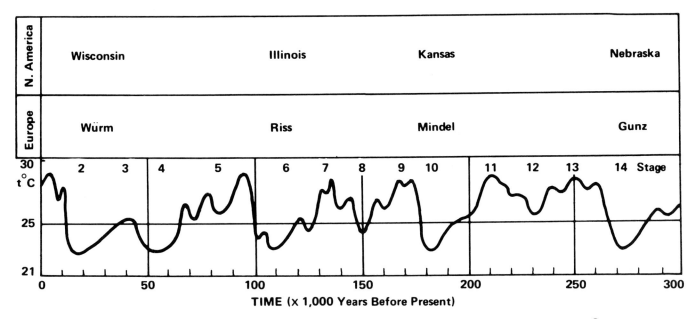

FIGURE 9 Shortened chronology of Pleistocene glacial stages. (After Emiliani, 1955; reprinted with permission; copyright © 1955, University of Chicago Press.)

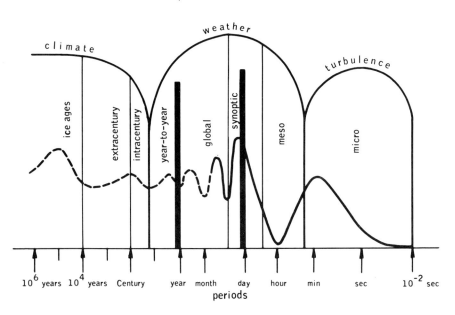
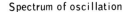

FIGURE 10 Spectrum of weather and climate oscillations. Hypothetical spectrum at longer periods represented by dashed line.

ble 1) is the nondimensional angular velocity of a planet's rotation, $2 \omega R/c$. If that parameter is small (Mercury, Venus), atmospheric motions are scarcely affected by the planet's rotation. Circulation may then be determined by temperature difference, not between the equator and the poles, but between the undersolar and antisolar points. If that parameter is of the order of unity (Earth, Mars, sun), circulation may be determined by both rotation and other factors (on the earth, for instance, by the difference be-

tween the oceans and the continents). If the parameter is large (big planets), rotation becomes decisive.

The other major parameter is the rate of production of kinetic energy (G. S. Golitsyn's formula). This can be used to evaluate both the level of kinetic energy and the typical velocity of atmospheric motions. Examples of such evaluations for Earth, Venus, and Mars are presented in Table 1.

Detailed calculation of a planet's climate from the pre-assigned "external parameters" requires physical-

TABLE 1 Thermodynamic Efficiency. Angular Velocity, Kinetic Energy, and Wind Velocity for Earth, Venus, and Mars[a]

	η	$\dfrac{2\omega R}{c}$	ϵ	U
Earth	0.2	2.8	4	10
Venus	0.15–0.3	0.01	0.1–0.2	4–6
Mars	0.2–0.4	1.7	200–400	30–50

[a]Symbols: ϵ = average rate of kinetic energy generation (and dissipation) (in unit mass); S = solar constant; A = albedo; R = radius of a planet; M = mass of the atmosphere; η = thermodynamic efficiency of an ideal heat engine; κ = numerical coefficient; L = typical horizontal scale of synoptic processes; c = sound velocity; ω = angular velocity of a planet's rotation; U = typical wind velocity.

NOTE: Energy production

$$\epsilon = \frac{\kappa\eta\pi R^2 S\,(1-A)}{M}$$

$$\kappa \sim 0.1$$

Energy level

$$U \sim (\epsilon L)^{1/3}$$

$$L \sim \begin{cases} \dfrac{c}{2\omega} & \text{if } \dfrac{2\omega R}{c} > 1 \\[2mm] R & \text{if } \dfrac{2\omega R}{c} < 1 \end{cases}$$

mathematical models of the planet's atmospheres. Work on the development of such models has been conducted by E. N. Blinova in the Soviet Union during the last 25 years.

In the last decade, the so-called numerical experiments on the general circulation of atmospheres have become very popular; these consist of the numerical integration of the equations of a model by steps in time for several months or even several years. Particular mention should be made of the experiments by Phillips (1956), Smagorinsky (1964), Smagorinsky et al. (1967), Mintz (1968), and Leith (1965).

Some of the results of the experiments are shown in Figures 11–15. Satisfactory agreement is obtained when actual data for the vertical distribution of zonal wind are compared with the results obtained by Phillips, Smagorinsky, and Mintz (Figure 11). Except for some details, actual data for surface pressure compare well with those obtained by Mintz (Figure 12). Vertical temperature profiles in different latitudes, as described by actual data and by Smagorinsky's results, give satisfactory agreement (Figure 13). Meridional temperature sections of Smagorinsky agree with those based on actual data (Figure 14). Finally, comparison of the rates of transformation of the zonal mean and fluctuating parts of potential energy and kinetic energy, as obtained experimentally by several authors and from actual data, shows agreement at least within an order of magnitude (Figure 15).

These results are promising and show that numerical experiments can be used for explaining the peculiarities of modern climate, for reconstructing paleoclimates, and for predicting future climatic changes, including the effects of man's influence.

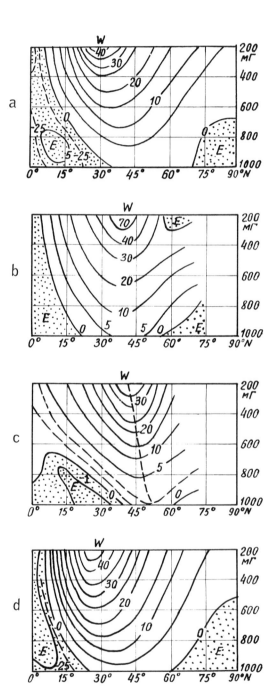

FIGURE 11 Mean zonal wind in Northern Hemisphere: (a) observation (after Mintz); (b) experiment of Phillips; (c) experiment of Smagorinsky; (d) experiment of Mintz.

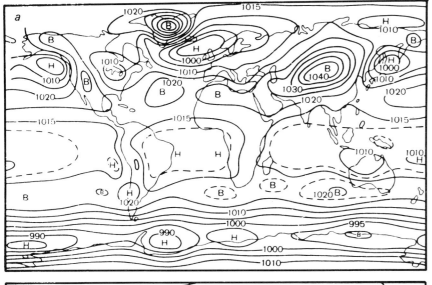

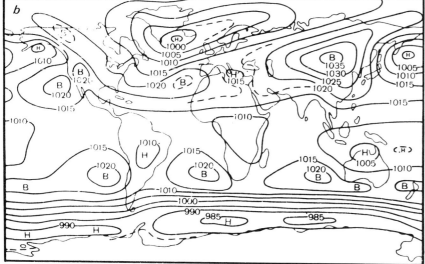

FIGURE 12   Mean surface pressure: (a) experiment of Mintz; (b) observation. (After Mintz, 1968; reprinted with permission.)

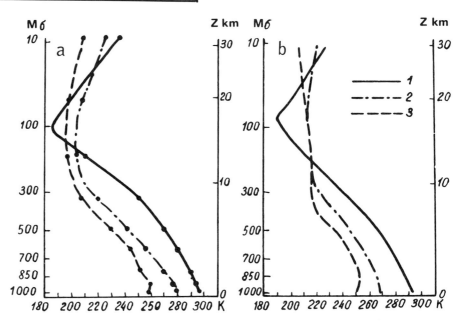

FIGURE 14   Meridional cross section of mean yearly zonal temperature field: (a) experiment of Smagorinsky; (b) observation.

FIGURE 13   Vertical profiles of mean yearly zonal temperature: (1) at 10°N; (2) at 50°N; (3) at 90°N; (a) experiment of Smagorinsky; (b) observation.

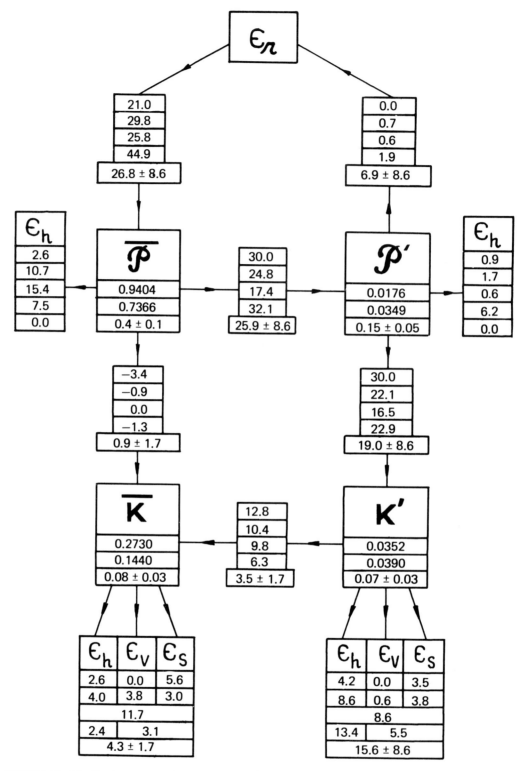

FIGURE 15 Distribution and transformation of zonal mean and fluctuating components of potential and kinetic energy.

$\bar{\epsilon}_r$ = external energy source

$\bar{P}$, P' = zonal mean and fluctuating components of potential energy

$\bar{K}$, K' = zonal mean and fluctuating components of kinetic energy

ϵ_h, ϵ_v, ϵ_s = rates of energy dissipation due to horizontal and vertical turbulent mixing and surface friction

Energy (joule per gram): Smagorinsky, 1964, 1967. Reality: Oort, 1964.

Energy transformation (10^{-3} joule per gram per day): Phillips, 1956; Smagorinsky, 1964, 1967; Gambo, 1967. Reality: Oort, 1964.

REFERENCES

Ahlmann, H. W. (1953) Glacier variations and climatic fluctuations. *Bowman Memorial Lecture.*, New York, American Geographical Society, 9 pp.

Brier, G. W. (1961) Some statistical aspects of long-term fluctuation in solar and atmospheric phenomena. *Ann. N.Y. Acad. Sci.*, *95*, 173–187.

Dzerdzievsky, B. L. (1956) Problem of variations in general atmospheric circulation and climate, pp. 109–122 in: A. I. Voeikov, ed., *Contemporary Problems in Climatology*, Gidrometizdat, Leningrad.

Emiliani, C. (1955) Pleistocene temperatures. *J. Geol.*, *63*, 538–578.

Gambo, K. (1967) Energy exchange between the zonal flow and the eddies, pp. 152–166 in: *Proceedings of the Dynamics of Large-Scale Atmospheric Processes*, editor A. S. Monin, Izdatvo Nauka, Moscow.

Gavrilin, B. L., and A. S. Monin (1967) A model of long-term interactions between the ocean and the atmosphere. *Dokl. Akad. Nauk. SSSR, 176*(4), 822–825.

Kolesnikova, V. N., and A. S. Monin (1965) Spectral variations of meteorological parameters. *Izv. Akad. Nauk. SSSR, Ser. Fiz. Atmos. Ok.*, *1*(7), 653–669.

Kolesnikova, V. N., and A. S. Monin (1966) On the year-to-year variations of meteorological elements. *Izv. Akad. Nauk. SSSR, Ser. Fiz. Atmos. Ok.*, *2*(2), 113–120.

Leith, C. E. (1965) Lagrangian advection in an atmospheric model. *WMO Tech. Note 66*, 168 pp.

Mintz, Y. (1968) Very long-term global integration of the primitive equations of atmospheric motion: An experiment in climate simulation. *Meteorol. Monogr.*, *8*(30), 20–36.

Mitchell, J. M. (1961) Recent secular changes of global temperature. *Ann. N.Y. Acad. Sci.*, *95*(1), 235–250.

Monin, A. S. (1956) Macroturbulent exchange in the earth's atmosphere. *Izv. Akad. Nauk. SSSR, Ser. Geofiz. 4*, 452–463.

Oort, A. H. (1964) On estimates of the atmospheric energy cycle. *Mon. Weath. Rev., Wash.*, *92*(11), 483–493.

Phillips, N. A. (1956) The general circulation of the atmosphere: A numerical experiment. *J. R. Meteorol. Soc.*, *82*(352), 123–164.

Smagorinsky, J. (1964) Some aspects of the general circulation. *J. R. Meteorol. Soc.*, *90*(983), 1–14.

Smagorinsky, J. *et al.* (1967) *Dynamics of Large-Scale Atmospheric Processes*. Nauka, Moscow, 70 pp.

Van der Hoven, J. (1957) Power spectrum of horizontal wind speed in the frequency range from 0.0007 to 900 cycles per hour. *J. Meteorol.*, *14*(2), 160–164.

Ward, F. (1964) General circulation of the solar atmosphere from observational evidence. *Pure Appl. Geophys.*, *58*, 157–180, plus 6 tables.

Colin S. Ramage

UNIVERSITY OF HAWAII,
DEPARTMENT OF GEOSCIENCES,
HONOLULU, HAWAII

METEOROLOGY OF THE SOUTH PACIFIC TROPICAL AND MIDDLE LATITUDES

My relief that the Scientific Committee on Oceanic Research (SCOR) did not ask me to present a paper similar to this one on the Indian Ocean in 1961 is erased by my trepidation at my present predicament. Anything I could have said then about Indian Ocean meteorology would before long have been rendered obsolete by the International Indian Ocean Expedition (IIOE); should the South Pacific now be explored, this paper will surely suffer an even worse fate.

Nevertheless, the effort is worth making, if only to chart the extent of our ignorance and to emphasize the promise of new technology, in particular satellites and super-pressure balloons, to delineate in unprecedented detail the clouds and winds of even the emptiest ocean.

The practice of synoptic meteorology is worldwide. However, except for daily analyses especially prepared from International Geophysical Year (IGY) observations for the period July 1957 through December 1958 (Weather Bureau, Republic of South Africa, 1966; Deutscher Wetterdienst, Seewetteramt, 1965), no synoptic weather charts are available for the region south of 25°S and between 90° and 140°W. Thus, the insights meteorologists develop from continual study of day-to-day weather are quite lacking over much of the South Pacific. In our ignorance, and influenced by the vast homogeneous surface of this uncharted area, we have assumed that weather systems there usually develop, intensify, decay, and move with a massive uninterrupted smoothness that is unknown over continents or smaller ocean reaches. I suspect that the truth is much more complex.

In the remainder of this paper, after listing my data sources, I depict average conditions in January and July, then treat atmospheric circulation components that are in some respects peculiar to the South Pacific, discussing in turn those features with insignificant annual variations, significant annual variations, and significant interannual variations (for area references, see Figure 1).

DATA SOURCES

Table 1 summarizes the data sources at my disposal. Such notable heterogeneity can scarcely be found anywhere else, and it must constantly be kept in mind as an essential modifier of any apparently dogmatic statements that follow. Rather than original sources, the most easily accessible sources have been listed.

JANUARY AND JULY

Average conditions in summer, typified by January, and in winter, typified by July, are depicted in Figures 2 through 9.

Only at two levels can the tropospheric circulation be described: at the surface, with averages computed from ships' wind observations, and at 200 millibars ($\approx$ 12.4 km), with averages based on balloon soundings, jet aircraft flying-level winds, and super-pressure balloon trajectories.

Experience has shown that the circulation as represented by mean resultant winds is closely related to average pressure distribution. Centers of average anticyclonic and cyclonic flow coincide with centers of average highs and lows, respectively, whereas clockwise-turning streamlines coincide with troughs, and anticlockwise-turning streamlines, with ridges. Of course, at the surface, friction causes streamlines to angle across isobars toward lower pressures, while in the immediate vicinity of the equator, steep latitudinal gradients of the coriolis force render pressure–wind relationships even more complex (Gordon and Taylor, 1968).

For delineation of weather, I depended on frequency of ship observations reporting precipitation — a long geographically inhomogeneous record — and on mean cloudiness determined from weather satellite data for a single January and a single July — a record whose brevity is compensated for somewhat by the density and geographical homogeneity of the observations.

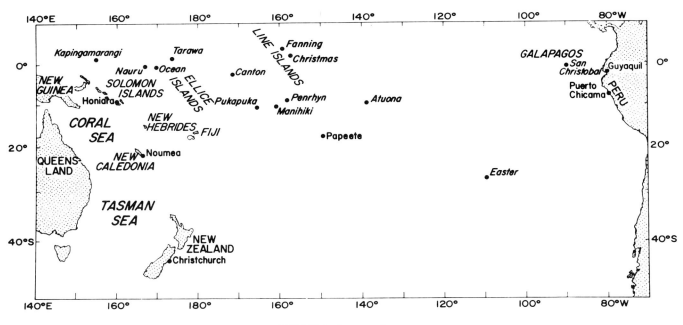

FIGURE 1 Locator chart.

TABLE 1 Sources of Published South Pacific Meteorological Data

Type of Data	Period	Source
A. SURFACE OBSERVATIONS		
1. Monthly climatological averages of ship observations	Approximately 100 years	McDonald (1938)
		Meteorological Office (1945)
		Meteorological Office (1950)
		U.S. Weather Bureau (1959)
2. Island climatology	1948–present	National Weather Records Center
	1952–present	Deutscher Wetterdienst, Seewetteramt (1952–)
	15–80 years	Various publications by meteorological services
3. Synoptic observations	January 1950–present	New Zealand Meteorological Service (1950–)
4. Synoptic analyses	July 1957–December 1958	Weather Bureau, Republic of South Africa (1966)
		Deutscher Wetterdienst, Seewetteramt (1965)
B. UPPER AIR OBSERVATIONS		
5. Monthly mean resultant winds, temperatures, and pressure heights at standard pressure levels	1948–present	National Weather Records Center (1948–)
6. Monthly mean resultant winds from aircraft measurements at 200, 250, and 300 millibars, Papeete–Honolulu; Papeete–Los Angeles	May 1961–April 1963	Pearson (1968)
7. Air trajectories at 200 millibars from super-pressure balloon flights	April 28, 1966–mid-August 1967	Solot (1967)
C. WEATHER SATELLITE OBSERVATIONS		
8. Once-daily "nephanalyses" from operational Tiros and Essa satellites	January 22, 1965–present	Environmental Science Services Administration
9. Monthly photoaverages for the tropics	February 1967–present	Department of Meteorology, University of Wisconsin
10. Cloud photos from ATS-1 synchronous satellite ($0°$; $155°$W). Variable frequency, 20 minutes to 24 hours	January 1967–present	Department of Meteorology, University of Wisconsin
11. 15-day photoaverages from the ATS-1	February 1967–present	Department of Meteorology, University of Wisconsin
12. Mean monthly cloudiness ($30°$N to $30°$S) derived from nephanalyses	February 1965–January 1967	Sadler (1968)

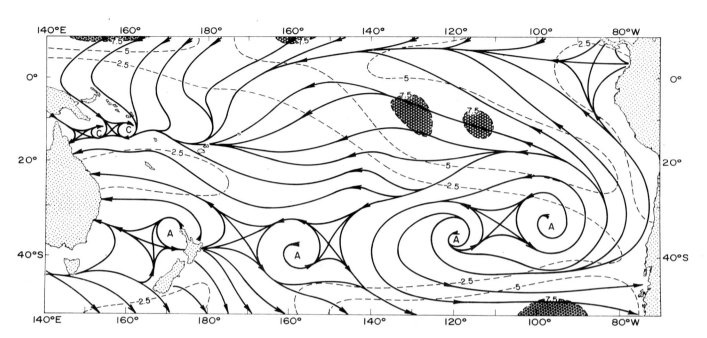

FIGURE 2   January. Mean resultant surface winds. Solid lines are streamlines; dashed lines are isotachs labeled in m sec^{-1}. (Reprinted with permission from McDonald, 1938.)

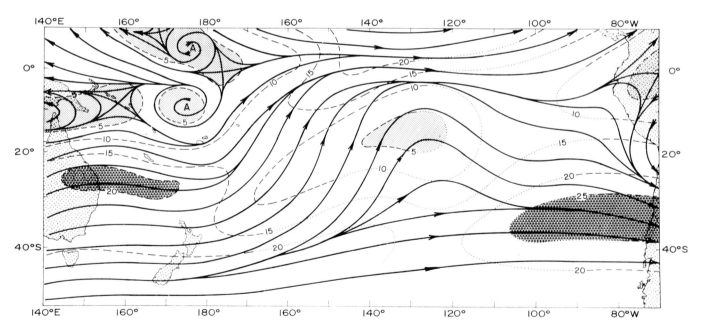

FIGURE 3   January. Mean resultant winds at 200 millibars ($\approx$ 12.4 km). Solid lines are streamlines; dashed lines (dotted where doubtful) are isotachs labeled in m sec^{-1}

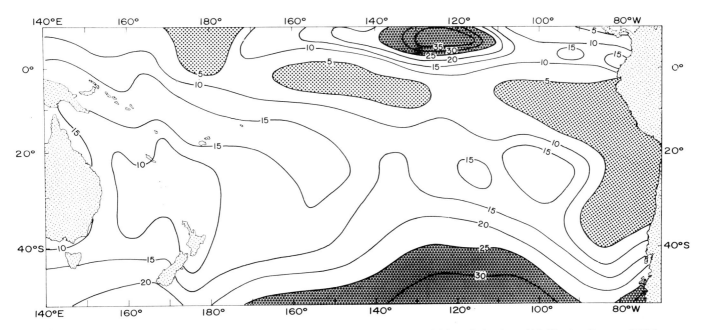

FIGURE 4   January. Percentage of ship observations reporting rain. (Reprinted with permission from U.S. Weather Bureau, 1959.)

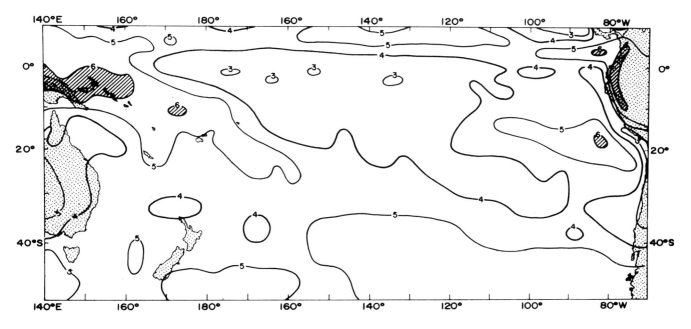

FIGURE 5   January 1967. Mean cloudiness determined from weather satellite observations. Isopleths are labeled in eighths of sky cover. (Reprinted with permission from Sadler, 1968.)

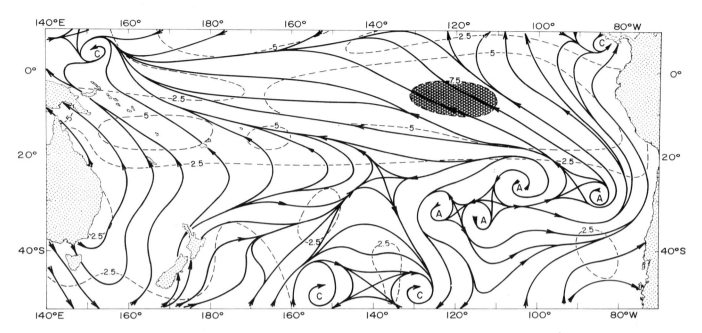

FIGURE 6 July. Mean resultant surface winds. Solid lines are streamlines; dashed lines are isotachs labeled in m sec^{-1}. (Reprinted with permission from McDonald, 1938.)

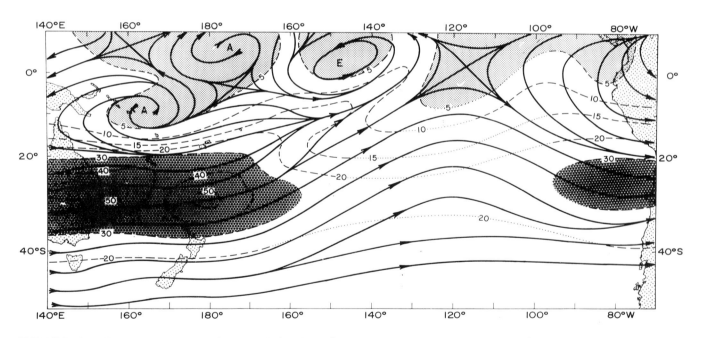

FIGURE 7 July. Mean resultant winds at 200 millibars ($\approx$ 12.4 km). Solid lines are streamlines; dashed lines (dotted where doubtful) are isotachs labeled in m sec^{-1}

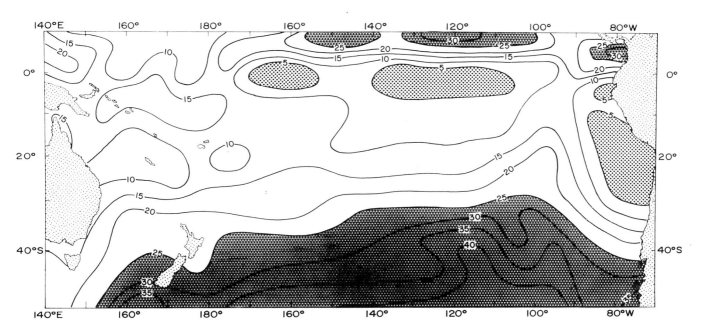

FIGURE 8   July. Percentage of ship observations reporting rain. (Reprinted with permission from U.S. Weather Bureau, 1959.)

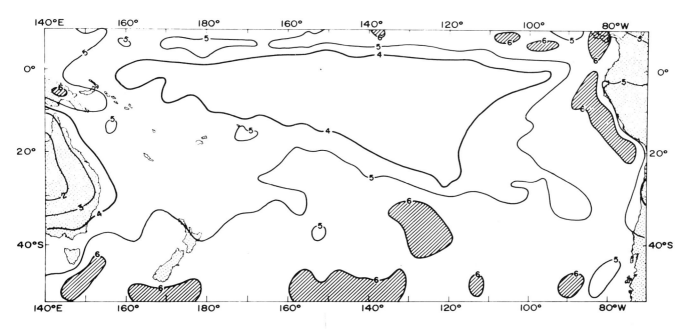

FIGURE 9   July 1967. Mean cloudiness determined from weather satellite observations. Isopleths are labeled in eighths of sky cover. (Reprinted with permission from Sadler, 1968.)

FEATURES THAT PERSIST THROUGHOUT THE YEAR

The Dry Zone off South America and in Equatorial Latitudes

In normal years, the southeast trades blow steadily around the eastern and northern periphery of the subtropical anticyclone.

Surface pressure falls toward a trough just north of the equator and toward the west along the equator. Thus, not only are the southeasterlies divergent, but easterly flow along the equator is also divergent (Gordon and Taylor, 1968). Throughout a normal year, then, cold water is upwelled along the coast of Peru (Gunther, 1936) and also along the equator as far west as 165°W (Austin, 1960). Subsidence resulting from the divergent surface winds prevents deep precipitating clouds from developing, while the relatively cold underlying waters further inhibit convection.

Bad Weather between New Guinea and Fiji/Ellice Islands

The northern part of the region between New Guinea and Fiji/Ellice Islands is monsoonal, with surface winds shifting from southeast in winter to north or northwest in summer. The summer bad weather—surface convergence and lifting in anticlockwise flow just south of the equator (Gordon and Taylor, 1968) and, farther south, a favored region for development and slow southward movement of tropical depressions or storms (see Figure 11) and subtropical cyclones (Taljaard, 1967)—is understandable.

The winter bad weather is much more complex. Hill (1964) describes frequent development of a thick persistent sheet of altostratus from which occasional rain falls. Although a cold front may be moving northeastward over the area, the cloud and rain seem not to be directly associated with it since the cold air behind the front is usually subsiding and stable. The weather appears, rather, to be produced by upward motion in the middle and upper troposphere, occurring to the east of, and induced by a vigorous trough in, the upper tropospheric westerlies. This is not unusual, for similar situations prevail over the eastern North Pacific in winter. What is unusual is the proclivity for development to recur over a single region. What is the anchor?

Although much work is needed to evaluate his hypothesis, Danielsen (personal communication, 1968) makes the plausible suggestion that the southwest monsoon of the northern Indian Ocean provides the anchor.

In the lower troposphere, strong southerlies prevail along the equator west of 60°E, veering over the Arabian Sea to sweep across India as the southwest monsoon. Compensating flow occurs in the upper troposphere—a northeast stream, which after crossing the equator, backs, sinks, and warms over Western Australia. There it meets cold air that has moved up from the south. A steep baroclinic gradient

and an extremely strong westerly jet stream result (Muffatti, 1964). Wave instability often develops, inducing upper tropospheric trough development downstream over eastern Australia or the Coral Sea. Since the Indian summer monsoon is anchored by ocean and continent, even its long-distance effects would tend to be anchored.

Mean resultant winds at 200 millibars between New Caledonia and western Queensland (Figure 7) fail to reflect the frequent presence of an upper tropospheric trough. However, careful analysis of the winds at Nouméa, New Caledonia, reveals the reason that mean resultant winds are indicators of synoptic climatology, as well as some of their shortcomings in doing so. The July 200-millibar mean resultant wind at Nouméa (274° 36 m sec^{-1}) is made up of a bimodal distribution of individual wind directions. The primary frequency maximum is for winds north of west (trough to the west), and the secondary frequency maximum is for winds south of west (no trough to the west).

Farther south, over the northern Tasman Sea, cutoff cyclones occasionally develop in the trough (Taljaard, 1967) and, accompanied by extensive cloud systems, drift slowly eastward.

Upper Tropospheric Trough in the Subtropics between 120° and 130°W

Figures 3 and 7 show a well-marked westerly trough in both January and July in the subtropics between 120° and 130°W. Pearson's charts (1968) confirm its persistence throughout the year as a climatological feature. Superpressure balloon trajectories (Solot, 1967), largely confined to middle and high latitudes, do not reveal any preferred longitudes for trough location (Table 2).

Generally, air east of a climatological trough in the westerlies rises; to the west it sinks. Rain should therefore fall more often east of the trough than west of the trough. Figure 10 seems to confirm this.

In the Northern Hemisphere summer a NE-SW oriented upper tropospheric trough persists across the central North Pacific. Ramage (1959) suggests that it is anchored by heat-flow circulations over Asia and North America. No such climatological trough is found in winter. Presumably, the Southern Hemisphere trough is anchored through continent-ocean configurations, but extensive observations are needed before one can attempt to understand the mechanism.

Middle-Latitude Systems

Van Loon (1965 and 1967) and Taljaard (1967), using IGY analyses (Weather Bureau, Republic of South Africa, 1966), conclude that occurrence, tracks, and speeds of extratropical cyclones vary little with the seasons (Table 2) and that apart from relatively common cyclogenesis over the Tasman Sea, geographical distribution is likely to be rather uniform.

Even during the IGY, no upper-air observations were

TABLE 2 Frequency of 200-Millibar Troughs Determined from 119 Super-Pressure Balloon Circuits of the Southern Hemisphere in 1966 and 1967 (from Solot, 1967)

Longitude			
160°E-170°W	170°W-140°W	140°W-110°W	110°W-80°W
39½	42	37½	45

TABLE 3 Average Speeds of Movement of South Pacific Surface Depressions in July, August, and September 1957 (Winter) and in January, February, and March 1958 (Summer) (from Van Loon, 1967)

	Speed (m sec⁻¹)		
	30-40°S	40-50°S	50-60°S
Winter	9.2	13.8	15.4
Summer	9.8	13.5	14.2

made in middle latitudes between 180° and South America, and thus it was impossible to delineate any of the large-scale circulation features that are best developed in the upper westerlies.

The National Center for Atmospheric Research (NCAR) Global Horizontal Sounding Technique (GHOST) (Lally *et al.*, 1966) has begun to fill this observational void. Super-pressure balloons, adjusted to fly at 200 millibars, are being released from Christchurch, New Zealand. In 119 circuits of the Southern Hemisphere polar westerlies during 1966 and 1967 (Solot, 1967), 164 troughs were encountered over the South Pacific, an average of 1.4 per circuit. This corresponds to a wavelength of about 85° longitude. At 45° latitude (about the average followed by the balloons), the critical zonal wind speed for such waves to be stationary is 19 m sec⁻¹ (Rossby, 1939). Since the actual zonal wind speed (as determined from the balloon trajectories) is only slightly higher, "blocking"* must occur at times, with consequent distortions in local weather. The baloon trajectories reveal no favored longitudes for blocking (Table 2); the converse is true in the Northern Hemisphere (Rex, 1950).

The GHOST results reinforce the impression, derived from IGY studies, of small annual and geographic variability.

FEATURES WITH SIGNIFICANT ANNUAL VARIATION

The Australian Monsoon

At *low levels*, the circulation between northern Queensland and the dateline reverses from winter to summer.

During winter a cold anticyclone centered over Australia forms part of the hemisphere-girding subtropical ridge.

---

*The obstructing on a large scale of the normal west-to-east progress of migratory cyclones and anticyclones.

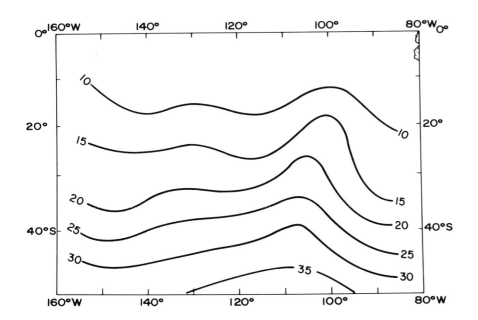

FIGURE 10 Year. Percentage of ship observations reporting rain. (Reprinted with permission from U.S. Weather Bureau, 1959.)

Thus, southeast tradewinds extend across the Coral Sea to the Queensland coast and northward to a weak trough just south of the equator (Figure 6).

By summer, a heat low has developed near 20°S over Australia. It forms part of the near-equatorial trough, which has moved south and intensified and now extends along about 12°S to the dateline. Winds blow from north or northwest north of the trough (Figure 2).

Axes of maximum rain frequency, maximum cloudiness, and maximum condensational heating do not shift much between winter and summer (Figures 4, 5, 8, 9), probably being anchored to the mountainous Indonesian islands. In summer, frequent thunderstorms over Indonesia and the Solomons convert large quantities of latent heat to sensible heat and transport it to the high troposphere. The consequent pressure rise at these levels prevents the subtropical ridge from moving poleward with the march of the sun (Figures 3 and 7) (in contrast to other regions) (Ramage, 1968b).

Tropical Cyclones

Tropical cyclones (Figure 11) are confined to the western part of the South Pacific and to the months of December through March. An average of four cyclones per season develops winds of at least Beaufort force 9. Frequencies range from 0 to 9 (Gabites, 1956).

The western North Pacific spawns many more tropical cyclones, and their development (referred to a unit area) occurs nearly four times more often than it does over the western South Pacific.

One peculiarity of South Pacific tropical cyclones is their very low median recurvature latitude, 15°, which is notable compared to that of 24°, north of the equator. In general, tropical cyclones lying equatorward of the upper tropospheric subtropical ridge move toward the west and poleward. As they cross the ridge, the zonal component of movement shifts to the east.

Over the western North Pacific, the 200-millibar subtropical ridge moves 2,000 km poleward between winter and summer, but over the South Pacific the poleward shift is insignificant, and tropical cyclones recurve abnormally soon after development.

FEATURES WITH SIGNIFICANT INTERANNUAL VARIATION

Significant interannual variations occur either when intensities of dry and wet regimes change, but their locations do not (a rather common occurrence over most of the tropics), or when significant circulation changes accompany major rearrangement of dry and wet areas, epitomized by development of persistent El Niño conditions off South America and over the Eastern Pacific equatorial zone.

Intensity of Tropical Weather Regimes

Early in 1965, weather satellites began photographing every part of the earth at least once daily. Sadler (1968), using National Satellite Center analyses of cloudiness, has derived charts of average cloudiness for the tropics for 2 years of individual months, commencing with February 1965. Figures 5, 9, and 12 are based on his work.

For the global tropics, every month from February to July 1965 was slightly less cloudy than the corresponding month in 1966. However, Sadler found that during the 1965 months normally cloudy areas were cloudier and normally fine areas were finer than in the 1966 months. In other words, *gradients* of average cloudiness were steeper in 1965.

Figure 12, showing the difference between July 1965 and July 1967, typifies what is apparently a rather common interannual variation in low latitudes. Areas of maximum and minimum cloudiness occupied the same locations in both months, but the 1965 gradients were steeper.

Perhaps large-scale vertical circulations were stronger

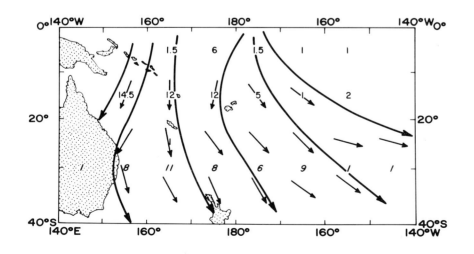

FIGURE 11 Tropical cyclones, December through April, 1940–1956. Streamlines delineate median tracks; upright numerals denote frequency of first detection by 5-degree squares; sloping numerals denote frequency of crossing 30°S, by 5-degree intervals. (Based on Gabites, 1956; reprinted with permission.)

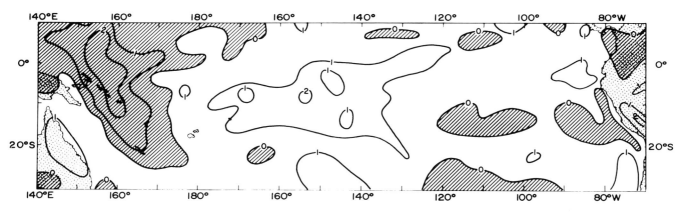

FIGURE 12 Difference in cloudiness between July 1965 and July 1967. Shaded areas: July 1965 cloudier.

over the western South Pacific in July 1965 than in July 1967. The tradewinds and upper tropospheric westerlies combine in an extensive vertical circulation with an upward cloudy branch over New Guinea–New Hebrides and a downward fine-weather branch east of the dateline. This circulation was probably more vigorous in July 1965 than in July 1967, with greater upward motion increasing cloudiness in the west and greater subsidence improving normally fair weather in the east (Figure 13 and Table 4).

Equatorial El Niño

Wyrtki (1966) lists many papers on El Niño, the abnormal warming of Peru coastal waters during the Southern Hemisphere summer. A significant weakening of the southeast trades results in disappearance of cool upwelled water. Bjerknes (1966) expands the discussion to include the equatorial eastern Pacific. He postulates that for a year or two prior to El Niño a recognizable trend sets in toward weaker southeast trades.

Bjerknes describes the following sequence of ocean and atmospheric changes along the equator: weaker southeast trades → upwelling stops, surface waters warm → increased heating and convection → precipitation increases. Investigations by Doberitz *et al.* (1967) throw some doubt on the validity of this description. I have reproduced two figures from their report: Figure 14, which shows that interannual variability in rainfall is greater over the central equatorial Pacific than farther east, and Figure 15, which reveals a

complexity of correlations between the temperatures of Peru coastal waters and Pacific islands rainfall. Changes throughout the region that is normally dominated by upwelling appear to occur at about the same time. Finally, the authors report that at Canton Island over a period of 16 years (1950–1965) mean monthly sea-surface temperature is most strongly correlated with rainfall of the *preceding* month.

Figures 14 and 15 suggest that over both the central and eastern Pacific, roughly simultaneous changes accompany development of El Niño and equatorial heating.

Perhaps the sequence is as follows:

1. Weakening of easterly flow along the equator results in reduced upwelling and in rising sea-surface and air temperatures.

2. With the temperature and pressure distributions now beginning to resemble those over other equatorial regions, well-marked near-equatorial troughs in both hemispheres extend eastward from east of the dateline, and depressions develop in them.

3. Although individual depressions may move westward initially, they develop successively farther east, and, as a consequence, the normal direction of the pressure gradient along the equator is reversed (Ichiye and Petersen, 1963).

4. Westerlies set in progressively from the west (Table 5), and since along the equator these *must* be convergent (Gordon and Taylor, 1968), the air rises and unusually heavy rains begin.

5. The equatorial westerlies wax and wane in unison with depression activity to northeast and southeast.

From November 1957 through March 1958 and from December 1958 through February 1959—during the strongest El Niño of recent years—rainfall at Canton Island averaged 15 mm on days when surface winds blew from the westerly quarter, but only 4 mm on days with winds from the easterly quarter.

6. Once the westerlies arrive, all trace of upwelling finally ceases, partially accounting for the slight lag of sea tempera-

TABLE 4 Rain at Honiara and Papeete in July 1965 and July 1967

	Amount (mm)		Number of Rain Days	
	July 1965	July 1967	July 1965	July 1967
Honiara	311	115	16	9
Papeete	70	38	10	13

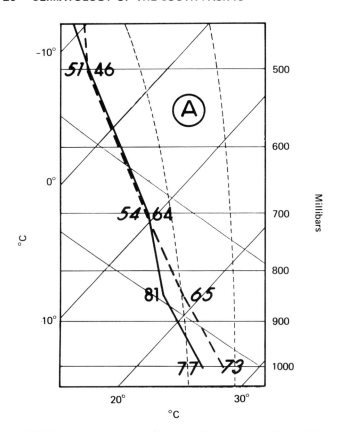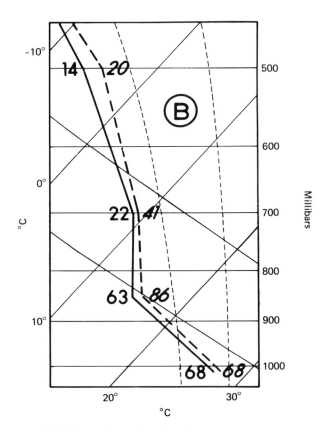

FIGURE 13 Average aerological soundings in July 1965 (solid lines) and July 1967 (dashed lines). Numbers denote percentage relative humidities. *A.* Honiara; *B.* Papeete.

tures behind rainfall reported by Doberitz *et al.* (1967) (see page 25).

Ichiye and Petersen (1963) state that the considerable excess of sea temperature over air temperature during El Niño enhances convection and rain.

However, Table 6 shows that in January 1958, at the height of El Niño, the sea around Canton Island gave up insignificantly more heat to the air than it did in January 1960, a typically dry upwelling month.

For the period November 1957 through March 1958, Canton Island had a mean sea-surface temperature of $30°C$, a mean minimum temperature on rainless days (maximum radiational cooling) of $27°C$, and a mean minimum temperature on days with more than 12 mm of rainfall of $24°C$. Such low values can stem only from rain evaporating into and cooling the air (Palmer *et al.*, 1955), which explains the apparent anomaly of mean surface-air temperatures at Canton Island showing only minor fluctuations between 1957 and 1960, while mean sea-surface temperatures ranged over $3°C$.

Simple cause-and-effect postulates are insufficient to account for El Niño. Evidence begins to suggest that the December–February rainfall regime over Indonesia might significantly reflect interactions with events far to the east

TABLE 5 Onset Dates of Surface Westerly Winds during El Niño Peaks of Late 1957 and Late 1958

	1957	1958
Tarawa (01°21′N; 172°55′E)	1 November	26 November
Canton (02°46′S; 171°43′W)	17 November	4 December

TABLE 6 Averages at Canton Island for January 1958 and January 1960. (Computations based on Budyko, 1956.)

Average	1958	1960
Air temperature ($°C$)	28.5	29.1
Sea-surface temperature ($°C$)	29.7	27.6
Vapor pressure of air (millibars)	30.87	27.09
Saturation vapor pressure at sea-surface temperature (millibars)	41.71	36.92
Wind speed (m sec^{-1})	6.4	7.5
Rate of cooling at sea surface from evaporation (cal cm^2 day^{-1})	359	381
Rate of sensible heat loss from sea surface to air (cal cm^2 day^{-1})	26	−38
Rainfall (mm)	416	11

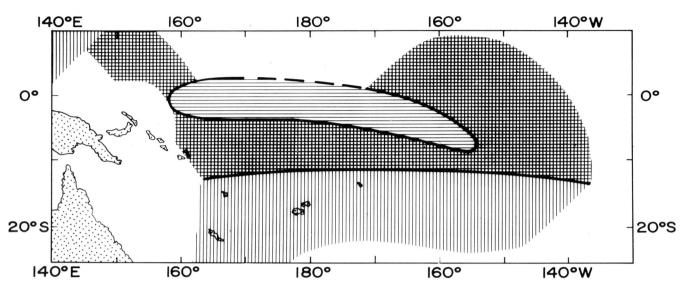

FIGURE 14 Rainfall regimes. Vertical hatching: marked annual cycle; horizontal hatching: strongly developed longer period variations; cross hatching: mixed. (Reprinted with permission from Doberitz *et al.*, 1967.)

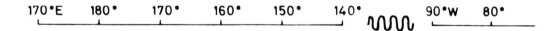

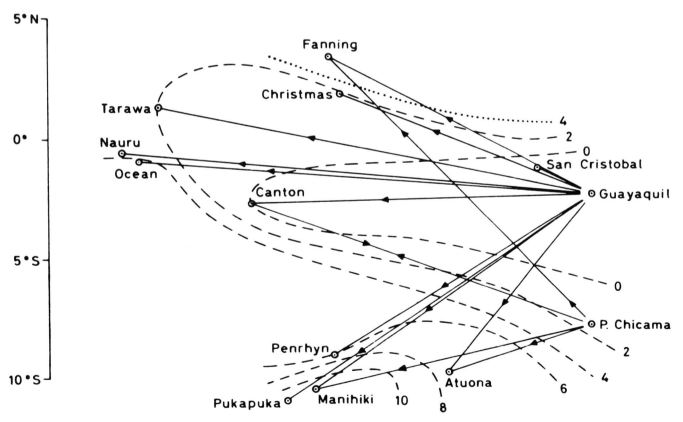

FIGURE 15 Time lag of maximum correlations between monthly values on the western coast of South America (Guayaquil, rainfall; Puerto Chicama, sea-surface temperature) and monthly rainfalls at Pacific islands. Dashed lines are isochrones labeled in units of months; arrows indicate lag directions. (Reprinted with permission from Doberitz *et al.*, 1967.)

(Ramage, 1968a). The problem demands a concerted attack by oceanographers and meteorologists.

CONCLUDING REMARKS

I am most interested in those aspects of its meteorology—El Niño and the winter rains of the Coral Sea–Fiji region—that distinguish the South Pacific from other oceans.

Both of these phenomena are telemorphically related to meteorological events elsewhere in the tropics and subtropics, and their *interannual* changes are significant. Therefore, any detailed studies would have to be undertaken as part of a more general simultaneous study of the whole tropics and should be pursued for at least 5 years.

Before long, the French National Center of Scientific Research plans to release super-pressure balloons from tropical South American stations (Morel, personal communication, 1968). The GHOST program is also likely to include low latitudes. Trajectories determined therefrom, as well as improved weather-satellite observations, will certainly enhance our knowledge of tropical South Pacific weather. To be quantitatively useful, the data thus obtained must be tied into routine rawinsoundings and oceanographic measurements from research vessels and from Kapingamarangi Atoll, Tarawa Island, the Line Islands, the Galápagos Islands, and Easter Island. Most important, the recently abandoned Canton Island weather station, with its unequaled long-period record, must be reactivated. These stations, once established, could be permanently maintained as part of the World Weather Watch network (World Meteorological Organization, 1967).

Two ambitious micro- and mesoscale investigations of air–sea interaction processes are scheduled for the Barbados region during 1968 and 1969 (Garstang, 1967; Bomex Project Office, 1968). Should they prove successful, the results could be used to parameterize routine larger scale observations made in the South Pacific.

This paper is Hawaii Institute of Geophysics Contribution No. 224, supported by the Atmospheric Sciences Section, U.S. National Science Foundation, NSF Grant GA-1009.

REFERENCES

Austin, T. S. (1960) Oceanography of the east central equatorial Pacific as observed during Expedition Eastropac. U.S. Fish and Wildlife Service, *Fishery Bull.*, *60*, 257–282.

Bjerknes, J. (1966) Survey of El Niño 1957–58 in its relation to tropical Pacific meteorology. *Bull. Inter-Amer. Trop. Tuna Comm.*, *12* (2), 1–62.

Bomex Project Office (1968–   ) Barbados oceanographic and meteorological experiment—*Bulletins*, 1– , Bomex Project Office, Rockville, Maryland.

Budyko, M. I. (1956) *The heat balance of the earth's surface.* Gidrometeorologicheskoe izdatel'stvo. Leningrad, 255 pp.

(Translated by Nina A. Stepanova and distributed by the U.S. Weather Bureau, Washington, D.C., 1958.)

Deutscher Wetterdienst, Seewetteramt (1952–   ) *Die Witterung in übersee.* Einzelveroffentlichungen. Hamburg.

Deutscher Wetterdienst, Seewetteramt (1965) *International Geophysical Year (1957–58) world weather maps. Part II: Tropical zone 25°N to 25°S. Daily sea level and 500 mb charts.* Einzelveröffentlichungen. Hamburg.

Doberitz, R., H. Flohn and K. Schütte (1967) Statistical investigations of the climatic anomalies of the equatorial Pacific. *Bonn. Meteorol. Abhand.*, *7*, 76 pp.

Gabites, J. F. (1956) A survey of tropical cyclones in the South Pacific, pp. 19–24 in: *Proceedings of the tropical cyclone symposium, Brisbane, December 1956*, Director of Meteorology, Melbourne, 436 pp.

Garstang, M. (1967) The Barbados meteorological experiment (abstract). *Bull. Amer. Meteorol. Soc.*, *48*, 634.

Gordon, A. H., and R. C. Taylor (1968) Numerical steady-state friction layer trajectories over the oceanic tropics as related to weather. *International Indian Ocean Expedition Meteorol. Monogr.* 7, East-West Center Press, Honolulu.

Gunther, E. R. (1936) A report on oceanographical investigations in the Peru coastal current. *Discovery Rep.*, *13*, 107–206.

Hill, H. W. (1964) The weather in lower latitudes of the southwest Pacific associated with the passage of disturbances in the middle latitude westerlies, pp. 352–365 in: *Proceedings of the Symposium on Tropical Meteorology, Rotorua*, editor J. W. Hutchings, N. Z. Meteorological Service, 737 pp.

Ichiye, T., and J. R. Petersen (1963) The anomalous rainfall of the 1957–58 winter in the equatorial central Pacific arid area. *J. Meteorol. Soc., Japan, Ser. II*, *41*, 172–182.

Lally, V. E., E. W. Lichfield, and S. B. Solot (1966) The Southern Hemisphere GHOST experiment. *World Meteorol. Org. Bull.*, *15*, 124–128.

McDonald, W. F. (1938) *Atlas of the Climatic Charts of the Oceans*, U.S. Government Printing Office, Washington, D.C., 130 charts.

Meteorological Office (1945) *Monthly Meteorological Charts of the Western Pacific Ocean*, H. M. Stationery Office, London, 120 pp.

Meteorological Office (1950) *Monthly Meteorological Charts of the Eastern Pacific Ocean*, H. M. Stationery Office, London, 122 pp.

Muffatti, A. H. J. (1964) Aspects of the subtropical jet stream over Australia, pp. 72–88 in: *Proceedings of the Symposium on Tropical Meteorology, Rotorua*, editor J. W. Hutchings, N. Z. Meteorological Service, 737 pp.

National Weather Records Center (1948–   ) *Monthly Climatic Data for the World*, National Weather Records Center, Asheville, North Carolina.

New Zealand Meteorological Service (1950–   ) *Daily Weather Bulletins*, Wellington, N.Z.

Palmer, C. E., C. W. Wise, L. J. Stempson, and G. H. Duncan (1955) The practical aspect of tropical meteorology. *Air Force Surveys in Geophys.*, *76*, 195 pp.

Pearson, A. D. (1968) *The Upper Tropospheric Wind and Weather Patterns of the Tropical Eastern Pacific*, M.S. Thesis, 61 pp. Department of Geosciences, University of Hawaii.

Ramage, C. S. (1959) Hurricane development. *J. Meteorol.*, *16*, 227–237.

Ramage, C. S. (1968a) Role of a tropical "maritime continent" in the atmospheric circulation. *Mon. Weather Rev.*, Wash., 96.

Ramage, C. S. (in press) Climate of the Indian Ocean in: *Climates of the Oceans*, editor H. Thomsen (Vol. 15 of *World Survey of Climatology*). American Elsevier, New York.

Rex, D. F. (1950) Blocking action in the middle troposphere and its effect upon regional climate; Part II: The climatology of blocking action. *Tellus*, *2*, 275–301.

Rossby, C.-G. (1939) Relation between variations in the intensity of the zonal circulation of the atmosphere and the displacements of the semipermanent centers of action. *J. Mar. Res.*, *2*, 38–55.

Sadler, J. C. (1968) Average cloudiness in the tropics from satellite observations. *International Indian Ocean Expedition, Meteorol. Monogr. 3*, East-West Center Press, Honolulu.

Solot, S. B. (1967) *GHOST Atlas of the Southern Hemisphere*. NCAR Global Atmospheric Measurement Program, Boulder, Colorado, 10 plates.

Taljaard, J. J. (1967) Development, distribution and movement of cyclones and anticyclones in the Southern Hemisphere during the IGY. *J. Appl. Meteorol.*, *6*, 973–987.

U.S. Weather Bureau (1959) *U.S. Navy Marine Climatic Atlas of the World, Volume 5, South Pacific Ocean*, Chief of Naval Operations, Washington, D.C., 267 charts.

Van Loon, H. (1965) A climatological study of the atmospheric circulation in the Southern Hemisphere during the IGY, Part I: 1 July 1957–31 March 1958. *J. Appl. Meteorol.*, *4*, 479–491.

Van Loon, H. (1967) A climatological study of the atmospheric circulation in the Southern Hemisphere during the IGY, Part II. *J. Appl. Meteorol.*, *6*, 803–815.

Weather Bureau, Republic of South Africa (1966) *International Geophysical Year (1957–58) World Weather Maps, Part III, Southern Hemisphere South of 20°S. Daily Sea Level and 500 mb Charts*, Pretoria, South Africa.

World Meteorological Organization (1967) *World Weather Watch, the Plan and Implementation Programme*. World Meteorological Organization, Geneva. 56 pp.

Wyrtki, K. (1966) Oceanography of the eastern equatorial Pacific Ocean. *Ocean. Mar. Biol. Ann. Rev.*, *4*, 33–68.

II
CIRCULATION OF
THE SOUTH PACIFIC

Bruce A. Warren

WOODS HOLE OCEANOGRAPHIC INSTITUTION,
WOODS HOLE, MASSACHUSETTS

GENERAL CIRCULATION OF THE SOUTH PACIFIC

INTRODUCTION

In 1853, the British geographer Alexander Findlay published a chart of surface currents of the Atlantic and Pacific oceans in connection with some proposals then being made for cutting canals across Central America. It is a very good surface current chart—to my knowledge, the best of those of comparable scope that were drawn up during the nineteenth century. Especially for the South Pacific, with the exception of the southern Tasman Sea, Findlay's chart is in remarkable accord with present ideas of the distribution of surface currents. Thus, except for the coastal regions off Australia and South America, our knowledge of the surface currents in the South Pacific is actually not much better than Findlay's was over 100 years ago, and he had very little information to go on. There is a great need, therefore, for physical-oceanographic exploration of the pattern and intensity of the surface circulation, and a still greater need to study the deep flow, because much less is known about it.

This paper reviews some of the principal things that have been learned about the general circulation of the South Pacific, with emphasis on ambiguities requiring investigation, and with reference to circulation theory. Plainly, the observational coverage of the South Pacific, particularly of its central regions, has been so limited that even observations made at random or on an arbitrarily selected geometric grid will have value. It is more rewarding, however, to carry out observational programs designed to answer questions or to resolve problems of more than local interest, and a theoretical framework provides a useful basis for defining such problems.

This will be a two-part discussion, including, first, some considerations of the oceanwide circulation in the upper kilometer, and then, some speculation and fragmentary evidence concerning deep-water movements.

NEAR-SURFACE CIRCULATION

Figure 1 is Reid's chart of dynamic topography at the surface of the Pacific, with reference to the topography of presumably much smaller amplitude at a depth of 1,000 m; it illustrates well the gross characteristics of the subtropical gyre in the South Pacific. In high latitudes, we see the eastward-flowing West Wind Drift, or Circumpolar Current, and in tropical latitudes, the westward-flowing South Equatorial Current, broken near 10°S by the eastward-flowing South Equatorial Countercurrent (Reid, 1959). The equatorward flow that connects the high- and low-latitude zonal flows is concentrated toward the eastern side of the South Pacific, near South America. In this respect, the subtropical gyre of the South Pacific differs markedly from those of other oceans, where the equatorward flow is distributed more uniformly over the full breadth of the ocean. As in other oceans, however, the subtropical gyre in the South Pacific is completed by a relatively narrow western boundary current, the East Australian Current, which flows southward against the coast of Australia. Here, at the sea surface, the center of the gyre seems to lie in fairly low latitudes, between 20° and 30°S.

What is known of the general flow pattern in the thermocline has been described by Reid (1965) in his monograph on intermediate waters of the Pacific Ocean. Figure 2 is his chart of the distribution of salinity on the surface of thermosteric anomaly equal to 125 centiliters per ton (cl/t) (σ_t = 26.8). The depth of this surface in the central South Pacific is 400–500 m, though it does intersect the sea surface close to 50°S. The most striking feature here is the broad tongue of low-salinity water in the eastern South Pacific; this tongue represents northward flow from the West Wind Drift, which subsequently turns and flows westward close to 20°S. The pattern is very much like that of the dynamic

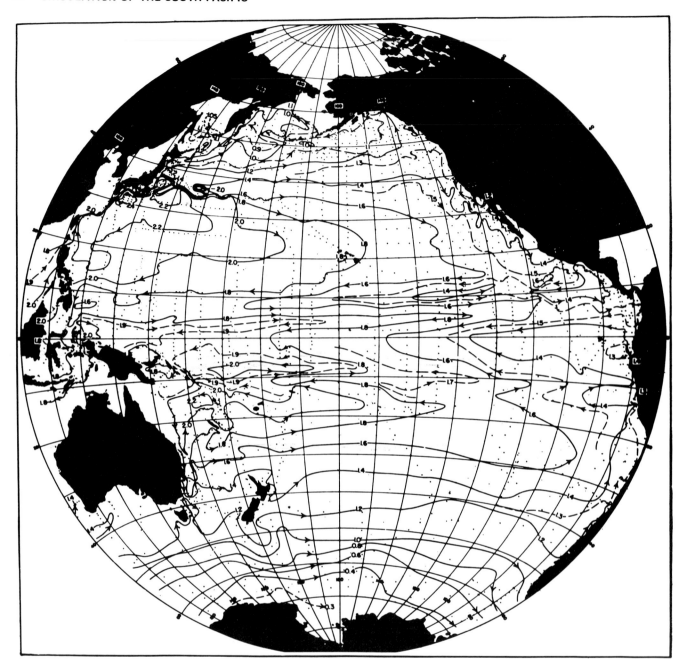

FIGURE 1   Dynamic topography (in dynamic meters) of the sea surface in the Pacific Ocean with respect to the 1,000-decibar surface. (Reprinted with permission from Reid, 1961.)

topography of the sea surface, although the central latitude of the gyre, the latitude of zero zonal velocity, seems to lie close to 30°S, some 5 degrees or so farther south than at the surface. Right against the coast of South America there is some evidence of the southward-flowing Peru-Chile Undercurrent, but its core is 100–200 m shallower than the 125-cl/t surface. Close to the coast of Australia, the low-salinity water appears to extend southward as a very narrow tongue, which is indicative of the East Australian Current;

it is a very irregular feature, however, and probably one that is not to be relied upon in detail as a definition of flow. It may be worth noting that although surface current charts based on ship set (e.g., Wyrtki, 1960) often show northeastward flow into the Tasman Sea from the West Wind Drift, there is no indication at all of such movement in Figure 2.

Figure 3 is also from Reid's monograph (1965), and it gives the salinity distribution on the surface of thermosteric anomaly equal to 80 cl/t (σ_t = 27.3). This surface ranges in

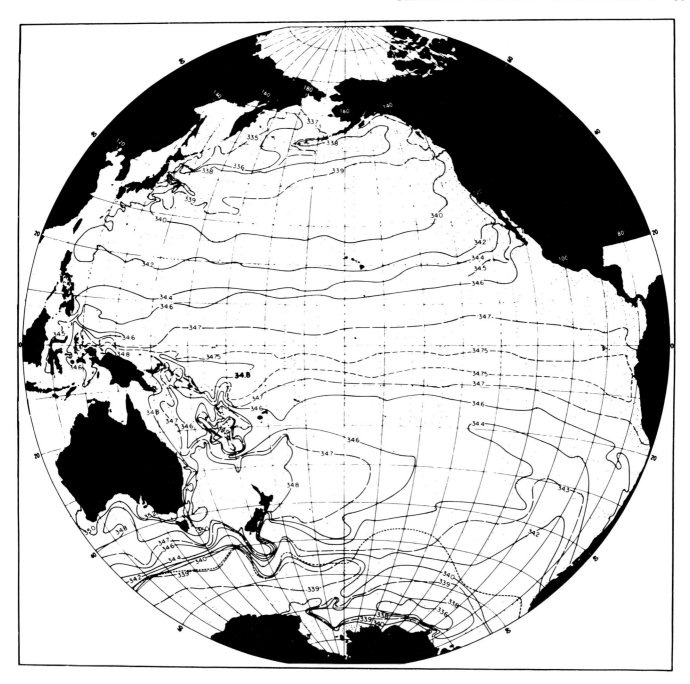

FIGURE 2   Distribution of salinity (⁰⁄₀₀) on the surface of thermosteric anomaly equal to 125 cl/t in the Pacific Ocean. South of the dashed line the quantity mapped is sea-surface salinity. (Reprinted with permission from Reid, 1965.)

depth from 800 to 1,200 m in the central South Pacific, and it intersects the sea surface between 60° and 70°S. Again, we see the broad northward flow in the eastern South Pacific, turning and flowing westward in latitudes of 20–25°S, slightly to the south of its core on the 125-cl/t surface. Neither the East Australian Current nor the Peru–Chile Undercurrent shows up very clearly here.

In summary, then, these three figures give a rather con-

sistent picture of the great subtropical gyre in the upper kilometer of the South Pacific. The West Wind Drift flows eastward in high latitudes, and part of it turns northward into the South Pacific; this northward flow is almost wholly confined to the eastern part of the ocean, and north of 25°S, it swings around to the west and becomes part of the South Equatorial Current. The gyre is completed by a narrow western boundary current, the East Australian Current,

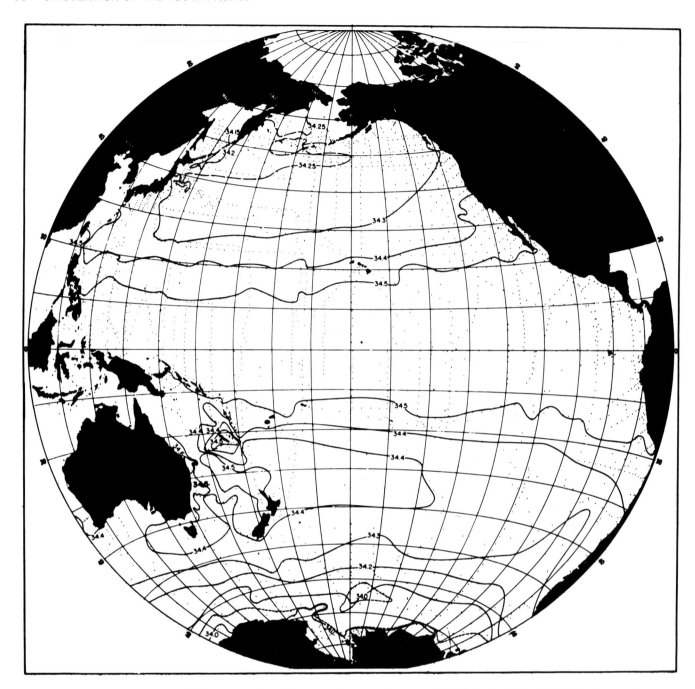

FIGURE 3 Distribution of salinity (⁰/₀₀) on the surface of thermosteric anomaly equal to 80 cl/t. South of the dashed line the quantity mapped is sea-surface salinity. (Reprinted with permission from Reid, 1965.)

flowing southward along the coast of Australia. It is in the upper kilometer that the effects of wind systems should be most evident, and I wish to suggest that this circulatory pattern is described remarkably well by the simple theory of Sverdrup transport; of all oceans, moreover, the South Pacific may be the one that is best described by the theory of wind-driven circulation. It remains to be seen, of course, how much of this apparent good agreement between theory

and nature is real and how much is only a reflection of our imperfect knowledge of the winds and currents.

Figure 4 is a calculation by Welander (1959) of the vertically integrated wind-driven volume transport in the World Ocean. Except at the western boundaries of the oceans, the transport picture given is the familiar Sverdrup transport, in which the meridional component of the vertically integrated velocity is calculated from the curl of the local average

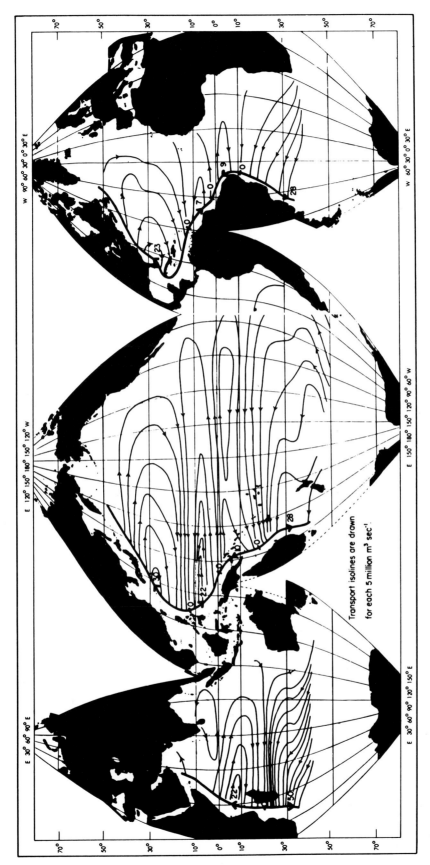

FIGURE 4 The Sverdrup transport and the western boundary currents in the oceans corresponding to the annual mean wind-stress field. The figures represent the volume transports (in millions of $m^3 sec^{-1}$) in different parts of the boundary currents. (Reprinted with permission from Welander, 1959.)

Transport isolines are drawn for each 5 million $m^3 sec^{-1}$

wind-stress vector. The zonal components are then derived by applying the equation of mass continuity to these calculated meridional components, with a boundary condition of zero zonal velocity at the eastern sides of the oceans. Such calculations do not, in general, lead to vanishing zonal velocities at the western boundaries as well; departures are permitted from the strict Sverdrup relation to the extent that narrow currents of unspecified higher order dynamics are introduced along the western boundaries in order to preserve continuity and to satisfy the western boundary condition.

Thus, the heavy lines in Figure 4 represent these western boundary currents, which patch up the incomplete field of Sverdrup transport, and the numbers printed beside them are their requisite volume transports expressed in millions of m^3 sec^{-1}. The lighter lines are the streamlines of the Sverdrup transport itself, each streamline representing a transport of 5×10^6 m^3 sec^{-1}. (The break in streamlines in the eastern Pacific is a result of binding of the original publication.)

The striking feature of the Sverdrup transport pattern in the South Pacific is that the northward flow is almost wholly confined to the eastern Pacific, quite in contrast to the patterns calculated for the other oceans, and very much in agreement with the actual flow pattern suggested by the distribution of properties in the upper kilometer. The central latitude of the calculated gyre lies 10–15° south of the corresponding feature indicated by the observed property distributions, but the theoretical central latitude is determined by the latitude of maximum wind-stress curl, which, as Hamon (1965) has pointed out, is poorly defined for the South Pacific because of the sparseness of data. This discrepancy may be removed with the accumulation of improved wind data, and it should probably not be taken too seriously yet.

If we think of the confinement of northward flow to the eastern Pacific as a simple effect of the local wind, then from a dynamical point of view, it would be improper to regard this flow as a *boundary* current, because its existence is not tied dynamically to any physical boundary. There are, of course, current phenomena of much smaller horizontal scale whose existence does seem to depend intimately on the physical boundary, and on descriptive grounds Wooster and Reid (1963) have found it convenient to combine these features with the much wider northward flow into an eastern boundary current system. In the construction of mathematical models of the South Pacific circulation, however, Gunther's earlier distinction (1936) between coastal currents and offshore oceanic currents will probably prove a more useful point of view.

One might note that in Figure 4 there is no indication of these coastal currents, such as the Peru-Chile Undercurrent. This omission results in part from the fact that Welander's calculations are of vertically integrated transport and in part

from the fact that these coastal currents are probably phenomena of higher order dynamics, perhaps related to the density stratification and not, in any case, included in this simple model.

Also missing from the picture is the South Equatorial Countercurrent reported by Reid (1959). Inasmuch as this feature is not more than two degrees wide, however, and the Sverdrup transports shown in Figure 4 are based on wind vectors averaged over five-degree squares, one would not expect to see such a feature here.

The theoretical intensity of the subtropical gyre is represented by the transport of the western boundary current introduced to complete it: 28×10^6 m^3 sec^{-1}. Hamon (1965) has listed 12 calculations of the volume transport of the East Australian Current, with reference to 1,300 m, and the average of these values is also 28×10^6 m^3 sec^{-1}. Such close agreement is certainly fortuitous, but it demonstrates clearly the potential of the simple Sverdrup transport theory to deduce correctly not only the overall pattern of the near-surface circulation, but also its approximate magnitude.

Welander's chart (Figure 4) shows, moreover, that this simple theory appears to give a more realistic circulation picture for the South Pacific than for any other ocean. One wonders why this is so, and I have no suggestions. To determine at the very least whether all this is really true, however, and not just chance agreement among sparse data, it is plain that more extensive wind observations are just as important as better coverage with hydrographic and current measurements.

From the pattern of the isolines in Figure 4, it is evident that Welander ignored the existence of New Zealand in his calculations, and I should like to suggest that New Zealand may have a rather profound effect on the course of the western boundary current. The whole New Zealand platform is strongly reminiscent of the partial meridional barriers modeled in laboratory experiments by Faller (1960) and discussed theoretically by Stommel and Arons (1960a) in their outline theory of the abyssal circulation. There is no difficulty, incidentally, in considering shallow wind-driven circulations in terms of the abyssal-circulation model: in the latter, horizontal flow is forced by vertical velocities at the base of the thermocline, while in the former, circulations are forced by vertical velocities at the base of the Ekman layer; the situations are quite analogous. Indeed, the formalism of Stommel and Arons (1960a) is much the same as that of Welander (1959): i.e., interior meridional motion is forced by an externally imposed divergence of the horizontal velocity field, normal velocity components are made to vanish at the eastern boundaries, and continuity and the western boundary condition are satisfied by adding narrow swift currents along the western boundaries. Eastern boundary currents are excluded as a means of patching up the strictly geostrophic flow, at least from the barotropic models, on the grounds that they would have to have anticyclonic absolute vorticity, which happens seldom if at all in nature. The flow regime defined by these rules has been realized in various forms by Faller (1960) in experiments with rotating basins.

Let us consider the implications of this model for the near-surface circulation in the vicinity of the Tasman Sea. Suppose that east of New Zealand the flow is eastward, as implied by the dynamic topography, rather than westward, as suggested by the badly defined latitude of maximum wind-stress curl. Not all of this eastward flow derives directly from the West Wind Drift, and part of it, in this dynamical scheme, must be fed from some sort of boundary current along the east coast of New Zealand. This current must, in turn, be connected to the boundary current along the east coast of Australia, and the only connection allowed here is a zonal jet because the meridional velocities are all specified by the wind-stress curl. The latitude of this jet will then be determined by the relative positions of New Zealand and Australia. The farthest south that the East Australian Current could conceivably flow would be the southern extremity of Tasmania (about 44°S) because there is no physical boundary any farther south. But New Zealand extends into higher latitudes than Australia (nearly to 50°S), if one considers the 200-m depth contour as its boundary. Thus, if the East Australian Current were to turn eastward as a zonal jet in the latitudes of Tasmania, it would be obstructed by South Island, and could get around it only as a boundary current along the Tasman Sea coast of New Zealand. But eastern boundary currents are excluded. The only way, in fact, for a boundary current along eastern Australia to join a western boundary current off New Zealand is through a zonal jet at the latitude of the northern extremity of New Zealand. This latitude is about 34°S, which is the same latitude, according to Hamon (1965), at which the East Australian Current turns abruptly eastward from the Australian coast.

Perhaps, then, it is the presence of New Zealand as a partial meridional barrier that is responsible for the separation of the East Australian Current from the coast. This explanation supposes the existence of both a fairly well-defined zonal jet across the Tasman Sea and some semblance of a southward western boundary current along the North Island of New Zealand, and I do not know to what extent either exists. The only chart of dynamic topography I have seen that covers the full Tasman Sea in any detail is one prepared by Wyrtki (1962); it shows a broad eastward flow from Australia in the appropriate latitudes but contains no suggestion of a southward boundary current off New Zealand. On the other hand, some station data (Garner and Ridgway, 1965) do indicate some sort of boundary current in this area, but detailed information is lacking. If it were possible to track the East Australian Current by some means, one might follow it across the Tasman Sea to the meridians of New Zealand to see what becomes of it and to test whether this simple hypothetical flow scheme makes any sense at all in the light of observations. It might be noted parenthetically that a *strictly* zonal current in the Tasman Sea should not be anticipated, because such large topographic features as Lord Howe Rise and Norfolk Ridge would necessarily force extensive meandering.

DEEP CIRCULATION

In the deep water, the circulation has been far less well delineated than the circulation at the surface. Until very recently, the major accomplishment from the limited data available was a description of the layering and origins of the water masses that make up the deep South Pacific. This type of description is conveniently done with reference to meridional sections; as an example, Reid's profile (1965) of the dissolved-oxygen concentration along 160°W is shown in Figure 5. The high-oxygen Antarctic Intermediate Water at depths of about 500–1,200 m, and the high-oxygen bottom water, at depths greater than 3,000–4,000 m, both clearly derive from the Antarctic, whereas the oxygen-minimum water in between seems to owe its characteristics to conditions in the tropical and North Pacific.

Although the *origins* of water masses are indicated well enough on a profile such as this, one cannot infer very much about the local *direction* of flow. At first glance, for example, the bottom water appears to be seeping northward here, but, in fact, it may have moved northward from its place of origin at some quite different longitude, and here at 160°W it may be moving eastward or westward, or even southward back toward the Antarctic. Indeed, the only dynamical theory of deep-water circulation, that of Stommel (1958) and Stommel and Arons (1960b), requires that the meridional component of the deep flow in the interior of the South Pacific be directed *toward* the pole and that all the northward flow from the Circumpolar Current be confined to a narrow current along the western boundary. To learn about the *pattern* of deep flow, therefore, one needs fairly detailed observational coverage in the zonal direction as well as broad meridional coverage.

Wooster and Volkmann (1960) studied the distribution of properties at a depth of 5 km, but because of the scarcity of data, they were unable to find evidence for zonal differences in the South Pacific. On the other hand, as noted in his classic paper on the deep Pacific, Knauss (1962) did see some marginal indication at shallower levels of a westward concentration of the northward flow. Figure 6 is Knauss's picture of the temperature distribution at a depth of 4,500 m; the colder water moving northward does seem to be confined to the western half of the central South Pacific basin, but because of the sparseness of the data, it is difficult to say much more about the flow. The temperature field at 3,500 m is qualitatively similar to that at 4,500 m (Knauss,

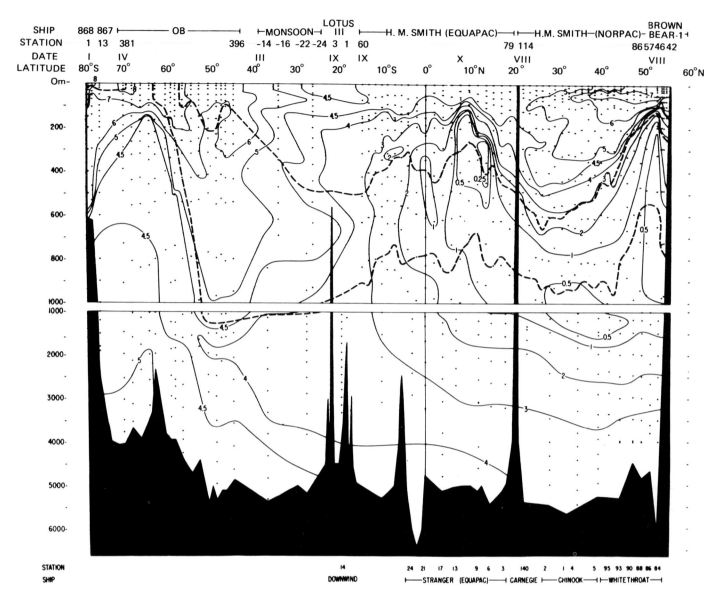

FIGURE 5   Dissolved-oxygen concentration (ml/l) along approximately 160°W from Antarctica to Alaska. (Reprinted with permission from Reid, 1965.)

1962). It might be noted in passing that the western boundary for the deep South Pacific is New Zealand and the Tonga-Kermadec Ridge, rather than Australia, because the Tasman Sea is closed off to the north at depths greater than 3,000 m.

In part to obtain more detailed deep-water observations, two transpacific hydrographic sections were made in 1967 on cruises of the USNS *Eltanin*, one at 43°S and the other at 28°S. With respect to the hydrographic stations, these two cruises have been called the Scorpio Expedition.* The most notable accomplishment of the expedition was a clear

delineation of the theoretically anticipated deep western boundary current. This current has already been reported briefly by Reid *et al.* (1968). Since the existence of this current has far-reaching implications in terms of the relevance of the abyssal-circulation theory to the deep South Pacific, I should like to describe the profiles of temperature and salinity across it.

The positions of the *Eltanin* hydrographic stations are shown in Figure 7. We are concerned at this point only with stations in the vicinity of the deep western boundary, that is, stations 23–45 at 43°S and stations 134–156 at 28°S.

Figure 8 is the western boundary portion of the temperature profile at 43°S, from the Chatham Rise, on the left, to the base of the East Pacific Rise, on the right. Plainly,

*Reports in preparation by J. L. Reid, Jr., H. Stommel, E. D. Stroup, and B. A. Warren.

near the boundary, the deep isotherms slope downward to the east, increasing in depth by about 500 m. The slope is very gradual, however, and occupies a band about 900–1,000 km wide. With a surface of zero horizontal velocity at some mid-depth, this slope is consistent geostrophically with northward flow of the deep water.

Figure 9 is the corresponding salinity profile, contoured above 2,000 m at intervals of 0.1‰ and in deeper water at intervals of 0.01‰. The striking feature in the deep water is the layer of maximum salinity at depths of 3–4 km; it is most strongly developed near the western boundary at about the same depth where the sloping isotherms are seen in Figure 8. This high-salinity water must have come from the Antarctic, because there is no water so saline in the Tropical or North Pacific; its confinement to the boundary current proves, therefore, that the deep current is indeed

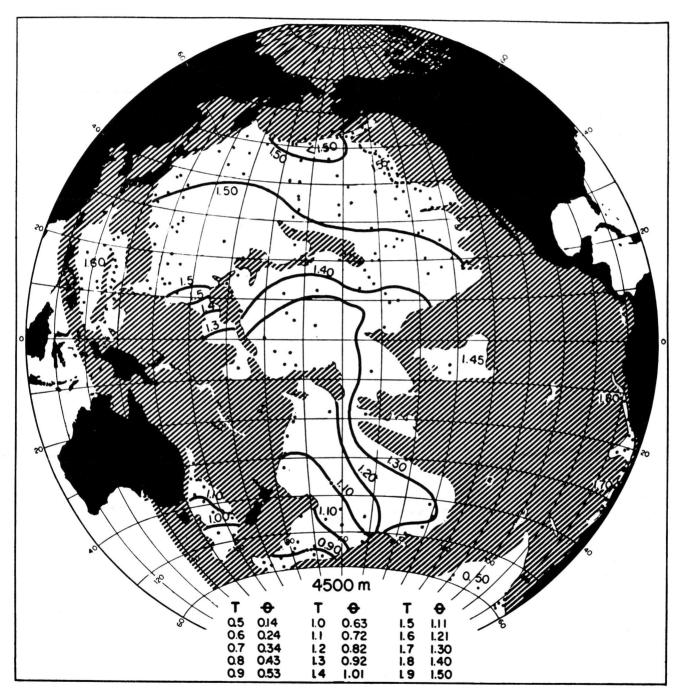

T	θ	T	θ	T	θ
0.5	0.14	1.0	0.63	1.5	1.11
0.6	0.24	1.1	0.72	1.6	1.21
0.7	0.34	1.2	0.82	1.7	1.30
0.8	0.43	1.3	0.92	1.8	1.40
0.9	0.53	1.4	1.01	1.9	1.50

FIGURE 6 Distribution of temperature *in situ* at a depth of 4,500 m in the Pacific Ocean. The corresponding potential temperature is indicated in the table. (Reprinted with permission from Knauss, 1962.)

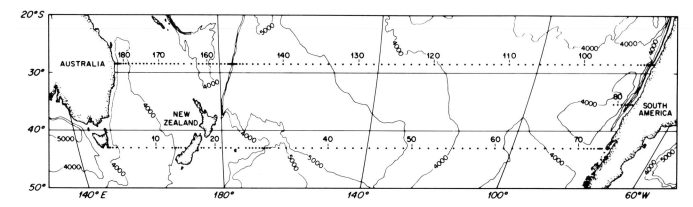

FIGURE 7   Positions of hydrographic stations occupied during Cruises 28 and 29 of USNS *Eltanin* (Scorpio Expedition).

flowing northward, as predicted by the abyssal-circulation theory (Stommel, 1958). The existence of the salinity maximum is due to the last traces of North Atlantic Deep Water carried eastward around Antarctica by the Circumpolar Current. One cannot exaggerate the need to use only the most sensitive conductivity methods in measuring deep-water salinities in the Pacific, because the total salinity difference across the boundary current is only about 0.03‰, too small a difference to be determined by the Knudsen titration method.

At 28°S, conditions appear similar to those at 43°S, although there are complexities that make detailed interpretation more difficult. Figure 10 is again the western boundary portion of the temperature profile. The Tonga-Kermadec Ridge is on the left, the foot of the East Pacific Rise is on the right, and the Tonga-Kermadec Trench is continued in the inset. As at 43°S, the deep isotherms descend toward the east, with the greatest change in level, about 500 m, near the 4-km depth. Within the broad region of sloping isotherms (about 900 km wide), there is a narrow zone close to the Tonga-Kermadec Ridge, where the isotherm slope is increased and then reverses across the trench. It is not yet clear whether this is a feature intrinsic to the large-scale circulation, or, perhaps, a local effect of the Tonga-Kermadec Trench on geostrophic motion.

The associated salinity profile is given in Figure 11. Once more, we see the deep layer of maximum salinity most strongly developed in the region where isotherms were observed to slope downward from the western boundary. As one would expect, the salinities here are somewhat lower than 43°S: The band of water with salinity greater than 34.72‰ is about three fourths as thick here as it is at 43°S, and the water of salinity greater than 34.73‰ is wholly confined to the relatively intense portion of the current lying between the Tonga-Kermadec Trench and Ridge.

The volume transport of the deep boundary current may be calculated from the station data, with an arbitrary assumption as to the depth of a surface of zero horizontal

velocity. If we choose 2,500 m, lying just above the salinity-maximum layer, as the reference level, the geostrophic transport of the current below this depth is $13\frac{1}{2} \times 10^6$ m^3 sec^{-1} at 43°S and nearly 17×10^6 m^3 sec^{-1} at 28°S. Both figures are fairly large for abyssal currents, though only about half as large as originally conjectured by Stommel and Arons (1960b).

The bare existence of the boundary current does not by any means prove in its entirety the theoretical deep-circulation scheme, i.e., a broad cyclonic gyre with an upward flux of water into the thermocline, but it does provide strong supporting evidence for it. It suggests, also, that this theory will be a fruitful starting point from which to plan deep-observation programs.

I have described this recent discovery in considerable detail because some have supposed that the deep South Pacific is a truly homogeneous water mass, lacking physical contrasts by which to infer its circulation. Plainly this is not so: With sufficiently dense observations of modern accuracy, significant differences in temperature–salinity characteristics are readily apparent, and there certainly does exist a circulatory pattern detectable by standard hydrographic stations.

Equally clear is the need for further extensive survey work in this area. Two sections are insufficient to map the deep boundary current; several more traverses are desirable for this purpose. Also, it would be useful to know how much farther north the current flows, or at least can be detected. Between 10° and 20°S, for example, the central basin of the deep South Pacific becomes severely constricted, and it is important to determine what fraction of the boundary current, if any, flows through this passage into the North Pacific. The extreme intensification of the current on the flank of the Tonga-Kermadec Ridge is a puzzling feature, and it would be informative to see how closely it correlates geographically with the Tonga-Kermadec Trench. All these matters could be resolved through carefully planned and executed hydrographic surveys. Deep current measurements, in addition, would remove the discomforting need to assume arbitrary motionless surfaces in estimating volume transports.

On the dynamical side, the circulatory pattern implied by the Scorpio data does not agree in detail with the simple model hypothesized by Stommel and Arons (1960b), and one would like to see a more refined construction based on more realistic topography, perhaps even taking some account of stratification effects. Furthermore, the width of the deep current, although small compared with the total

breadth of the South Pacific, is still considerably greater than that envisioned in the usual boundary layer analysis; moreover, it is about twice as great as the width of the corresponding boundary current in the South Atlantic (Wüst, 1955). These are both mystifying points and worthy subjects for theoretical inquiry.

There is one additional feature of the deep water in the South Pacific from which information about the circulation could probably be inferred. Knauss (1962) has pointed out

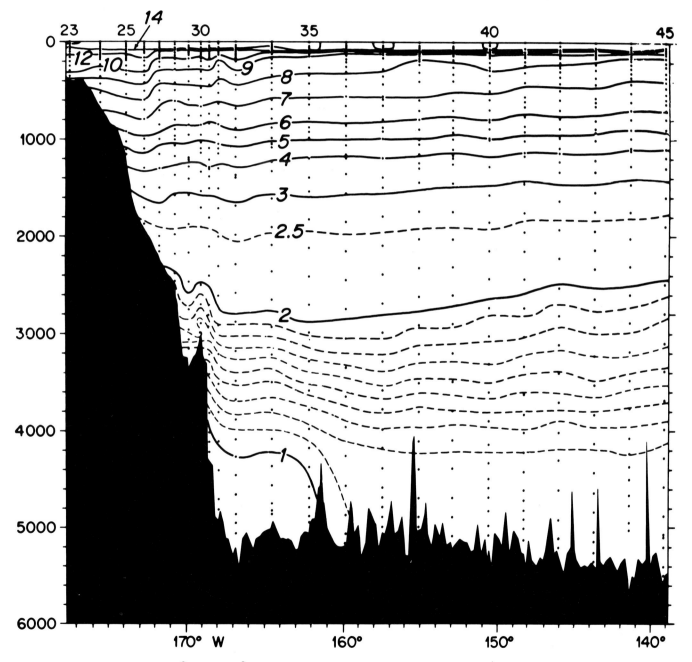

FIGURE 8 Profile of temperature (°C) along 43°S near western boundary of deep South Pacific. Scorpio Stations 23–45, April 1–15, 1967. Depth given in meters.

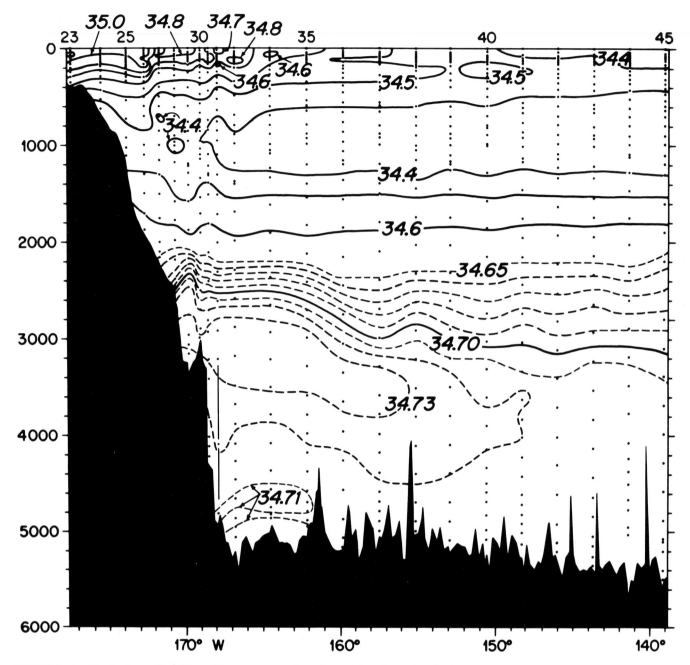

FIGURE 9   Profile of salinity (‰) along 43°S near western boundary of deep South Pacific. Scorpio Stations 23–45, April 1–15, 1967. Vertical line beneath Station 32 indicates data taken from STD trace. Depth given in meters.

that deep-water temperatures appear to be slightly high over the crest of the East Pacific Rise, near a zone of anomalously large heat flow from the ocean floor (von Herzen, 1959). The local situation is complicated; the data are too few, and our theoretical understanding is inadequate to use this heat-flow phenomenon to advantage in describing the deep circulation, but since it offers promise as a future source of significant information, it is a subject very much worth investigating.

Figure 12 is a compilation by Bullard (1963) of heat-flow measurements in the eastern Pacific. The shading indicates values greater than 2μ cal cm^2 sec^{-1}, the hatching indicates values less than 1μ cal cm^2 sec^{-1}. The point of interest here is the extraordinarily high values, reaching a maximum of 8μ cal cm^2 sec^{-1}, on the crest of the East Pacific Rise. The Scorpio sections previously discussed for the western South Pacific cross the East Pacific Rise well south of the heat-flow maximum, at 28° and 43°S. Nevertheless, in the vicinity of 100°–110°W, effects of the high heat flow can be seen fairly clearly in the distribution of potential temperature and salinity at 28°S. (Some effect is apparent at 43°S, but it is not strong enough to warrant discussion here.)

Figure 13 shows profiles of salinity and potential temperature at depths greater than 2,000 m at 28°S. The profile is limited to the East Pacific Rise, and is constructed from *Eltanin* stations 96–125 (see Figure 7 for station positions). The interpretation of these profiles is somewhat ambiguous

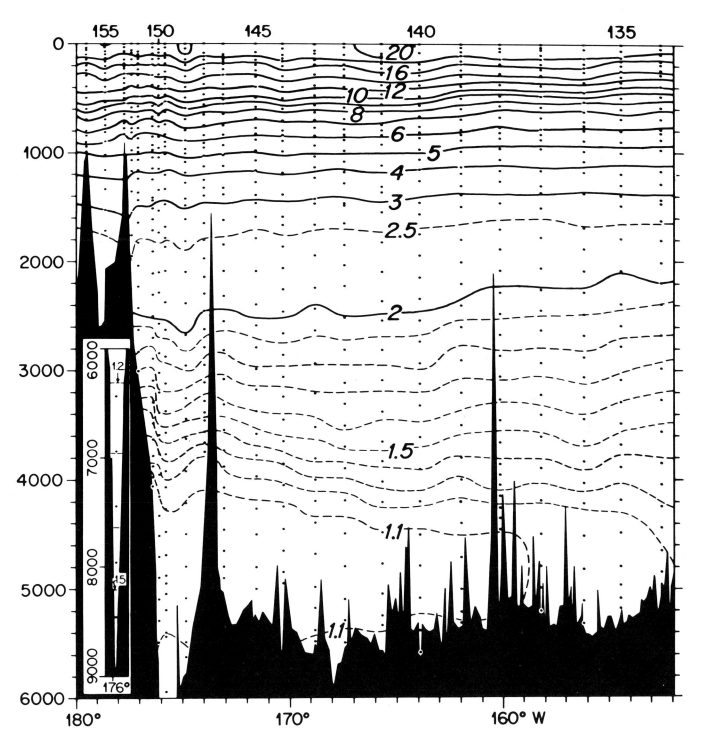

FIGURE 10   Profile of temperature (°C) along 28°S near western boundary of deep South Pacific. Scorpio Stations 134–156, July 3–18, 1967. Depth given in meters. Tonga-Kermadec Trench continued in inset.

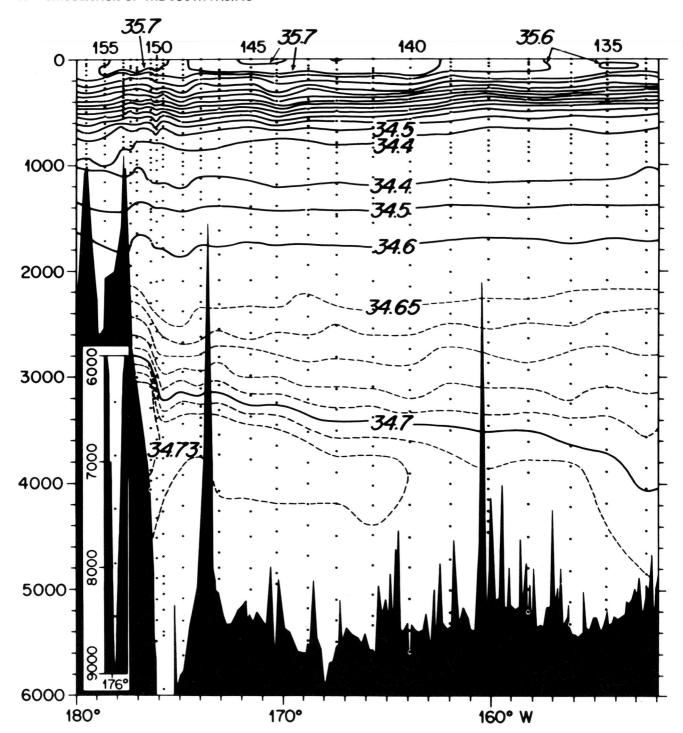

FIGURE 11 Profile of salinity (‰) along 28°S near western boundary of deep South Pacific. Scorpio Stations 134–156, July 3–18, 1967. Depth given in meters. Tonga-Kermadec Trench continued in inset.

because of a regional variation in deep temperature–salinity characteristics: It seems that, at a given salinity, deep water on the eastern side of the East Pacific Rise is 0.1–0.2°C warmer than deep water on the western side. Despite the confusion introduced by this regional difference, the potential isotherms show a pronounced downward dip in the sev-

eral hundred meters of water overlying the crest of the East Pacific Rise at Station 110. This temperature anomaly of 0.05–0.10°C lies just within the longitudes of high heat flow, shown in Figure 12, and must be attributed to this heat source.

The mechanism of heat transfer in this water is not at all

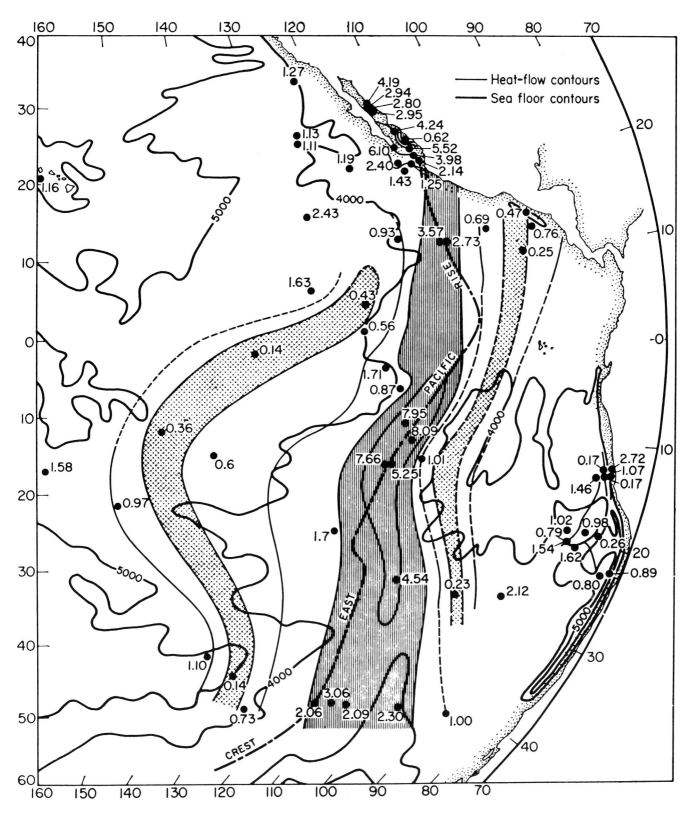

FIGURE 12   Heat-flow measurements in the eastern Pacific. (Reprinted with permission from Bullard, 1963.)

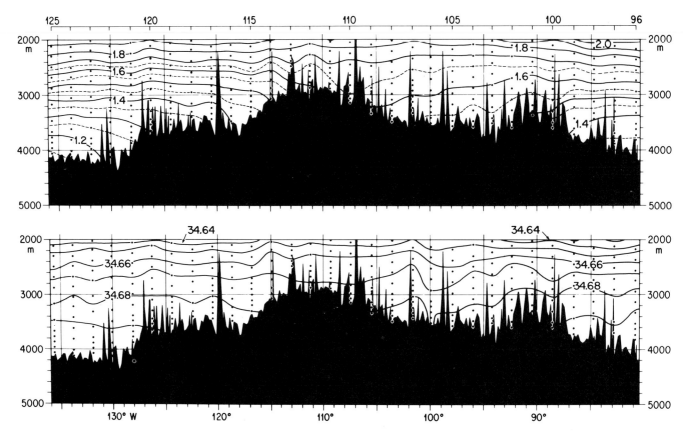

FIGURE 13 Profiles of potential temperature (°C, top) and salinity (‰, bottom) at depths greater than 2,000 m across the East Pacific Rise along 28°S. Scorpio Stations 96–125, June 8–28, 1967. Depth given in meters.

clear, however. If the water were actively convecting in response to the heat source, one would expect to see the salinity field nearly as well mixed as the temperature field, with little distortion of the temperature–salinity relation. On the other hand, if the heat transfer were by eddy conduction, only the temperature field should be affected; the isohalines should be undisturbed, and the temperature–salinity relation near the bottom should be markedly distorted. On the western flank of the rise (Stations 115–123), the potential isotherms do dip downward abruptly as they approach the rise, with no corresponding distortion of the isohalines, as if conductive heat transfer were taking place there, but on the eastern flank of the rise (Stations 100–109), the deep isotherms and the isohalines are essentially parallel as they approach the crest of the rise, as if convective mixing were occurring there. The whole situation is complicated by the regional temperature–salinity difference, and it seems impossible to determine exactly what is happening to the water at the crest of the rise from the limited data now available.

High values of heat flow appear to be a common feature of midocean ridges throughout the World Ocean (Bullard, 1963), but to date it is only on the East Pacific Rise that this abnormal heat flow has been shown to modify the deep water. This fact alone makes the deep eastern Pacific an area of special interest to physical oceanography, and it will be a particularly challenging task to disentangle the various complexities of this area and find out what is really happening to the bottom water there.

This problem is relevant to deep circulation studies because the magnitude and pattern of the temperature anomalies must be strongly affected by the distribution of currents; the temperature anomalies could, therefore, serve as a tool for estimating the direction and intensity of the deep-water motion near the East Pacific Rise. At present, however, observational coverage of both heat flow and water characteristics is too sparse, and our understanding of the effect of nonuniform heat flow on stratified fluids is too imperfect to put this tool to any use.

CONCLUSION

I have not attempted in these general comments to present a comprehensive description of the overall circulation in the South Pacific, because we really do not know it very well. I have tried, rather, to focus attention on specific problems within the general circulation—problems with which substantial progress can be made with present facilities. My selection of problems is admittedly biased: They are problems that intrigue me, both as being amenable to solution and as having a wide bearing on the circulation of the South Pacific as a whole.

From time to time, one hears uninformed remarks to the effect that the era of the hydrographic survey is over. Most people at work in the field know that this is nonsense, and it should be clear that the South Pacific, especially, is an ocean where for some time to come the major contributions to physical-oceanographic knowledge will come from carefully designed hydrographic surveys making full use of the most accurate modern instrumentation. I have tried to suggest—or at least imply—some ways in which detailed hydrographic work might be used to good advantage in investigating the circulation of this ocean.

This paper is Contribution No. 2460 from the Woods Hole Oceanographic Institution, Woods Hole, Massachusetts.

REFERENCES

Bullard, E. C. (1963) The flow of heat through the floor of the ocean, pp. 218–232 in: *The Sea*, editor M. N. Hill, vol. 3, Ch. 11, 218–232. Wiley-Interscience, New York, London, Sydney.

Faller, A. J. (1960) Further examples of stationary planetary flow patterns in bounded basins. *Tellus*, *12* (2), 159–171.

Findlay, A. G. (1853) Oceanic currents, and their connection with the proposed Central America canals. *J. R. Geogr. Soc.*, *23*, 217–240.

Garner, D. M., and N. M. Ridgway (1965) Hydrology of New Zealand off-shore waters. *Mem. N.Z. Oceanogr. Inst.*, *12*, 62 pp.

Gunther, E. R. (1936) A report on oceanographical investigations in the Peru Coastal Current. *Discovery Rep.*, *13*, 107–276.

Hamon, B. V. (1965) The East Australian Current, 1960–1964. *Deep-Sea Res.*, *12* (6), 899–922.

Knauss, J. A. (1962) On some aspects of the deep circulation of the Pacific. *J. Geophys. Res.*, *67* (10), 3943–3954.

Reid, J. L., Jr. (1959) Evidence of a South Equatorial Countercurrent in the Pacific Ocean. *Nature, 184*, 209–210.

Reid, J. L., Jr. (1961) On the geostrophic flow at the surface of the Pacific Ocean with respect to the 1,000 decibar surface. *Tellus, 13* (4), 489–502.

Reid, J. L., Jr. (1965) Intermediate waters of the Pacific Ocean. *Johns Hopk. Oceanogr. Stud.*, *2*, Johns Hopkins Press, Baltimore, 85 pp.

Reid, J. L., Jr., H. Stommel, E. D. Stroup, and B. A. Warren (1968) Detection of a deep boundary current in the western South Pacific. *Nature*, vol. *217*, 937.

Stommel, H. (1958) The abyssal circulation. *Deep-Sea Res.*, *5* (1), 80–82.

Stommel, H., and A. B. Arons (1960a) On the abyssal circulation of the world ocean: I. Stationary planetary flow patterns on a sphere. *Deep-Sea Res.*, *6* (2), 140–154.

Stommel, H., and A. B. Arons (1960b) On the abyssal circulation of the world ocean: II. An idealized model of the circulation pattern and amplitude in oceanic basins. *Deep-Sea Res.*, *6* (3), 217–233.

Von Herzen, R. (1959) Heat-flow values from the southeastern Pacific. *Nature, 183*, 882–883.

Welander, P. (1959) On the vertically integrated mass transport in the oceans, pp. 95–101 in: *The Atmosphere and the Sea in Motion*, editor B. Bolin, Rockefeller Institute Press and Oxford University Press, New York.

Wooster, W. S., and J. L. Reid, Jr. (1963) Eastern boundary currents, pp. 253–280 in: *The Sea*, editor M. Hill, vol. 2, Ch. 11, Wiley-Interscience, New York, London, Sydney.

Wooster, W. S., and G. H. Volkmann (1960) Indications of deep Pacific circulation from the distribution of properties at five kilometers. *J. Geophys. Res.*, *65* (4), 1239–1249.

Wüst, G. (1955) Stromgeschwindigkeiten im Tiefen–und Bodenwasser des Atlantischen Ozeans auf Grund dynamischer Berechnung der *Meteor*–Profile der Deutschen Atlantischen Expedition 1925-27. *Pap. Mar. Biol. and Oceanogr., Deep-Sea Res., Suppl., 3*, 373–397.

Wyrtki, K. (1960) Surface circulation in the Coral and Tasman seas. *CSIRO Div. Fish. Oceanogr. Tech. Pap., 8.*

Wyrtki, K. (1962) Geopotential topographies and associated circulation in the western South Pacific Ocean. *Aust. J. Mar. Freshwater Res., 13*, 89–105.

Bruce V. Hamon

DIVISION OF FISHERIES AND
OCEANOGRAPHY, C.S.I.R.O.,
CRONULLA, NEW SOUTH WALES

WESTERN BOUNDARY CURRENTS IN THE SOUTH PACIFIC

INTRODUCTION

The East Australian Current is the only feature in the South Pacific large enough to be classed as a western boundary current. South-flowing currents off the north island of New Zealand, the East Auckland Current and the East Cape Current (Garner and Ridgway, 1965), appear to be well defined, but are relatively weak. Off the south island of New Zealand, the near-shore currents are to the north (the Canterbury Current). No volume transports for these currents have been published, nor is there any evidence of their seasonal variation.

Study of the East Australian Current began in 1950. Physical results up to the end of 1964 have been published (Hamon, 1961, 1965; Wyrtki, 1962b). Water masses and circulation in the area have been studied by Rochford (1957, 1958, 1959), Wyrtki (1961, 1962a), and Garner (1959). A recent review of work in the area is given by Highley (1967), and the hydrology of the southwest Pacific Ocean is reviewed by Rotschi and Lemasson (1967). Wyrtki (1960) gives monthly diagrams of surface currents and surface streamlines for the area, derived from current atlases. Since 1964, there have been eight cruises in this area; their results are to be published (Boland and Aamon, in press).

The present paper reviews the main features of the East Australian Current, with emphasis on physics rather than water mass analysis; suggestions for future work will also be discussed.

MAIN FEATURES OF THE EAST AUSTRALIAN CURRENT

Area and Bathymetry

As shown on navigation charts, the East Australian Current is a strong, narrow, south-flowing current near the edge of the continental shelf between 27° and 37°S. From evidence of oceanographic cruises, however, it is clear that we should think of a complex current system instead of a simple south-flowing current. This system extends at least 300 miles offshore. Figure 1 is a simplified bathymetric chart of the Tasman Sea and part of the Coral Sea. The area in which the East Australian Current system has already been investigated on oceanographic cruises is outlined, but it is not suggested that this outline be regarded as a natural boundary of the system.

The main features of the bathymetry are the restricted channels between Australia and New Caledonia (~ 20°S) and the Lord Howe Rise. A chain of seamounts appears along 155–156°E; the most northerly one might influence the dynamics of the current, but the others probably do not.

Surface Currents and Eddies

Surface currents are complex, variable, and strong (up to 4.0 knots). Figure 2 shows the results (0/1,300-decibar topography) obtained on four cruises at 2-month intervals, September 1963–March 1964. The growth, or entrance from the north, of an anticyclonic eddy off Sydney and its subsequent movement farther south can be clearly seen. Note that on Cruise G1/64 (Figure 2c) there was no strong south-flowing current near the shelf edge at 30°S. The same condition was noted in November 1965.

Large anticyclonic eddies (~ 250 km in diameter) appear to be frequent east and south of Sydney. Eddies were observed in November 1960, January 1961, March 1961, March 1965, and November 1965 (closure of the eddy on the seaward side of the eddy of November 1965 was not proven). The sequence shown in Figure 2 suggests southward movement of the eddies. This suggestion is supported by hydrological evidence, which shows that water of Coral Sea origin may be found as far south as eastern Bass Strait

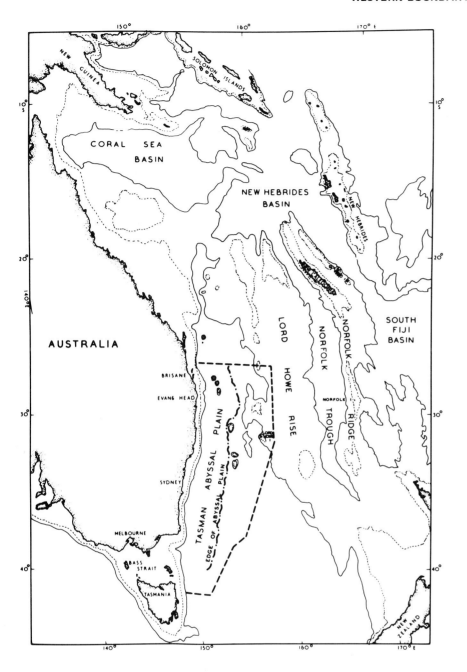

FIGURE 1 Simplified bathymetry of the Tasman and Coral seas. The area in which most cruises to study the East Australian Current have taken place is enclosed by the dashed line. The thin dotted and continuous lines are, respectively, the 1,000-m and 3,000-m contours.

(40°S). Since no cruise has yet shown a continuous current along the coast at any one time (this is a current atlas fiction), it seems likely that the water is carried south by eddies.

The strong, variable circulation pattern is by no means confined to summer months, when, according to current atlas evidence, the East Australian Current is best developed. As an example, Figure 3 shows the pattern in July 1965. A strongly developed pattern was also observed in August 1964. Note, in Figure 3, the sharp bend away from the coast and back again at about 32°30'S.

The Formation of Eddies

It seems very likely that the frequent anticyclonic eddies are formed when the "main" current bulges toward the south and becomes unstable, causing the bulge to separate as an eddy. It is difficult to understand how such large features could form by any other mechanism. Cruise results that appear to represent the early stages of formation of an eddy are shown in Figure 4. An eddy with its center at 36°S was present 6 months later, but there is, of course, no evidence of continuity over this period.

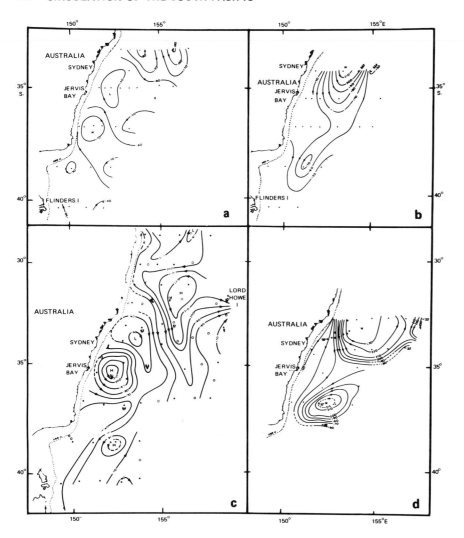

FIGURE 2 Surface dynamic topographies relative to 1,300 decibars, for four cruises at 2-month intervals. (a) G4/63 (September 9–18, 1963). (b) G5/63 (November 7–15, 1963. (c) G1/64 (January 13, 1963–February 6, 1964). (d) G3/64 (March 18–25, 1964). (Reprinted with permission from Hamon, 1965.)

The above hypothesis concerning eddy formation seems reasonable, except for the nature of the main current. As shown in Figure 2 (Cruise G1/64), and confirmed by the findings of other cruises, there is no simply structured south-flowing near-shore boundary current that then turns east in such a way as to enclose a large area of relatively motionless warm water: i.e., there is no local equivalent of the Sargasso Sea.

Subsurface Currents and Structure

Geostrophic currents relative to 1,300 decibars decrease to half the surface value at a depth of about 250 m and then decrease gradually to zero at the reference depth. The computed current profiles are smooth and uniform in shape.

The vertical temperature structure is of the usual shape, except for the fairly frequent occurrence of very deep (> 200 m) and very shallow (< 20 m) mixed layers. These layers are believed to result from convergence and divergence associated with the flow pattern, rather than from local wind or seasonal effects (Hamon, 1968a). (An unusual vertical temperature distribution, with an isothermal and isohaline layer between depths of 270 and 500 m, was found at a station inside an anticyclonic eddy in November 1960, but similar structure was not found in later eddies.)

Sampling was restricted to depths above 1,500 m on most cruises. At 1,300 m, the temperature distribution is remarkably similar to the surface dynamic topography. The temperature range at 1,300 m observed on all cruises was 3.2° to 4.8° C. The similarity between temperature distribution at 1,300 m and surface dynamic topography suggests that the current structure extends much deeper than the assumed reference level of 1,300 m (see the following section).

Direct Measurements of Current

Eight direct measurements of current have been made using Swallow floats. Figure 5 shows the results for two floats in one area, fitted in the usual way to geostrophic currents from pairs of stations at the corners of a square surrounding the pinger station. The pingers were located below a strong surface current (Cruise Dm2/67; 33°15′S, 153°36′E; April

1967). The most interesting feature of Figure 5 is that the 1,200-m pinger moved in the same direction as the surface current, tending to confirm the deeper current structure suggested above. The 3,200-m pinger moved slowly (2 cm sec^{-1}) to the east at right angles to the surface current.

On a later cruise (Dm4/67, September 1967) Swallow-float measurements were again made below a moderate surface current (approximately 85 cm sec^{-1} toward the southeast). Three floats were tracked, at depths of 1,300, 2,300, and 3,400 m, respectively. The main results are shown in Figure 6, again fitted to geostrophic currents from a network of four stations. Figure 6 indicates that at least the east–west component of the current extends to a depth of 3,500 m or deeper.

Volume Transports

Above a depth of 1,300 m, volume transports are in the range 12–43 $\times$ 10^6 m^3 sec^{-1}. If the direct measurements of current shown in Figures 5 and 6 are typical, the maximum surface-to-bottom transport would be much greater than 43 $\times$ 10^6 m^3 sec^{-1}. On cruise Dm4/67, a transport to the south of 57 $\times$ 10^6 m^3 sec^{-1}, relative to the 3,500-m level, was estimated.

Short-term Variability

Evidence of appreciable short-term variability is accumulating, but more time-series observations are needed.

Geomagnetic electro-kinetograph (GEK) measurements were made continuously in one area while pinger tracking was conducted. For the area to which Figure 5 applies, the surface current increased from 1.8 to 3.0 knots in 1.5 days, followed by a decline at a slower rate over the next few days. (The local acceleration corresponding to this increase is only 5 percent of the Coriolis term.) Similar fairly rapid changes in surface current were observed during cruise Dm4/67.

Time Scales of Motion Deduced from Sea Levels

Monthly mean sea levels at Lord Howe Island have a range of 67 cm as a result of changes in dynamic height near the island. Recent spectrum analyses (Hamon, 1968b) show an interesting broad peak in the sea-level spectrum at Lord Howe Island for periods between 50 and about 1,000 days, the maximum being at about 170 days. This peak is believed to provide some measure of the time scales of movement of baroclinic circulation patterns near the island, but it is not clear at present how such a measure can be interpreted.

In the frequency range of the spectral peak, the sea-level spectrum at Sydney is much lower than at Lord Howe Island and does not show a maximum (except for the usual annual and semiannual terms). This is understandable, since the mechanism operating at an island, which is sometimes on the left and sometimes on the right of a strong current, obviously cannot operate at a coastal station.

For periods between 2 and 7 weeks, the sea-level spectra at both Sydney and Lord Howe Island are larger than one might expect, and too large to be caused by atmospheric pressure effects. It is suggested that sea-level changes in this part of the spectrum might still be associated with the strong

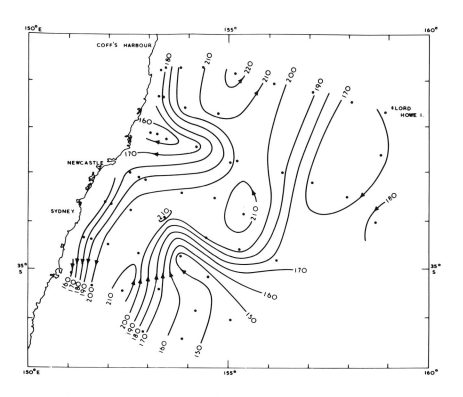

FIGURE 3 Surface dynamic topography relative to 1,300 decibars, for a winter cruise (G6/65, July 12–25, 1965).

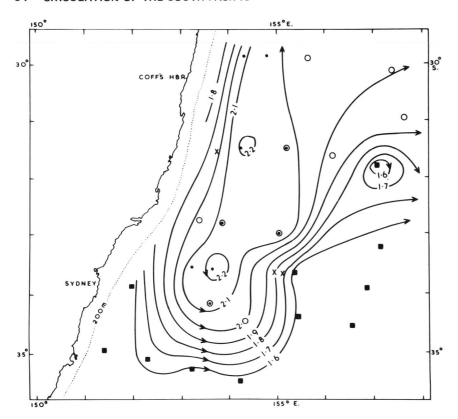

FIGURE 4 Surface dynamic topography relative to 1,300 decibars, for cruise G6/64 (September 16–27, 1964). (Reprinted with permission from Hamon, 1965.)

circulation in the area, but with a different mechanism (perhaps barotropic) that could be almost equally effective at coastal and island stations.

Sea-level differences over relatively short distances on the east Australian coast (Sydney-Port Kembla, 40 miles apart) show large variations (about 10 cm) that might be associated with the offshore circulation, but more theoretical work is needed before these differences can be interpreted. There is also some evidence that large dynamic height changes sometimes occur near the shelf edge but do not appear in mean sea levels at nearby tide stations (Hamon, 1969).

Horizontal Scales

Most information on horizontal scales comes from GEK results. These show that the total width of the current ("main" current, or half of an eddy) is as much as 150 km, with the stronger flowing "core" (> 50 cm sec^{-1}, perhaps) 40 to 80 km wide. Average horizontal shear ($\partial u/\partial y$) of the surface velocity is 9×10^{-5} sec^{-1} on the right-hand side of the current and 2.6×10^{-5} sec^{-1} on the left-hand side, facing downstream. Some fine structure is superimposed, with scales of the order 5–10 km (see Hamon, 1965, Figure 13).

Near-shore Measurements

1. A section 45 miles seaward from Evans Head (29°S) has shown, at weekly intervals, rapid changes in hydrological conditions and current strengths. There is slight evidence of a periodicity of about 6 weeks in the temperature structure over the continental shelf.

2. Analyses of surface currents from ships' navigation logs between Cape Moreton (27°S) and Sugarloaf Point (32°S) show reasonable persistence of current anomalies (i.e., departures from the mean current) over distances of about 100 km, but persistence is poor over greater distances (Hamon and Kerr, 1968). These results are for currents approximately at the 100-fathom depth. From the same data, time persistence at fixed latitudes was estimated to be 15–20 days, with slight evidence of a periodicity of about 70 days. These estimates are very imprecise because the data are imprecise (east–west variation of ships' tracks is probably the worst feature), but the results do tend to confirm the time-and-space scales found by other means.

Hydrology

No attempt is made to review the hydrology of this region in detail, but the following points may be of interest with respect to dynamics.

In the region of the East Australian Current (outlined in Figure 1) the temperature-salinity diagram is usually simple. A slight subsurface salinity maximum (depth < 100 m) appears only in the northern part of the area. It is caused by Subtropical Lower Water (Wyrtki, 1962b), which in turn is derived from Subtropical Surface Water formed in the South Pacific to the east of New Zealand.

A feature of the area is the constancy of the salinity

minimum of the Antarctic Intermediate Water; the mean value is 34.456‰, and most values lie within ±0.005‰ of this. Wyrtki (1962b) has suggested, on the basis of core-layer analysis, that Antarctic Intermediate Water enters the region from the east, between 10° and 30°S, rather than by the alternative, more direct route from the south.

At greater depths (~ 3,000-m-deep salinity maximum), again on the basis of core-layer analysis, Wyrtki (1961) deduced a northward movement of water in the East Australian basin.

To a good approximation, there is no change in shape of the temperature-salinity (T-S) curve across a current in this area. The only change is that particular given combinations of temperature and salinity are found at different depths.

Continuous records of salinity and temperature as functions of depth in this area rarely show measurable medium-scale structure below a depth of 300 m, but at some stations this structure extends to about 600 m. Figures 7 and 8 are T-S diagrams from TSD recorder results for two stations, which show exceptional medium-scale structure. The salinity maximum at 520 m in Figure 7 is believed to have been caused by water of Bass Strait origin.

DISCUSSION

The 16 major cruises that have been carried out in the region of the East Australian Current permit an adequate description of the main circulation features, but we do not yet have an understanding of these main features.

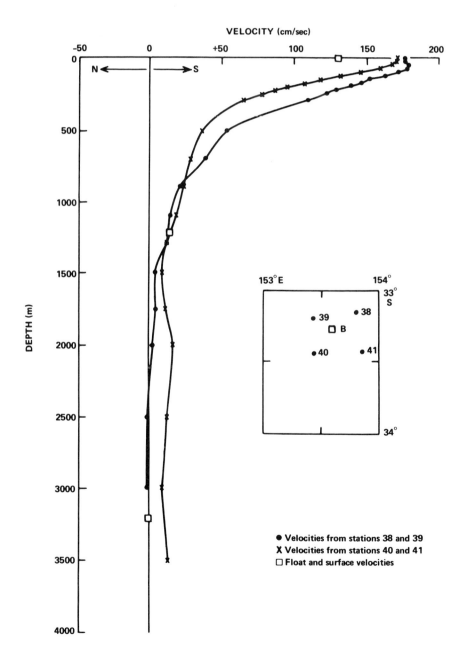

● Velocities from stations 38 and 39
X Velocities from stations 40 and 41
□ Float and surface velocities

FIGURE 5 North-south geostrophic velocities as functions of depth, neutrally buoyant float results, and surface velocity, cruise Dm2/67 (March 29-April 18, 1967). Both geostrophic velocity curves were calculated relative to 1,300 m and then fitted to the 1,200-m float result. The east-west velocities were negligible. The station positions relative to the floats are shown inset.

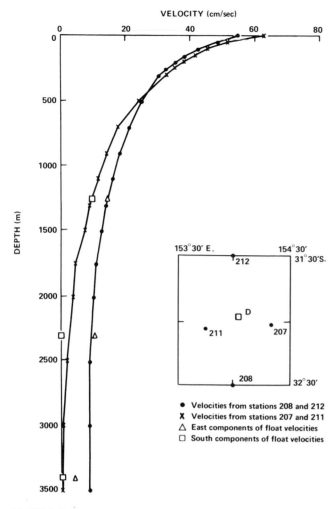

FIGURE 6   North–south and east–west velocities as functions of depth, and neutrally buoyant float results, cruise Dm4/67 (August 14–September 18, 1967). The geostrophic velocities have been matched to the corresponding components of the float velocity at 1,250 m. The station positions relative to the floats are shown inset.

The view that there is a general anticyclonic wind-driven circulation pattern at midlatitudes in the South Pacific can be accepted (Wyrtki, 1962b; Reid, 1961, 1965). Water moves westward into the Coral Sea in the South Equatorial Current and generally moves eastward at latitudes $\geqslant 30°$S. The East Australian Current appears at first to be a western boundary current, required by continuity to join these two zonal flows along the Australian coast. But this explanation does not account for the frequently observed northward-flowing currents to the east of the East Australian Current. There is very little net transport when the northward and southward transports are added together. Nor does it account for the frequent large eddies or for the occasional absence of a strong southward current seen in east–west sections between 28° and 30°S.

Instead of speaking of an East Australian Current as the

boundary phenomenon, it would not be unreasonable to discuss the existing observations as representing a continuing succession of south-moving eddies by which Coral Sea water is carried south and eventually mixed with water of more southern origin. On the basis of this hypothesis, the marked southerly sets shown on current atlases just seaward of the 100-fathom line would be interpreted to be merely the near-shore flows of anticyclonic eddies. The general predominance of southerly sets in the area would simply reflect the fact that most shipping is coastwise and therefore would never encounter the northward sets on the eastern sides of eddies. The volume transports computed from sections across "the current" would then be meaningless in relation to water budgets in the Tasman and Coral seas. Instead, we would need to know the rates of movement, spacings, and sizes of complete eddies.

From a different point of view, Boland (personal communication) has suggested that a wide spectrum of water movements might be present in the northern part of the area. Some of these might be damped out, while others grow to form eddies or sharp meanders as they move south. Topography might determine which part of the spectrum grows.

FUTURE WORK

In our laboratory, it is felt that we already have enough of the kind of information that can be obtained from single-ship hydrological survey cruises, at least for the area outlined in Figure 1. The information is certainly not enough to define climatological mean conditions, but the work necessary to do this would be unwarranted.

The following sections outline some ideas for future work. These ideas are little more than guesses suggested by a phenomenological approach to the existing data.

There are two important barriers to further progress: lack of money (including lack of ship facilities) and lack of theoretical guidance in what features to look for. Of these, lack of money is the more obvious, but probably not the more important. Given unlimited resources, we would, no doubt, stumble across many interesting new facts, but we would still not understand the system unless theory and observation could be brought more closely together.

Extension of Area

The area investigated has been limited by ship availability and the need to work at fairly closely spaced stations. (Other cruises have been made in the Tasman and Coral seas for biology and hydrology studies, but the station spacing is too great for useful physical interpretation unless results of several cruises are combined.) We think it desirable to look at the adjoining areas to the north, northeast, and south, but that this would be of little value unless the area outlined in Figure 1 were surveyed at the same time. This would require the use of more than one ship, or at least the use of one ship for about 2 months. It might be possible to reduce the amount of ship time by making use of coastal tide gauges and of anchored current and temperature recorders, but at present it would be hard to interpret the results.

Extension of the area of study to the north and northeast would throw light on the "source region" and on the existence or nonexistence of eddies in these latitudes. Extension to the south would enable us to observe the decay of an eddy, but this would involve several cruises over a period of 6–12 months. On one cruise, there was evidence of appreciable temperature structure at a depth of 1,300 m east of Tasmania; this structure was not paralleled in the dynamic heights. This suggests that the near-surface structure of an eddy might decay first. It would be interesting to investigate this possibility.

Direct Measurements of Currents

A few more Swallow-float measurements should probably be made at the 1,300-m level below a strong surface current in order to check the significance of the results reported

above. An extensive program of current measurements at greater depths does not seem warranted until there is more chance of interpreting the results. Deep current measurements in the narrow channel at 22°S, 155°E would, however, be interesting.

Time and Space Variability—Use of Aircraft

The East Australian Current area offers excellent scope for the use of aircraft to study the development of a circulation system. Flying conditions are usually good, and there are adequate airport facilities. An airborne infrared radiometer is already being used in part of this area for tuna research and is showing interesting rapid changes in surface-temperature pattern. Unfortunately, these changes cannot be interpreted physically, since surface temperature is not a good enough indicator of the current structure in this area.

A proposal was made a few years ago to use expendable bathythermographs (B/T) dropped from aircraft for this work. The design of a suitable B/T was started, but later abandoned. Apart from the use of a large number of anchored buoys, or full-time use of two or three ships, the air-dropped B/T scheme seems the only way in which nearly continuous coverage of the area can be maintained. It was

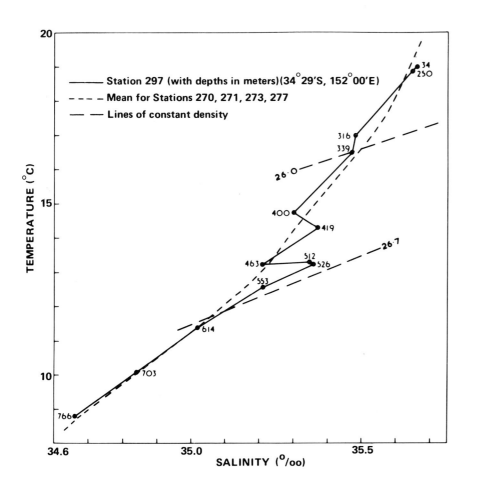

FIGURE 7 Temperature–salinity curves, cruise G6/65 (July 12–25, 1965), from TSD recorder.

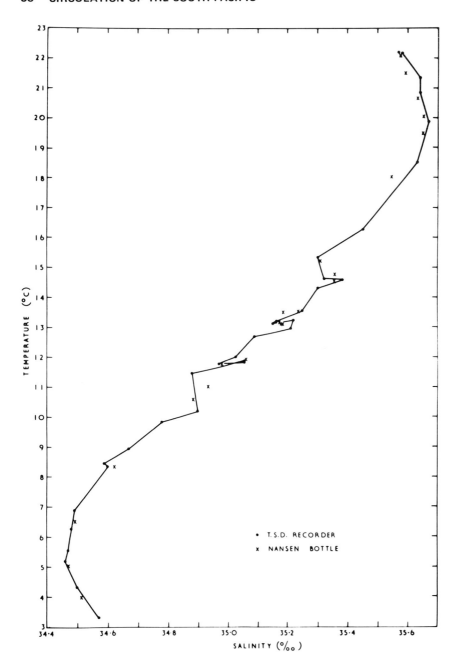

FIGURE 8 Temperature–salinity curves, Station 424, G9/65 (29°33'S, 155°29'E; November 19, 1965).

estimated that the project might cost nearly $90,000 (U.S.) for 20 surveys of an area of 600 miles by 200 miles.

Use of Recording Current Meters

A project is under way to use recording current meters and thermographs on the continental shelf and slope, probably off Evans Head (29°S). The aims are to study the connection between shelf, slope, and (possibly) offshore currents in order to find out more about their time scales and to determine whether coastal sea levels are in any way connected with the offshore or shelf current patterns.

Nonlinearity

If nonlinearity is important in the dynamics of the area, we would like to be able to measure or estimate the nonlinear terms, but at present we cannot see how this can be done.

Numerical Models

It is likely that the formation of an eddy, or its later movement and gradual dissipation, can be studied by means of numerical models, but no serious attempt has yet been made to work along these lines. Initial and boundary conditions would be difficult to specify, but if this problem could be overcome, the numerical approach offers good prospects for finding out at least whether nonlinear effects and bottom topography are important.

REFERENCES

Boland, F. M., and B. V. Hamon (in press) The East Australian Current, 1965–1968. *Deep Sea Res.*

Garner, D. M. (1959) Nomenclature of water masses in the Tasman Sea. *Aust. J. Mar. Freshwater Res.*, *10*, 1–6.

Garner, D. M., and N. M. Ridgway (1965) Hydrology of New Zealand offshore waters. *Mem. N.Z. Oceanogr. Inst.*, No. 12, 62 pp.

Hamon, B. V. (1961) Structure of the East Australian Current. *CSIRO Div. Fish Oceanogr. Tech. Pap.*, No. 11.

Hamon, B. V. (1965) The East Australian Current, 1960–1964. *Deep-Sea Res.*, *12*, 899–921.

Hamon, B. V. (1968a) Temperature structure in the upper 250 metres in the East Australian Current area. *Aust. J. Mar. Freshwater Res.*, *19*, 91–99.

Hamon, B. V. (1968b) Spectrum of sea level at Lord Howe Island in relation to circulation. *J. Geophys. Res.*, *73*, 6925–6927.

Hamon, B. V. (1969) Review papers: I "Current-effect" on sea level. II Oceanic eddies. *CSIRO Div. Fish. Oceanogr. Rep. 46.*

Hamon, B. V., and J. D. Kerr (1968) Time and space scales of variations in the East Australian Current, from merchant ship data. *Aust. J. Mar. Freshwater Res.*, *19*, 101–106.

Highley, E. (1967) Oceanic circulation patterns off the east coast of Australia. *CSIRO Div. Fish. Oceanogr. Tech. Pap.*, No. 23.

Reid, J. L. (1961) On the geostrophic flow at the surface of the Pacific Ocean with respect to the 1,000-decibar surface. *Tellus*, *4*, 489–502.

Reid, J. L. (1965) Intermediate waters of the Pacific Ocean. *Johns Hopk. Oceanogr. Stud.*, *2*, 85 pp.

Rochford, D. J. (1957) Identification and nomenclature of the surface water masses in the Tasman Sea (data to the end of 1954). *Aust. J. Mar. Freshwater Res.*, *8*, 369–413.

Rochford, D. J. (1958) Total phosphorus as a means of identifying East Australian water masses. *Deep-Sea Res.*, *5*, 89–110.

Rochford, D. J. (1959) The primary external water masses of the Tasman and Coral seas. *CSIRO Div. Fish. Oceanogr. Tech. Pap.*, No. 7.

Rotschi, H., and L. Lemasson (1967) Oceanography of the Coral and Tasman seas. *Oceanogr. Mar. Biol. Ann. Rev.*, *5*, 49–97.

Wyrtki, K. (1960) The surface circulation in the Coral and Tasman seas. *CSIRO Div. Fish. Oceanogr. Tech. Pap.*, No. 8.

Wyrtki, K. (1961) The flow of water into the deep sea basins of the western South Pacific Ocean. *Aust. J. Mar. Freshwater Res.*, *12*, 1–16.

Wyrtki, K. (1962a) The subsurface water masses in the western South Pacific Ocean. *Aust. J. Mar. Freshwater Res.*, *13*, 18–47.

Wyrtki, K. (1962b) Geopotential topographies and associated circulation in the western South Pacific Ocean. *Aust. J. Mar. Freshwater Res.*, *13*, 89–105.

Warren S. Wooster

SCRIPPS INSTITUTION OF OCEANOGRAPHY,
LA JOLLA, CALIFORNIA

EASTERN BOUNDARY CURRENTS IN THE SOUTH PACIFIC

The differences in the oceanography of the eastern and western sides of the oceans are by now well known. The intensification of horizontal flow in the west is matched by an intensification of oceanographic research activity in the Gulf Stream and the Kuroshio. On the eastern side, where oceanographic conditions are somewhat less dramatic, only the California Current has been studied with comparable vigor, and the North Atlantic analog, the Canary Current, remains little known to this day.

Modern oceanographic investigations of the eastern South Pacific began about 40 years ago with the cruises of the *Carnegie* (1928-1929) and the *William Scoresby* (1931). But more than 400 years ago, the Spanish explorers were familiar with the cold northward flow along the coasts of Chile and Peru (Gunther, 1936). Humboldt observed in 1802 that the ocean was colder than the air along these coasts and postulated that the low sea temperatures were due to the Antarctic origin of the coastal current. By the middle of the nineteenth century, the fact that coastal temperatures did not increase monotonically from southern Chile to northern Peru had been noted, and the upwelling of subsurface water had been invoked as an explanation. The controlling influence of the trade wind was also recognized. Thus, the major elements of information needed for an interpretation of this eastern boundary current have been available for a long time.

Yet, for a number of reasons, knowledge of the oceanography of the eastern South Pacific has remained inadequate. A resurgence of modern investigation began in 1952 with the *Shellback* Expedition and has continued to this day. Since 1958, Peruvian investigators have been making routine cruises designed to monitor seasonal and yearly changes in the circulation. The eventual synthesis of all existing data should permit a most interesting and useful analysis. However, it is not yet possible to make such a synthesis, and I shall restrict myself to commenting on some interesting features of the circulation that are perhaps not generally known and that may suggest possibilities for further work in the region.

The general characteristics of eastern boundary currents (Wooster and Reid, 1963) should be kept in mind while looking at the situation in the South Pacific. Such currents tend to flow slowly toward the equator in a broad shallow drift. Average speeds are less than 25 cm sec^{-1}, and the transport appears to be of the order of 10-20 $\times$ 10^6 m^3 sec^{-1}. The surface waters of these currents are relatively cold, largely because of coastal upwelling, which also weakens the thermocline and brings nutrient elements into the surface layer. Commonly, a layer of deoxygenated water lies below the surface current, and next to the coast there is often a poleward undercurrent in the midthermocline.

In the South Pacific, the eastern boundary current system appears to extend approximately from 50$°$ to 5$°$S. Beyond the southern limits of the system, waters of the Antarctic Circumpolar Current flow toward the Drake Passage and the Atlantic; north of 5$°$S, the current swings west to feed the South Equatorial Current. In between, through some 45 degrees of latitude, lies what is often called the Humboldt Current, but which I prefer to call the Chile-Peru Current system.

A superficial impression of this system can be gained from a look at the mean surface temperature along the coast (Figure 1). The seasonal rhythm is apparent, with low temperatures in the winter and high temperatures in the summer. Note the large gradients north of 5$°$S, where the equatorial front or north boundary of the system lies. Note also the region of temperature minimum at 15$°$S, where, throughout the year, temperatures are lower than to the north or south. In a different summary of the same data (Figure 2), the steady equatorward increase of temperature along the Chilean coast appears to be interrupted at about 20$°$S, where a major change in coastal orientation can be

TIME OF YEAR

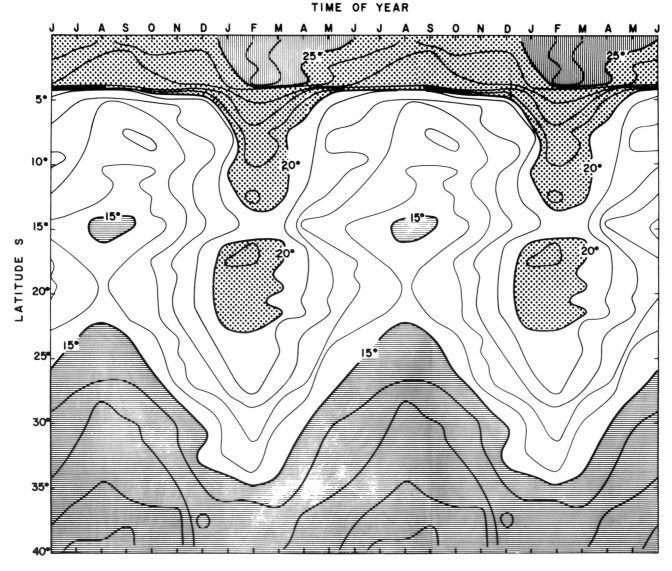

FIGURE 1 Average surface temperature, by one-degree squares and months, along west coast of South America.

observed. North of 15°S, the same rate of equatorward temperature increase is offset, by several degrees, toward colder values.

These thermal features and their lateral extent are apparent in a chart of July surface temperatures (Figure 3). Surface isotherms bend northward along the coast; cold regions off Peru and off Chile are separated by a warm belt at 20°S. Low temperatures extend along the equator far to the west. In January, on the other hand (Figure 4), the equatorial temperature minimum is much weaker, coastal temperatures are higher, and the warm water at 20°S appears to extend southward from the equator (with some additional warming locally), driving a wedge between the colder re-

gions. This January picture suggests the possibility of an actual separation between the Peruvian and Chilean components of the coastal circulation.

Available surface-current charts are not adequate to establish whether this is indeed the fact. No significant southward flow from the equator appears, and equatorward flow along the coast is more-or-less continuous. Subsurface circulation, however, is probably quite different, and geostrophic studies (Wyrtki, 1963) suggest that significant poleward transport of equatorial water occurs offshore. This transport may be masked at the surface by a shallow wind-blown drift. Further evidence for the separation can be seen in the subsurface distributions of salinity and dissolved oxy-

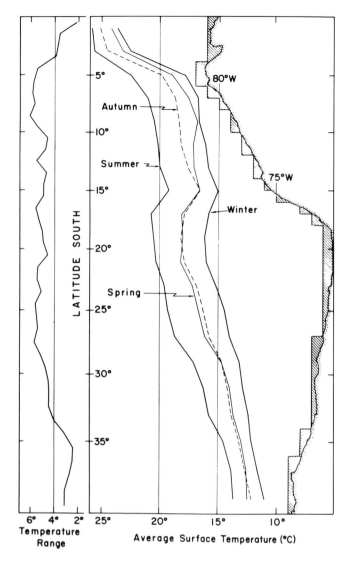

FIGURE 2 Average surface temperature, by two-degree squares and seasons, and annual temperature range, along west coast of South America.

gen. For example, the waters off the Chilean coast characteristically show two salinity minima, for subantarctic and Antarctic intermediate water, that are not found off Peru.

One can then speculate that most of the Chile Current goes off to the west before reaching 20°S, that the Peru Current arises along the southern coast of Peru, that this current is fed largely from the west and from below, and that the 18°–22° zone off northern Chile represents a quiet backwater between the two circulations. To demonstrate whether this is indeed the case, and if so, how and why the separation occurs, is one of the intriguing scientific problems of the eastern South Pacific that remains unresolved.

Investigations of the Chilean portion of the system are being pursued by Chilean scientists and will be discussed elsewhere. The remainder of the present paper is restricted to the Peruvian part of the system, extending from southern Peru north and west to the Galápagos Islands.

Characteristics of the Peru Current can be seen in a vertical section measured perpendicular to the coast at about 15°S. The temperature field (Figure 5) parallels the density field. Thus, the coastward shoaling of the thermocline is accompanied by a coastward shoaling of the isopycnals. The consequent slope of the sea surface is about 22 cm in 1,000 km, equivalent at this latitude to an average northward flow of about 5 cm sec^{-1}. Downward-sloping isotherms indicate the presence of an undercurrent near the coast and a countercurrent about 400 km offshore. Both of these are subsurface flows, and they transport water of equatorial origin toward the south. This can be seen in the salinity distribution (Figure 6), which indicates high-salinity water at depth in both places. Also, a shallow salinity minimum is indicated offshore. This is a vestige of the Chilean Current, which is far offshore at this latitude. The deeper salinity minimum is the Antarctic Intermediate Water referred to earlier. Finally, the oxygen distribution (Figure 7) shows a well-developed oxygen minimum near shore, where a large secondary maximum of nitrite nitrogen has been observed (Wooster et al., 1965); the higher oxygen content of the Antarctic Intermediate Water can also be seen.

Discussions of the circulation of this region mention not only countercurrents and undercurrents, but also coastal and oceanic currents. In my opinion, this multiplicity of current designations reflects our ignorance of a large and complex system of surface and subsurface circulation. The taxonomy of this system badly needs a careful and extensive revision.

Low temperatures along the eastern boundary coasts are usually attributed to upwelling. This must be the situation off Peru since the region is one of positive heat balance throughout the year (Wyrtki, 1966). The upwelling, in turn, is attributed to the action of the equatorward component of the wind stress parallel to the coastal boundary. In the simple example analyzed by Ekman, the offshore transport (and hence the transport of replacement water from below) is directly proportional to the wind-stress component and inversely proportional to the Coriolis parameter. Off the Peruvian coast, the low latitude and the trade-wind stress combine to produce notoriously vigorous and persistent upwelling. As a result, surface waters are both cold and nutrient-rich, with about ten times the open-ocean content of phosphate phosphorus and similar substances, and, of course, with the greatest single-species fishery in the world, as Dr. Kasahara points out in his paper, "Commercial Fisheries," p. 252.

The mechanisms of upwelling and the accompanying vertical mixing are poorly understood, and direct measurements of vertical speed are beyond the scope of present technology. A number of indirect estimates have been made. Wyrtki (1963) has examined divergence of flow in

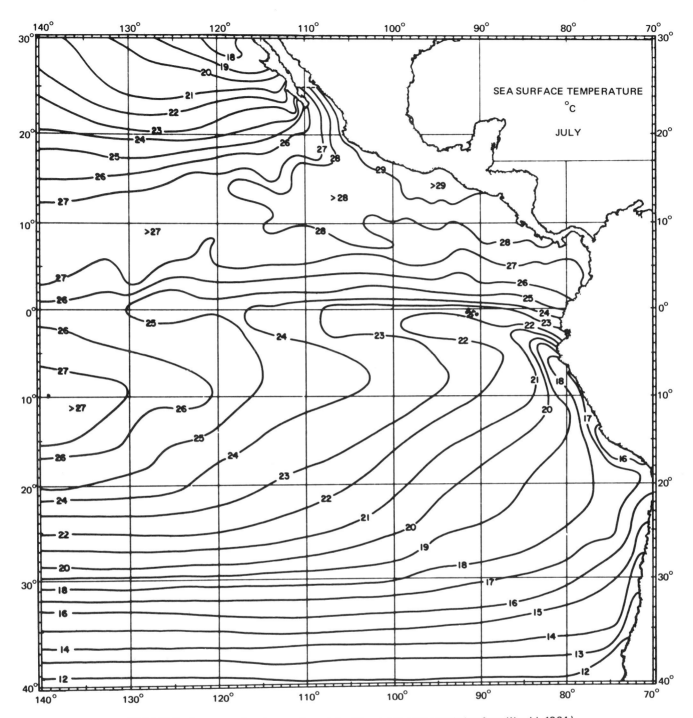

FIGURE 3   Average surface temperature, July. (Reprinted with permission from Wyrtki, 1964.)

the upper 100 m of the Peru Current and has estimated vertical speeds of 1–8 m per month. Much higher values, up to 100 m per month, were estimated by Wooster and Reid (1963) from mean-wind-stress data. An independent analysis has recently been made using the average offshore component of ship drift; vertical speeds of 20 to more than 100 m per month were obtained, the values depending on

location and season. Although these various estimates may have some utility for setting limits on the potential productivity of the region, they leave much to be desired from the physical point of view.

The Peru Current, like all near-surface features of the ocean, is highly variable, the variability occurring over a wide range of scales and frequencies. Of particular interest

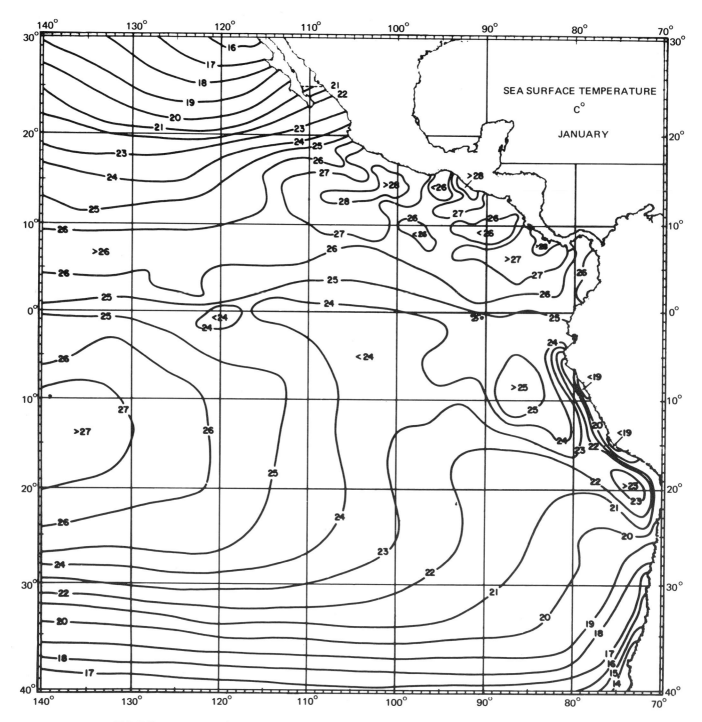

FIGURE 4 Average surface temperature, January. (Reprinted with permission from Wyrtki, 1964.)

are the periods of months to years during which the usual seasonal rhythm may be substantially altered and the Niño phenomenon can develop.

As Figure 8 shows, year-to-year changes in surface temperature along the Peruvian coast can be very large. These changes are determined by changes in the surface winds, acting directly through their effect on the intensity of upwelling and indirectly through their effect on the exchange

of latent and sensible heat. The weakening of the trades each summer is accompanied by a weakening in equatorward flow and upwelling and by higher temperatures along the coast. As Bjerknes (1966) has shown, the Niño phenomenon is associated with an extreme relaxation of the trades, permitting an invasion of warm surface water from the north and west and the virtual cessation of upwelling on the equator and along the Peruvian coast.

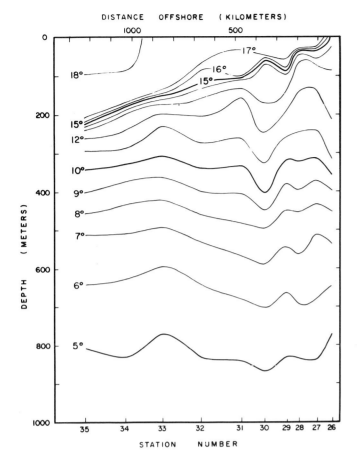

FIGURE 5   Distribution of temperature off Peruvian coast at 15°S, STEP-1 Expedition (October 28–November 2, 1960). (Reprinted with permission from University of California, 1961.)

Atmospheric and oceanic events related to the Niño are of very large dimensions, and the changes in heat storage in the surface layer of the ocean have substantial effects on the atmosphere. Although the catastrophic consequences of El Niño have been somewhat exaggerated, there is no doubt that the effects on the biota, including its human predators, are also large. Thus, studies of the phenomenon, and the collection of observational data from the region between the Galápagos and the mainland, have intensified in recent years, with the ultimate goal of successful prediction.

In this connection, a feature of particular interest is the north boundary of the Peru Current. The current leaves the coast north of 6°S and turns west toward the Galápagos, where it feeds into the South Equatorial Current. To my knowledge, no one has explained why the current leaves the coast in this particular locality. There must be, of course, a transition from the eastern boundary current to the zonal equatorial circulation, but the factors controlling the location and nature of this transition have never been evaluated.

The differences between the equatorial surface waters and those of the Peru Current are so great as to justify speaking of an "equatorial front." To the north, waters are

warmer, fresher, contain less dissolved oxygen and fewer nutrients, and are separated from deeper waters by a shoaler and more-intense thermocline than to the south. Because of the simultaneous temperature decrease and salinity increase, there is a large change in density and specific-volume anomaly across the front. This change is confined to the surface layer of less than 100 m, below which no significant differences in the characteristics of the waters to the north and south are apparent (see Wooster, 1969).

From surface-temperature averages it can be seen that the front extends from about 4°S near the coast to perhaps 1°N near the Galápagos. The position of the front and the gradients across it vary seasonally. The front appears to be farthest north and most fully developed during the southern winter.

During a cruise in November 1955, detailed observations were made from the research vessels *Baird*, *Horizon*, and *Bondy*. Velocity measurements by GEK showed that the front was strongly convergent; i.e., from north to south, the meridional component changed from southward to northward, and the zonal (westward) component increased. The difference in velocity across the front decreased steadily from east to west.

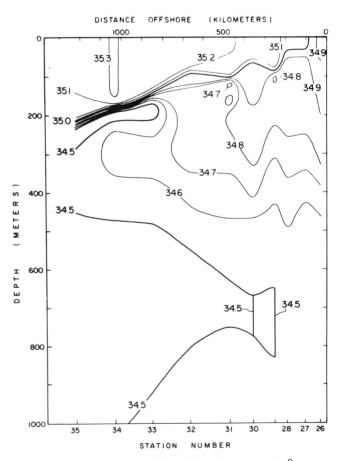

FIGURE 6   Distribution of salinity off Peruvian coast at 15°S, STEP-1 Expedition (October 28–November 2, 1960). (Reprinted with permission from University of California, 1961.)

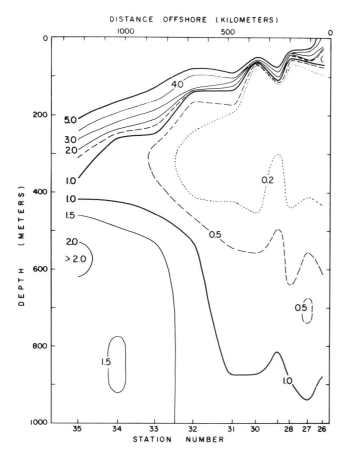

FIGURE 7 Distribution of dissolved oxygen off Peruvian coast at 15°S, STEP-1 Expedition (October 28–November 2, 1960). (Reprinted with permission from University of California, 1961.)

By chance, the *Baird* crossed the front about one week before the *Horizon*. During this interval (173 hours), the front moved southward a distance of 118 km at an average speed of 19 cm^1 sec^{-1}. This can be compared with a speed of 20–22 cm^1 sec^{-1}, which was calculated by Fedorov (1963) using the observed pressure gradient and an estimate of friction. It appears that this southward motion, which is against the wind, is of thermohaline character and is related to the southward pressure gradient resulting from the differences of temperature and salinity, and therefore of density, across the front.

It is possible to map the surface position of the front from changes in surface velocity, temperature, and salinity; the subsurface position, from changes in the intensity of the thermocline. The horizontal temperature gradient across the front was as great as 1.6°C per 10 km; the vertical temperature gradient in the thermocline decreased from 3.5°C per 10 m in the north to about 1.5°C per 10 m in the south. Figure 9 shows the apparent location of the front. West of 84°W, the front extended generally east–west, with latitude decreasing gradually to the west; east of 84°W, the line dipped toward the south. At about 84°W, the relative posi-

tions of the surface and subsurface fronts also changed, with the surface front lying north of the subsurface front in the west, and south of it in the east. The explanation for this east–west difference must be found in the dynamics of the front. In the west, the front appears to have been relatively stable, with the boundary of the warm surface waters being pushed to the north by the prevailing southerly winds. In the east, as we have already noted, tropical waters were moving rapidly to the south. In both places, the dynamic processes transforming the thermocline lagged behind.

This paper has discussed a sampling of features and phenomena in the eastern South Pacific. Work in the region during the past 10 or 15 years has permitted the recognition of a number of problems, the solutions of which will require additional work. I would like to close by commenting on what this additional work might include.

Exploration of the surface waters of the region within 1,000 km of the coast is relatively complete, or will be within the next year. Farther offshore, few oceanographic stations have been made. As in most parts of the ocean, the upper kilometer has been more intensively sampled than the deeper waters. Studies of the deep circulation will require many more deep stations, with high-quality observations down to the bottom. A suitable initial goal might be to have a grid of such stations on 500-km centers.

With regard to the upper kilometer, the pressing needs are for investigation and for monitoring. Certain features worthy of further investigation have already been mentioned, including the north boundary of the system, the center of permanent upwelling at about 15°S, the separation of the Chile and Peru Currents near 20°S, and the details of the system of surface and subsurface currents and countercurrents east of 90°W. These investigations should make use of modern *in situ* measuring equipment and of the direct methods now available for measuring velocity. Availability of precise satellite navigational equipment should greatly facilitate this work.

As in all eastern boundary current regions, a much more quantitative evaluation of vertical motion and mixing along the boundary is required. An investigation of the upwelling process may have considerable chance of success here, where upwelling is apparently so intense and persistent, where the boundary is so well defined, and where the response to changes in the driving forces can be so readily observed.

Finally, a system for monitoring the changes in atmospheric and oceanic circulation is needed. Eventually, this job may be done by buoys and satellites, but in the meantime much can be done with existing technology and resources. The long coastline, the offshore islands of Chile (Easter, Juan Fernandez, and San Ambrosio) and Ecuador (the Galápagos), and the coastal islands of Peru provide abundant sites for continuous recording of sea level and of

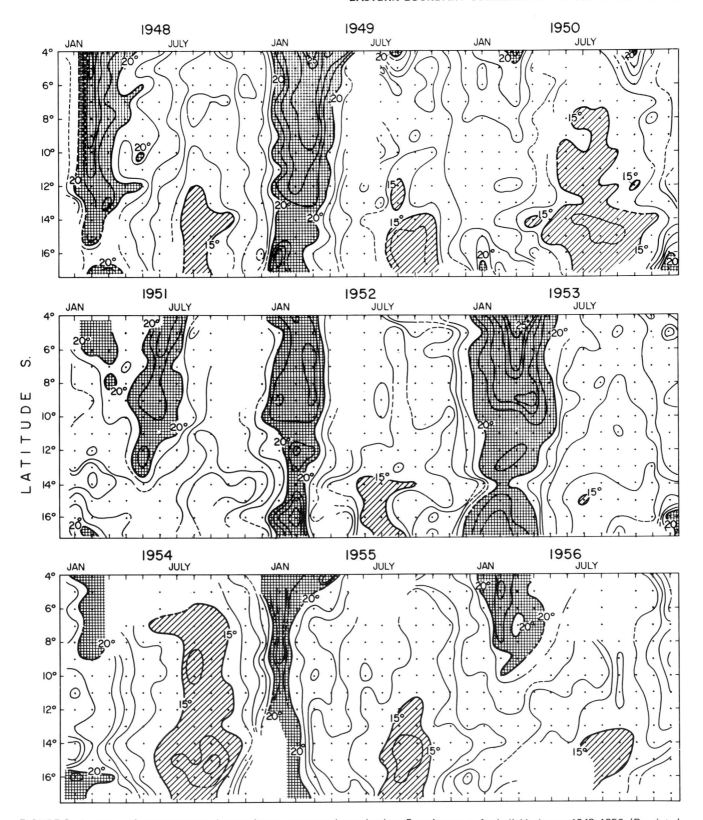

FIGURE 8 Average surface temperature, by one-degree squares and months along Peruvian coast, for individual years 1948–1956. (Reprinted with permission from Wooster, 1961.)

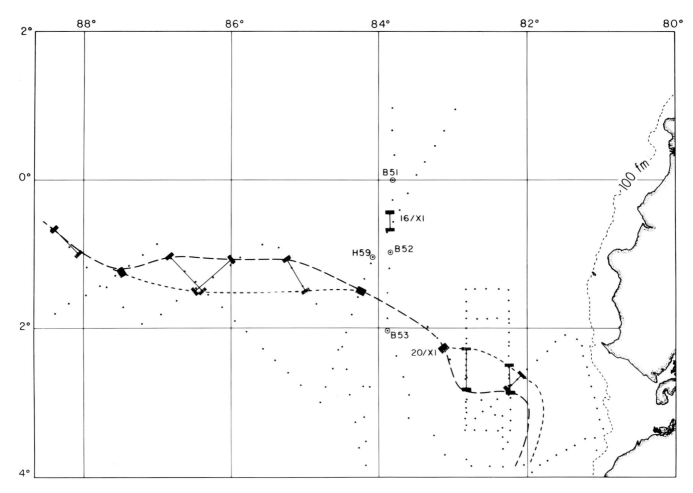

FIGURE 9 Position of equatorial front between Ecuador and Galápagos, November 1955 (EASTROPIC Expedition). Heavy line, position at surface; fine line, position in thermocline. (Reprinted with permission from Wooster, 1969.)

other pertinent oceanic and atmospheric parameters. Coastal shipping could be much more extensively used for monitoring changes in thermal structure, surface salinity, and surface weather. The coastal states could develop efficient monitoring cruises with their research vessels that could still leave time for more specific investigations on problems of concern.

The intensive study of the Chile-Peru Current system could lead to the solution of problems of considerable scientific interest. The vigorous cooperation of the countries of the region, required for successful prosecution of such a program, would also tend to strengthen marine research in each of the participating countries. Finally, direct and early benefits for the fisheries and shipping of the coastal states of the region could confidently be expected.

REFERENCES

Bjerknes, J. (1966) Survey of El Niño 1957–58 in its relation to tropical Pacific meteorology. *Bull. Inter-Amer. Trop. Tuna Comm.*, *12*(2), 25–86.

Fedorov, K. N. (1963) Some peculiarities of the currents and the ocean level at the equator. *Okeanologiia*, *1*, 3–12.

Gunther, E. R. (1936) A report on oceanographical investigations in the Peru Coastal Current. *Discovery Rep.*, *13*, 107–276.

University of California (1961) Preliminary report STEP-I Expedition 15 September–14 December 1960. Part I. Physical and chemical data. Mss. Rep. (SIO Ref. 61-9).

Wooster, W. S. (1961) Yearly changes in the Peru Current. *Limnol. Oceanog.*, *6*(2), 222–226.

Wooster, W. S. (1969) Equatorial front between Peru and Galápagos. *Deep Sea Res.*, Suppl. *16*, 407–419.

Wooster, W. S., and J. L. Reid, Jr. (1963) Eastern boundary currents, pp. 253–280 in: *The Sea: Ideas and Observations on Progress in the Study of the Seas*, editor M. N. Hill, Wiley-Interscience, New York. Vol. 2.

Wooster, W. S., T. J. Chow, and I. Barnett (1965) Nitrite distribution in Peru Current waters. *J. Marine Res.*, *23*(3), 210–221.

Wyrtki, K. (1963) The horizontal and vertical field of motion in the Peru Current. *Bull. Scripps Inst. Oceanogr.*, *8*(4), 313–346.

Wyrtki, K. (1964) The thermal structure of the eastern Pacific Ocean. *Dt. Hydrogr. Z., Erg. Reihe A* (8°), No. 6, 84 pp.

Wyrtki, K. (1966) Seasonal variation of heat exchange and surface temperature in the north Pacific Ocean. Hawaii Institute of Geophysics Rep. HIG-66-3.

Mizuki Tsuchiya

UNIVERSITY OF TOKYO,
TOKYO, JAPAN

EQUATORIAL CIRCULATION OF THE SOUTH PACIFIC

INTRODUCTION

This symposium is devoted to South Pacific oceanography, and, according to my understanding, the South Pacific is the area from the equator southward to the Antarctic Convergence. However, the Equatorial Undercurrent, which is an essential component of the equatorial current system, lies over both hemispheres. For this reason, I shall extend the area of interest slightly to the north of the equator so that the whole meridional extent of the Equatorial Undercurrent may be included.

Considerable progress has been made in Pacific equatorial oceanography during the last 20 years. I will try to summarize existing knowledge and to point out some problems that remain to be solved and that may be studied extensively during the proposed exploration of the South Pacific.

The equatorial currents are closely related to one another, but I shall discuss each separately, with emphasis on its association with other currents.

EQUATORIAL UNDERCURRENT

The Equatorial Undercurrent was discovered at 150°W in August 1952, and its existence is now established for almost the entire length of the equator in the Pacific Ocean.

Direct measurements made recently show how far west the undercurrent originates. Measurements from anchored buoys during *Vityaz* cruise 38 (January–April 1966) suggest that the undercurrent starts near 135°E (Kort *et al.*, 1966). Masuzawa (personal communication, 1967) observed strong eastward currents at depths between 50 m and 300 m at and north of the equator along 137°E in January 1967 and January 1968. This far west, it is usually difficult to distinguish clearly the Equatorial Undercurrent from the North Equatorial Countercurrent to the north.

In the far eastern Pacific, the undercurrent extends nearly to the coast of Ecuador. There is evidence that the undercurrent veers around the north side of the Galápagos Islands (Knauss, 1966).

The velocity core of the Equatorial Undercurrent lies at about the 300-centiliters-per-ton (cl/t) level, except in the far eastern Pacific, where the core shifts to a surface of lower thermosteric anomaly. The actual depth of the core is 200 m in the west and decreases eastward to 50 m near the Galápagos Islands; farther east the core deepens suddenly.

The distribution of oxygen on the 160-cl/t surface, which lies slightly below the thermocline bottom and well below the velocity core of the undercurrent, exhibits a pronounced tongue of high oxygen along the equator almost all the way across the Pacific Ocean (Tsuchiya, 1967). This high-oxygen tongue is related to the equatorial troughing of oxygen isopleths in the lower part of the thermocline. The 160-cl/t surface itself shows a trough along the equator, but the troughing of oxygen isopleths is even more intense. Thus, the 160-cl/t surface exhibits a maximum of oxygen along the equator. Oxygen in the high-oxygen tongue decreases from greater than 3 milliliters per liter (ml/l) at 150°E to 1 ml/l near the coast of Ecuador. This tongue indicates the importance of the eastward transport of oxygen by the Equatorial Undercurrent. The distributions of salinity and oxygen suggest that undercurrent water with the thermosteric anomaly approximately at the 160-cl/t level comes from the western South Pacific, probably through the Coral Sea.

On the map of salinity at the 300-cl/t surface, which lies near the velocity core of the undercurrent in the western and central Pacific, a high-salinity tongue is revealed just south of the equator from 150° to 118°W (Tsuchiya, 1968). This tongue represents the isolated high-salinity core that can be seen on some meridional sections with closely spaced samples (Montgomery and Stroup, 1962, p. 24; Knauss, 1966; Wyrtki, 1967b). The eastward decrease in salinity suggests that the saline water in the tongue is flowing east in the undercurrent. The source of this water is the southern Tropical Water, which is formed at the sea surface near the Tropic of Capricorn and spreads as a subsurface salinity maximum (at 300–400 cl/t) in the South Equatorial Current. The high-salinity tongue may start farther west, but the available data from the western equatorial Pacific fail to reveal it because of inadequate spacing of samples in the thermocline. Accordingly, how the saline Tropical Water enters the undercurrent from the South Equatorial Current is not known.

Although the saline water of Southern Hemisphere origin forms the core of the undercurrent, it also contains less saline water of Northern Hemisphere origin. This fact can be seen in the distribution of salinity on isanosteric surfaces in the upper layers of the intertropical Pacific Ocean (Tsuchiya, 1968) or on meridional sections crossing the equator. On the section along 140°W in September 1961 (Knauss, 1966, Figure 10), an isolated low-salinity core (enclosed by the isohaline for 34.8‰) centered at the 200-cl/t level is indicated at 1°N. This low-salinity water, flowing east in the undercurrent, is clearly of Northern Hemisphere origin and must have entered the undercurrent farther west than this longitude. Masuzawa (1967) observed isolated low-salinity water (less than 34.9‰ at the core) at about the 300-cl/t level between 0° and 2°N at 137°E in January 1967. Recent measurements with an *in situ* salinometer (STD) at 156°W and 157°30'W in February and March 1967 also show a distinct vertical minimum of salinity (34.8‰) at about the 300-cl/t level near 1°N, just north of the high-salinity core mentioned above (Wyrtki, 1967b). The low-salinity waters observed by both Masuzawa and Wyrtki are flowing east and are clearly of Northern Hemisphere origin.

As Montgomery and Stroup (1962, p. 57) have pointed out, an intense meridional gradient of salinity and an intense vertical gradient of thermosteric anomaly are characteristic of undercurrent water near the velocity core. Therefore, the water flux near the core is distributed over water classes with wide ranges in salinity and in thermosteric anomaly. In contrast to the nonhomogeneity of the core water, the water below about 200 cl/t is remarkably homogeneous, which is clearly demonstrated on the maps of salinity at the 160-cl/t surface and of the thickness of the layer between 155 cl/t and 165 cl/t (Tsuchiya, 1967). Water with temperature about 12.5° C, salinity about 34.9‰, and thermosteric anomaly about 160 cl/t dominates within about 5° of the equator east of 140°W. This distinct water type forms the flux mode of the undercurrent and has been named Equatorial 13-C Water. It is seen, on the thickness map, extending north and south beyond the region of the undercurrent. How this water is formed is not yet understood.

A study of downstream changes in water-flux characteristics may be necessary for a proper understanding of the undercurrent. On the basis of current-measurement and hydrographic data, the zonal flux of water near the equator is estimated for the four layers centered at 160, 200, 300, and 400 cl/t, each having a thickness of 10 cl/t. The estimated fluxes are summarized in Table 1 together with the total fluxes of the undercurrent.

For Swan Song data (August–December 1961), there is no significant change in the east flux within each layer from 140° to 118°W. At and east of 96°W, the flux of the lighter undercurrent water (295–305 cl/t), which contains the velocity core in the western and central Pacific, decreases to vanishingly small values. On the other hand, the flux of the heavier undercurrent water (155–165 cl/t), which forms the flux mode, remains uniform as far east as 96°W and then decreases farther east. In other words, the lighter undercurrent water near the velocity core begins to be lost

TABLE 1 Zonal Flux of Water near the Equator. East (Positive) Flux and West (Negative) Flux Are Tabulated Separately

| δ_T (cl/t) | Flux (km³/hr) Swan Song, August–December 1961 | | | | | Dolphin, April 1958 |
	140°W 2°N–2°S	118°W 2°N–2°S	96°W 1°N–2°S	93°W 1°N–1°S	87°W 1°N–1°S	140°W 2°N–2°S
395–405	{ + 0.1 { − 2.8	0.0 0.3	0.0 0.1	0.0 0.1	0.0 0.1	3.6 0.0
295–305	{ + 1.6 { − 0.0	2.9 0.0	0.0 0.4	0.4 0.2	0.0 0.2	3.1 0.0
195–205	{ + 5.6 { − 0.0	4.3 0.1	1.5 0.2	1.7 0.3	0.0 3.1	8.4 0.0
155–165	{ +11.7 { − 0.0	11.4 1.8	10.8 0.8	2.6 0.0	5.2 0.0	11.7 0.5
Total flux of the undercurrent	79	72	36	29	14	151

somewhere between 118° and 96°W, and the velocity core shifts to a lower thermosteric anomaly in the far eastern Pacific. The heavier undercurrent water begins to be lost somewhere between 96° and 93°W, 100 to 300 miles west of the Galápagos Islands.

East of 100°W, the oxygen content observed at 160 cl/t is higher to the north of the equator than south of it (Tsuchiya, 1967). This distribution may be taken as evidence that more undercurrent water with this thermosteric anomaly is discharged toward the north. During *Alaminos's* EASTROPAC cruise (January–April 1967), Cochrane (1967) found that an eastward current observed at 1°N, 84°W continued to flow east and then turned northward along the edge of Panama Bight. However, he also found evidence of a connection from the undercurrent south of the Galápagos Islands to the Peru-Chile Undercurrent.

The uniformity with longitude of the flux within the layer from 155 cl/t to 165 cl/t west of 96°W is perhaps surprising. This uniformity implies that there is no net meridional flow into or out of the undercurrent at the 160-cl/t level between 140° and 96°W. There is a possibility that the flux within this layer is also uniform with time. The estimated flux between the 155-cl/t and the 165-cl/t surfaces at 140°W on April 23-27, 1958 (*Horizon*, Dolphin cruise) is practically the same as that observed at the same longitude in September 1961, though the total flux of the undercurrent in 1958 was twice as large as that in 1961.

For the western equatorial Pacific, data adequate for this sort of study are not available. Precise measurements of the current velocity and water properties are very much needed.

SOUTH EQUATORIAL COUNTERCURRENT

Evidence of the existence of the Pacific South Equatorial Countercurrent was first presented by Reid (1959), and further evidence is being accumulated (Reid, 1961, 1965; Wooster, 1961; Koshliakov and Neiman, 1965; Rotschi and Lemasson, 1968). Most of the evidence available at present is indirect, and there have been only a few direct measurements (Burkov and Ovchinnikov, 1960; Koshliakov and Neiman, 1965). Thus, direct measurements of the South Equatorial Countercurrent, accompanied by water-property measurements, would be one of the interesting problems to be studied.

Geostrophic computation combined with data from various expeditions indicates that the countercurrent at the sea surface starts near the Solomon Islands and extends all the way east to the coast of South America (Reid, 1961). The axis lies near 10°S as far east as 120°W and then appears to shift southward. On the 125-cl/t isanosteric surface, which lies in the 300–400-m depth range in the area of present interest, the countercurrent is found along about 5°S in longitudes west of 120°W; east of 120°W, it is found

farther south (Reid, 1965, Figure 18). The estimated maximum speed is about 10 cm sec^{-1}. Both at the sea surface and at the 125-cl/t level, the countercurrent is better defined in the west than in the east. Wyrtki's surface-current charts (1965) for the intertropical Pacific east of 140°W, based on ship-drift data, do not show any eastward flow near these latitudes, and he suggests that the South Equatorial Countercurrent exists only in subsurface layers.

Although the eastward countercurrent is revealed rather well on each meridional section used for mapping, it is not easy to draw a zonally continuous picture all the way across the ocean. This difficulty may be due in part to use of data from different cruises, but there remains some doubt that the zonal continuity indicated by these maps is real. The southward shift of the axis in the eastern South Pacific seems to me open to question; Wooster (1961) found the countercurrent, with the maximum speed between 100 m and 200 m, at 6°S, 95°W, in November and December 1960, and Cochrane (1967) found evidence of the countercurrent, at 5°-6°S, 89°W, in March 1967 during the first survey cruise of EASTROPAC.

As stated before, the South Equatorial Countercurrent, as deduced from geostrophic computation, is found near 10°S at the sea surface and near 5°S in subsurface layers below about 250 m. At about the 200-cl/t level (about 200 m in the central South Pacific), the countercurrent is not clearly defined. This situation is seen on most meridional sections from the South Pacific Ocean. For instance, on Reid's section at 160°W (1965, Figure 6), the slope of isanosteres suggests an eastward current at and near the sea surface at about 8°S and a subsurface eastward current near 5°S, where the surface current is westward. These two eastward geostrophic currents are sometimes distinct from each other (Figure 1).

The deeper subsurface countercurrent near 5°S may be identified laterally by higher salinity and higher oxygen content. Wooster (1961) stated that the South Equatorial Countercurrent at 95°W in November and December 1960 contained slightly more salt and oxygen than waters to the north and south. Reid's map (1965, Figure 20) of oxygen content at the 125-cl/t level shows a possible tongue of high oxygen content at the latitudes of eastward geostrophic flow east of 150°W.

On the other hand, the shallower countercurrent near 10°S may be identified laterally by lower salinity and lower oxygen content, at least in the western South Pacific. The distribution of oxygen at the 400-cl/t surface indicates a band slightly lower in oxygen at about the position of eastward geostrophic flow west of 170°W (Tsuchiya, 1968). Recently, the New Caledonian research vessel *Coriolis* made nine cruises along 170°E between November 1965 and August 1967. Sections from these cruises show low surface salinity (sometimes as low as 34‰) near 10°S, where eastward geostrophic flow is indicated (Rotschi and Lemasson,

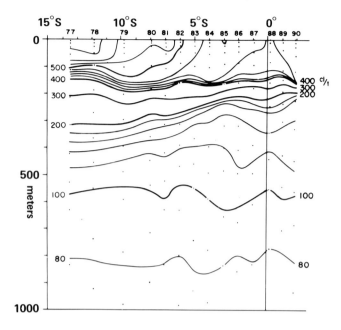

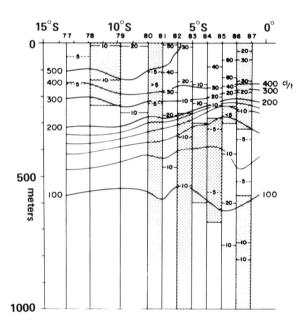

FIGURE 1   Thermosteric anomaly, in centiliters per ton (upper), and zonal geostrophic speed, in centimeters per second, relative to 1,000 decibars (lower), along 172°W. *Hugh M. Smith* cruise 8 (February–March 1951). Shaded areas denote eastward flow.

1968). Much of this low-salinity water, formed probably in the Southern Hemisphere by the excessive rainfall, is flowing east in the countercurrent. In the tongue of highly saline Tropical Water (δ_T about 350 cl/t) extending in the section from the south toward the equator, two cores of high salinity are revealed in almost all of the nine cruises. The slightly lower salinity (by approximately 0.1‰) be-

tween the two high-salinity cores is always associated with lower oxygen content. This low-salinity, low-oxygen water is found where the thermocline deepens toward the equator. Both salinity and oxygen on the isanosteric surface of 350 cl/t in the western South Pacific decreases westward. The low-salinity, low-oxygen water, therefore, suggests the importance of the eastward convection by the South Equatorial Countercurrent. It would be interesting to observe how far east the low-salinity, low-oxygen water could be traced.

Variations of the South Equatorial Countercurrent with longitude and with season remain to be studied. The countercurrent may be fed by the South Equatorial Current, but where and how it occurs is not yet clear. What happens to the countercurrent when it approaches the coast of South America remains obscure. Does it contribute to the Peru-Chile Undercurrent? Does it return to the west? These are all interesting problems to be investigated more thoroughly during the South Pacific expedition.

SOUTH EQUATORIAL CURRENT

The South Equatorial Current is the westward current bounded on the north by the North Equatorial Counter-current and extending southward across the equator to the axis of the South Pacific anticyclonic gyre. The Equatorial Undercurrent and the South Equatorial Countercurrent flow east in the region of the South Equatorial Current.

There are very few oceanographic sections covering the whole meridional extent of the South Equatorial Current, and, as far as I know, no detailed study of the current structure exists. For instance, there is little description of the South Equatorial Current in *The Oceans*, by Sverdrup *et al.* (1942), and, even now, the South Equatorial Current is the least investigated current of the Pacific equatorial current system.

Various surface-current charts indicate that the highest speed (exceeding 1 knot or 50 cm sec^{-1}) of the South Equatorial Current is found close to the equator in the eastern Pacific. Farther south, the current is weaker. There appear to be considerable seasonal variations in the current near the equator; the current is generally stronger in the northern summer than in the winter. According to Wyrtki (1965, Figures 1–12 and 16), the South Equatorial Current just north of the equator in the eastern Pacific is strongest in August and September and weakest in March. Puls (1895) noticed that the surface current at the equator between 115°W and the Galápagos Islands changes its direction to the east in March and April. This phenomenon may now be interpreted as a surfacing of the Equatorial Undercurrent during weakening of the easterly trade winds (Cromwell *et al.*, 1954).

The total flux of the South Equatorial Current is not

well known. Wyrtki (1967a) states that the flux north of 20°S and south of the North Equatorial Countercurrent is 160–220 km³ hr⁻¹. Montgomery and Stroup (1962, p. 56) have estimated the zonal geostrophic flux over the 300-decibar surface for the South Equatorial Current between 6°N and 7°S at 150°W in July and August 1952 and have obtained a westward flux of 227 km³ hr⁻¹. Most of this flux is concentrated in the surface homogeneous layer above the δ_T = 400 cl/t surface.

Water characteristics of the South Equatorial Current exhibit an interesting variety. Water above the 400-cl/t surface shows a large variation in salinity, ranging from 34.8 to 36‰. This variation arises from the meridional salinity gradient associated with an abrupt termination of the high-salinity tongue of southern Tropical Water near the equator. Between the 200-cl/t and 400-cl/t surfaces, although the flux is small, the South Equatorial Current north of the equator carries water with salinity less than 35‰, and that part of the current south of the equator carries water of much higher salinity. (The Equatorial Undercurrent carries water of intermediate salinity.) Below the 200-cl/t level, the southern part of the South Equatorial Current transports water of low salinity and high oxygen content, while the northern part transports water of relatively high salinity and extremely low oxygen content. In the upper layers, above the 300-cl/t surface, oxygen in the South Equatorial Current decreases downstream, but in the lower layers, below about the 200-cl/t surface, consumption of oxygen is exceeded by replenishment due to mixing, and oxygen content increases downstream.

A vertical minimum of salinity is found at about the 200-cl/t surface in the South Equatorial Current of the eastern South Pacific. This minimum is distinct from that of the Antarctic Intermediate Water, centered in the 80–100-cl/t zone, and may correspond to the shallower minimum (the deeper minimum, at the 125-cl/t surface, represents the North Pacific Intermediate Water) originating in the eastern North Pacific and extending in the California and North Equatorial currents. The southern minimum originates in the Peru Current and spreads from the coast of South America between 10° and 36°S to about 12°S, 121°W (Reid, 1965, p. 17). However, the distribution and characteristics of this water are obscure because data are insufficient.

How the Equatorial Undercurrent interacts with the South Equatorial Current just above it remains to be studied. Knauss (1966) assumed a linear interaction as a first approximation and explained the difference in transport of the undercurrent in different years. Variations of the South Equatorial Current may affect the upper part of the undercurrent and may, thus, result in variations in the total flux. However, the South Equatorial Current does not seem to affect the lower part of the undercurrent. The possibility that the flux of the undercurrent between the 155-cl/t and 165-cl/t surfaces is uniform with time has been stated before.

The South Equatorial Current feeds the East Australian Current, the South Equatorial Countercurrent, and part of the Equatorial Undercurrent. The detailed process of this feeding is an interesting and important problem that must be investigated more thoroughly.

CONCLUDING REMARKS AND RECOMMENDATIONS

This report summarizes our knowledge of the South Pacific equatorial circulation and points out what I think are exciting problems to be studied during the proposed expedition in the South Pacific Ocean.

There are some points that are worth mentioning in connection with the implementation of fieldwork. For studying the Equatorial Undercurrent, it is becoming a common practice to space both hydrographic stations and sampling depths more closely than in earlier observations. However, stations farther south are usually spaced at greater intervals. In view of the discovery of the South Equatorial Countercurrent, this station spacing seems to me unsatisfactory. The width of the countercurrent is limited to about 2° of latitude, and spacing stations at intervals greater than 1° is obviously not adequate for studying it. The differences in properties between countercurrent water and that to the north and south are slight. Therefore, measurements must be made in such a way that values of properties, particularly in the thermocline, are less ambiguous than they have been in the past. Close spacing of sampling bottles and use of the *in situ* salinometer (STD) are strongly recommended, even outside the region of the Equatorial Undercurrent.

Another point worth mentioning is the similarity between the Pacific and Atlantic equatorial circulations. I have examined some data from the intertropical Atlantic and found that the circulation system is essentially the same for the two oceans. Reference to Atlantic Ocean conditions would help to understand those in the Pacific. For instance, the source area of the Atlantic Equatorial Undercurrent has been investigated more thoroughly than the source area of the Pacific Ocean analogue (Metcalf and Stalcup, 1967), and examination of Atlantic data may facilitate the study of the Pacific source.

In concluding, I would like to emphasize the need for revision of the names of the equatorial currents. The discoveries of the Equatorial Undercurrent and the South Equatorial Countercurrent made the name South Equatorial Current quite ambiguous. What has been called the South Equatorial Current may be divided into two or more currents. I do not now propose any particular division; the best time to do that would be at the completion of the South Pacific expedition. By that time, we will have a better understanding of the distributions and characteristics of the equatorial currents.

REFERENCES

Burkov, V. A., and I. M. Ovchinnikov (1960) Structure of zonal streams and meridional circulation in the central Pacific during the Northern Hemisphere winter (in Russian). *Trud. Inst. Okeanol.*, *40*, 93–107.

Cochrane, J. D. (1967) Preliminary report on the Texas A & M EASTROPAC cruise, 21 January to 10 April 1967. Texas A & M University, Department of Oceanography, Texas A & M Reference 67-5-T, 21 pp., 5 figures.

Cromwell, T., R. B. Montgomery, and E. D. Stroup (1954) Equatorial Undercurrent in Pacific Ocean revealed by new methods. *Science*, *119*, 648–649.

Knauss, J. A. (1966) Further measurements and observations on the Cromwell Current. *J. Mar. Res.*, *24*, 205–240.

Kort, V. G., V. A. Burkov, and K. A. Chekotillo (1966) Recent data on the equatorial currents in the western part of the Pacific (in Russian). *Dokl. Akad. Nauk SSSR*, *171*, 337–339.

Koshliakov, M. N., and V. G. Neiman (1965) Some results of measurements and calculations of zonal currents in the equatorial part of the Pacific Ocean (in Russian). *Okeanologiia*, *5*, 235–249.

Masuzawa, J. (1967) An oceanographic section from Japan to New Guinea at 137°E in January 1967. *Oceanogr. Mag.*, *19*, 95–118.

Metcalf, W. G., and M. C. Stalcup (1967) Origin of the Atlantic Equatorial Undercurrent. *J. Geophys. Res.*, *72*, 4959–4975.

Montgomery, R. B., and E. D. Stroup (1962) Equatorial waters and currents at 150°W in July–August 1952. *Johns Hopk. Oceanogr. Stud.*, *1*, 68 pp.

Puls, C. (1895) Oberflächentemperaturen und Strömungsverhältnisse des Aequatorialgürtels des Stillen Ozeans. *Arch. Dtsch. Seewarte*, *18* (1), 38 pp., 4 Tafeln.

Reid, J. L., Jr. (1959) Evidence of a South Equatorial Countercurrent in the Pacific Ocean. *Nature*, *184*, 209–210.

Reid, J. L., Jr. (1961) On the geostrophic flow at the surface of the Pacific Ocean with respect to the 1,000-decibar surface. *Tellus*, *13*, 489–502.

Reid, J. L., Jr. (1965) Intermediate waters of the Pacific Ocean. *Johns Hopk. Oceanogr. Stud.*, *2*, 85 pp.

Rotschi, H., and L. Lemasson (1968) Variations observed during two years in the western equatorial Pacific. *Adv. Fish. Oceanogr.*, *Tokyo*, *2*, 13–15.

Sverdrup, H. U., M. W. Johnson, and R. H. Fleming (1942) *The oceans, Their Physics, Chemistry and General Biology*, Prentice-Hall, 1087 pp., 7 charts.

Tsuchiya, M. (1967) Distribution of salinity, oxygen content, and thickness at 160 cl/t of thermosteric anomaly in the intertropical Pacific Ocean. *Stud. Trop. Oceanogr.*, *5;* Proceedings of the International Conference on Tropical Oceanography, November 17–24, 1965, Miami Beach, pp. 37–41.

Tsuchiya, M. (1968) Upper waters of the intertropical Pacific Ocean. *Johns Hopk. Oceanogr. Stud.*, *4*.

Wooster, W. S. (1961) Further evidence of a Pacific South Equatorial Countercurrent. *Deep-Sea Res.*, *8*, 294–297.

Wyrtki, K. (1965) Surface currents of the eastern tropical Pacific Ocean. *Bull. Inter-Amer. Trop. Tuna Comm.*, *9*(5), 269–304.

Wyrtki, K. (1967a) Circulation and water masses in the eastern equatorial Pacific Ocean. *Int. J. Oceanol. Limnol.*, *1*, 117–147.

Wyrtki, K. (1967b) Oceanographic observations during the Line Islands Expedition, February–March, 1967. Hawaii Institute of Geophysics, University of Hawaii, HIG-67-17, 35 pp.

Henri Rotschi

CENTRE DE NOUMEA, NEW CALEDONIA,
OFFICE DE LA RECHERCHE SCIENTIFIQUE ET
TECHNIQUE OUTRE-MER (O.R.S.T.O.M.)

VARIATIONS OF EQUATORIAL CURRENTS

From October 1965 to May 1968 the research vessel *Coriolis* of the Centre O.R.S.T.O.M. of Nouméa, New Caledonia, made 10 cruises along 170°E, from 20°S to 4°N. On seven of these cruises, direct measurements were made of the equatorial current system. Some of the most significant preliminary results of these cruises are presented in this paper.

CURRENT MEASUREMENTS

Direct current measurements were made with either one or two self-recording Hydro-Products current meters that were attached to the hydrographic cable. The measurements were made immediately after hydrological stations during which the ship was maneuvered to find the proper speed and bearing for keeping the hydrographic cable vertical.

The first current meter was attached to the cable after 1,000 m of it had been paid out. The second meter was attached 1,000 m above the first, and the measurements were made keeping the shallow meter in the depth range of 0–500 m, which put the deep meter in the range of 1,000–1,500 m below the surface. One continuous measurement lasting 4 minutes was made every 20 m in the 0–300-m depth range and every 50 m at depths from 300 to 500 m. Measurements were made again at these intervals while the cable was being retrieved. During the whole series of measurements, the speed and bearing of the ship were kept constant so that the angle of the cable would be minimal. If one current meter was out of order, the same technique was used with the remaining meter, which was sent at regular intervals to depths of 500 or 1,000 m in order to give a deep reference. During the first two cruises, the reference level was 500 m.

In this technique, it is considered that the water in the depth range 1,000–1,500 m is motionless and that the meter in it records the drift of the ship. A vector subtraction gives the true current. In any case, it must be kept in mind that the current velocities in the upper layers are relative to the velocities in the 1,000–1,500-m layer (or at 500 m for the first two cruises). Obviously, there can be no guarantee that the drift of the ship remained constant during the station or that the evaluation of surface drift is reliable.

A study of the current at 500 m relative to that at 1,000 m during two prolonged stations at the equator, at 170°E and 169°E (Figure 1), shows some variation in velocity of both the east–west and the north–south components, but with an average that is the same for both stations: approximately 0.4 knot (positive to the east) for the east–west component, and –0.2 knot (positive to the north) for the north–south component. Similarly, the comparison of the results obtained 5 weeks later at another prolonged station at the equator and 170°E, shows a current at 500 m relative to 1,000 m with the same east–west component of 0.4 knot and a north–south component of –0.2 knot.

Thus, at prolonged stations at two different locations on the equator and at 5-week intervals, the current at 500 m relative to that at 1,000 m was found to be the same. Assuming a stable deep current, evaluation of the drift of the ship by the above method appears to be acceptable. The measurements of the east–west component of the current are computed to within ± 0.1 knot.

Five cruises were made at 5-week intervals, in March, April, June, July, and August 1967. The eastward flux of the Equatorial Undercurrent was computed using the 0 cm sec^{-1} isotach as the northern boundary, 4°S as the southern boundary, 400 m as the lower boundary, and assuming a narrowing of the flow to the east, which has always been found toward 3°N (Figure 2). The flux computed in Figure 2 showed considerable variation over a 3-month period, from April to July, increasing about fourfold; within a 1-month period, from March to April or from July to August, it decreased by about half (Table 1).

It must be pointed out that during all the cruises but one

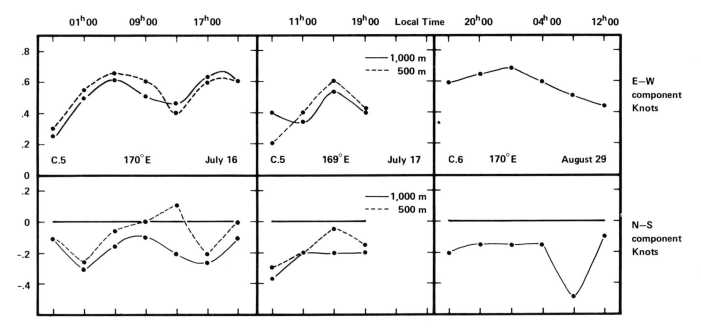

FIGURE 1   East–west and north–south component of the current at 500 m relative to 1,000 m at 0° and 170°E and 169°E, respectively, during cruise Cyclone 5, and at 0° and 170°E 5 weeks later, during cruise Cyclone 6. (Reprinted with permission from B. Warren, ed., *Progress in Oceanography*, vol. 6. Copyright © 1968, Pergamon Press Ltd.)

(cruise Cyclone 3), the Equatorial Undercurrent existed at least between the 500-cl/t and 150-cl/t isanosteric levels (Figure 3); when the flow was at a minimum during this cruise, the undercurrent existed only between the 350-cl/t and 150-cl/t levels.

Another point worth noting is that there is an obvious continuity between the "North Equatorial Countercurrent,"[*] the Equatorial Undercurrent, and a subsurface flow to the east that extends the undercurrent to the south and to greater depth (Figure 2). The deeper part of this third flow, below the 150-cl/t isanostere shows some stability; between this isanosteric level and the 400-m depth, a flux of about 4×10^6 m³ sec⁻¹ has been measured (Table 2).

Moreover, on three cruises, Cyclone 2, 5, and 6, the eastward flow had two cores, the upper one at the 450–500-cl/t level and the deeper one close to the 150–250-cl/t level (Table 2). The minimum flux is close to the 350–450-cl/t level. It appears also (Figure 3) that the salinity maximum of the Subtropical Lower Water of the South Pacific is distinct from the lower core of the undercurrent. Thus, one can consider that the flow of the undercurrent is vertically stratified, with, from top to bottom, an upper core in which the thermosteric anomaly is greater than 450 cl/t, a layer of

[*]As suggested by Tsuchiya in this publication, the existence of a pattern of zonal currents more complicated than the classical three-current system (which is actually in use) makes it necessary to revise the terminology. The term "North Equatorial Countercurrent," used here, refers to the so-called Equatorial Countercurrent and allows for the existence of a "South Equatorial Countercurrent," the permanent presence of which has now been demonstrated.

TABLE 1   Variations of the East Flux of the Equatorial Undercurrent Observed during Five Cruises of the Research Vessel *Coriolis* in 1967

Cyclone Cruise	Month	Depth of Reference (m)	Flux (10⁶ m³ sec⁻¹)
2	March	500	20
3	April	500	12
4	June	1,000	34
5	July	1,000	54
6	August	1,000	28

TABLE 2   Variations of the Zonal Flux in Various Layers as Observed during Five Cruises of the Research Vessel *Coriolis* in 1967

Level (cl/t)	Flux (10⁶ m³ sec⁻¹) Cyclone 2	Cyclone 3	Cyclone 4	Cyclone 5	Cyclone 6
>500	0.1	0	2.0	3.9	1.3
450–500	3.3	0	1.0	8.1	2.8
400–450	0.4	0	2.0	6.1	1.7
350–400	0.7	0	3.3	5.1	1.5
300–350	1.5	0.8	4.9	5.9	1.5
250–300	2.1	2.0	4.8	4.9	2.5
200–250	4.8	2.3	5.4	4.6	5.0
150–200	3.0	3.2	6.8	8.7	7.2
<150	4.4	3.4	4.0	6.1	4.0

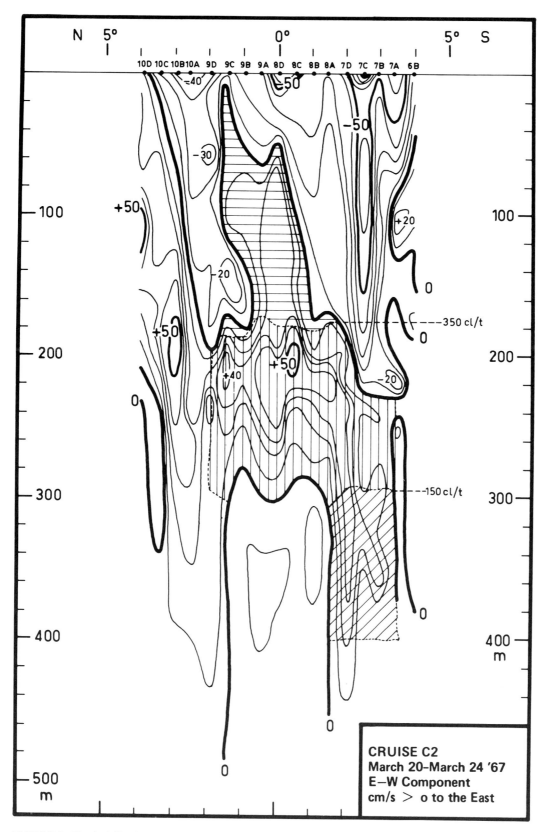

FIGURE 2   Vertical distribution of the east–west component of the currents at 170°E, between 4°S and 4°N, in March 1967, cruise Cyclone 2. (Reprinted with permission from B. Warren, ed., *Progress in Oceanography*, vol. 6. Copyright © 1968, Pergamon Press Ltd.)

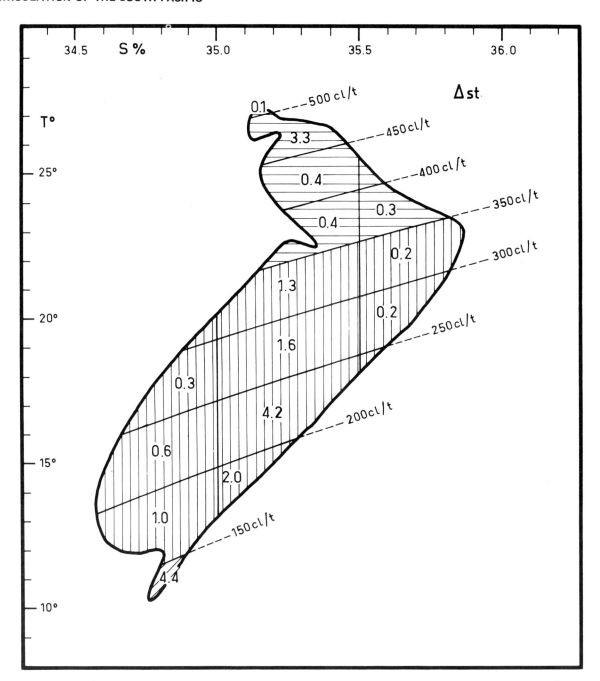

FIGURE 3   Fluxes in the Equatorial Undercurrent for various classes of salinity and thermosteric anomaly, at 170°E, in March 1967, cruise Cyclone 2 (10^6 m^3 sec^{-1}). Total transport = 20.

minimum flux with a thermosteric anomaly between 350 cl/t and 450 cl/t, a salinity maximum with a thermosteric anomaly equal to 250–350 cl/t, a lower core at 150–250 cl/t, and the deep flow mentioned earlier. The most stable part of the flux is the deepest part, which is formed of water originated at intermediate depth in the Southern Hemisphere, and also the lower core, which is a mixture of Equatorial Water and North Equatorial Countercurrent Water. The least stable part is the shallowest, composed of surface

and subsurface water of both the Northern and the Southern hemispheres (Table 3).

It is interesting to compare these measurements to those made farther east, at 150°W, by Montgomery and Stroup (1962), who computed a flux of 34 × 10^6 m^3 sec^{-1} relative to 300 dB; this flux is comparable to that found during cruise Cyclone 4 of the research vessel *Coriolis*.

It can be seen from Table 4 that the flux above the 350-cl/t isanosteric surface decreased from west to east; in the

TABLE 3 Variations of the Flux in the Five Main Layers of the Undercurrent

Layer	Thermosteric Anomaly (cl/t)	Flux (10^6 m^3 sec^{-1})				
		Cyclone 2	Cyclone 3	Cyclone 4	Cyclone 5	Cyclone 6
Upper core	>450	3.4	0	3.0	12.0	4.1
Minimum of flux	350–450	1.1	0	5.3	11.1	3.2
Salinity maximum	250–350	3.6	2.8	9.7	10.8	4.0
Lower core	150–250	7.8	5.5	12.2	13.3	12.2
Deep flux	<150	4.4	3.4	4.0	6.1	4.0

TABLE 4 Comparison of the Fluxes at 150°W and 170°E (10^6 m^3 sec^{-1})

Layer (cl/t)									
		Increase to the East		Unchanged		Decrease to the East			
Longitude	150	150–200	200–250	250–300	300–350	350–400	400–450	450–500	500–550
150°W	1.9	9.4	7.8	4.9	5.0	2.8	1.9	0.7	0.0
170°E	4.0	6.8	5.4	4.8	4.9	3.9	2.0	1.0	2.0

layer of the salinity maximum, the east and west fluxes were the same, and in the core layer and below there was an increase from west to east. The difference between the east and west fluxes below 150 cl/t is probably due to the fact that the flux obtained from data of cruise Cyclone 4 has been evaluated to 400 m. Finally, it seems that the undercurrent cools as it moves to the east.

EQUATORIAL UPWELLING

In the western part of the Pacific Ocean, cooling of the equatorial surface waters by upwelling has seldom been observed, except at 150°E at the beginning of the year. Nevertheless, the atlases of temperature distribution at the surface indicate that there could be an extension to 170°E of the equatorial upwelling, but at the beginning of the year only. Thus, it could be considered that enrichment of the upper layers of this region of the Pacific is unlikely, in spite of the fact that the average wind has a noticeable westward component.

During all the cruises of the research vessel *Coriolis* along 170°E, the wind observed in the equatorial region had a westward component stronger than 2 m sec^{-1}, except in December 1965 and in June 1966, when it blew for short periods from the west. The Ekman transport, as observed from the drift of the ship and from the direct current measurements, was in the direction of the wind, except in April

1967, when, as observed during cruise Cyclone 3, the surface current flowed toward the east, in spite of a rather strong westward wind. During June and September 1966, a tendency for a surface divergence to exist at the equator has been noted.

Thus, in most places and at most times, conditions were such as to induce a surface divergence, and a divergence was indeed observed during most of the cruises (Figure 4). The vertical distribution of temperature in the upper 100 m, between 4°S and 4°N, gives a good picture of the surface circulation; it indicates that there was upwelling during all cruises except those of December 1965 (cruise Bora 1) and of April 1967 (cruise Cyclone 3). During these two cruises, on the contrary, the eastward surface current induced a convergence, which was responsible for the low flux of the undercurrent measured during cruise Cyclone 3.

The vertical distribution of the nutrients phosphate and nitrate confirms that there is effectively an upwelling, with enrichment of the upper layer at the equator. During the two intervals when there was a convergence, the surface waters were poorer in nutrients and no enrichment at the equator relative to the adjacent waters was observed.

The flux minimum of the Equatorial Undercurrent, observed during cruise Cyclone 3 in April 1967, is directly bound to the particular structure of the currents during this period. On all the other cruises, the direct measurements indicated that the westward flow at the surface is very shallow and does not extend, at the equator, deeper than 60 m.

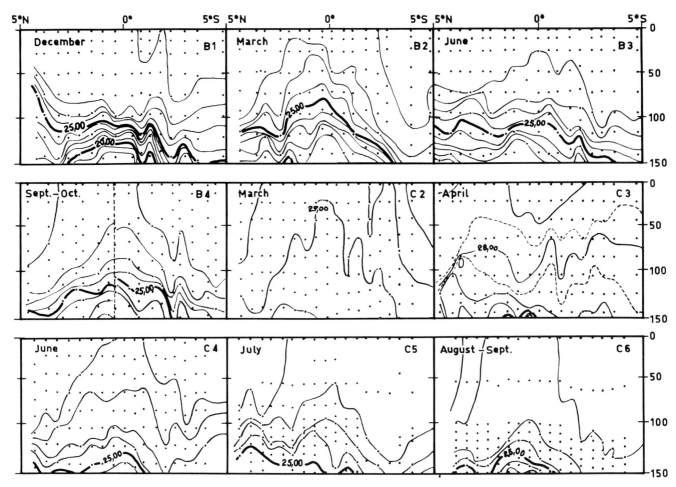

FIGURE 4 Vertical distribution of temperature (°C) in the upper 150 m, at 170°E, during various cruises of the research vessel *Coriolis*. (Reprinted with permission from Rotschi, 1968.)

Below this depth, there is a continuous flow to the east down to a depth of at least 300 m. Farther north and south, the eastward current extends much deeper. During cruise Cyclone 3, when the flux of the undercurrent was smallest, the vertical structure at the equator was different: The Equatorial Undercurrent proper existed only between 200 m and 300 m; in the upper 100 m, between 2°N and 2°S, there was a flow to the east; in the intermediate layer (100–200 m), the flow was to the west. Thus, in the presence of a surface convergence, the undercurrent was deprived of its shallower part, which is composed of warm and less saline water; this loss, added to the obvious slackening of the speed at the core, was responsible for the considerable decrease of the flux.

SOUTH EQUATORIAL COUNTERCURRENT

Jarrige (1968), in a preliminary study of the data of all the cruises of the research vessel *Coriolis* from the point of view

of the geostrophic circulation, found a permanent eastward component of the surface current relative to 1,000 dB. This flow, located near 10°S, corresponds to the South Equatorial Countercurrent, shown in various atlases of the surface circulation and reported by Reid (1959).

The characteristics of this flow are highly variable (Table 5). Its depth was found to be near 200 m on one cruise, 300 m on two others, and greater than 500 m on all others. Nevertheless, the velocity core, with a velocity of 10–30 cm sec^{-1}, has always been very close to the surface. The average width of this eastward component of the surface current is 560 km, varying between 330 km and 620 km; the southern boundary, which is bound to a permanent lessening of the westward component of the wind, does not vary much in latitude, whereas the northern boundary varies much more. Moreover, the upper 100 m of the current is always associated with a salinity minimum having salinity values between 34.00‰ and 34.80‰.

The existence of this current has been confirmed by the capture at 10°S, 170°E, of *Euphausia fallax*, which origi-

TABLE 5 Characteristics of the South Equatorial Countercurrent at 170° E (after Jarrige, 1968)

Cruise	Date	Extreme Latitude	Depth (m)	Maximum Velocity (cm/sec)	Latitude of the Velocity Maximum	Width (km)	Volume Transport (10^6 m^3 sec^{-1})	Salinity Minimum (‰)	Drift of the Ship
Bora 1	Dec. 1965	7°00′ S–12°15′ S	160	12	9°45′ S	580	2.2	34.6	—
Bora 2	March 1966	4°00′ S–10°15′ S	>500	31	7°50′ S	695	19.6	34.3	6°00′ S–5°00′ S 1 knot–ENE
Bora 3	June 1966	5°50′ S–12°05′ S	>500	24	6°50′ S	685	9.6	34.5	9°00′ S
Bora 4	Sept.–Oct. 1966	7°30′ S–10°25′ S	>500	29	8°00′ S	330	8.8	34.7	0.2 knot–E
Cyclone 2	March 1967	8°20′ S–12°10′ S	>500	31	9°40′ S	410	18.6	34.8	10°30′ S–9°00′ S 0.8–1.0knot–E
Cyclone 3	April 1967	5°00′ S–12.45′ S	>500	22	6°00′ S	930	14.7	34.2	9°00′ S 1.5–2.0 knot–SE
Cyclone 4	June 1967	8°45′ S–13°05′ S	>500	20	12°00′ S	480	12.4	34.0	11°30′S–11°00′S 1.0 knot–E
Cyclone 5	July 1967	9°20′ S–13°00′ S	280	21	12°00′ S	410	6.6	34.8	
Cyclone 6	August 1967	8°25′ S–13°10′ S	220	15	9°00′ S	540	3.4	34.8	—

nates north of New Guinea, and by the capture at about the same location of very young stomatopod larvae that could have originated only in the northern part of the New Hebrides archipelago. Recent GEK measurements indicate that in May 1968 there was an eastward drift at the latitude of the geostrophic South Equatorial Countercurrent. Finally, star fixes and dead reckoning always indicated an eastward drift of the ship at the latitude where the geostrophic eastward current exists.

THE HYDROLOGICAL STRUCTURE OF THE EQUATORIAL UNDERCURRENT

Knauss (1960) has noted that in the core of the Equatorial Undercurrent oxygen content did not vary and that this homogeneity can be considered to be an indication of a fairly strong vertical mixing process within the core of the undercurrent.

A close examination of the measurements made along the equator from the research vessel *Coriolis* during its Alize cruise (Rotschi *et al.*, 1967), however, shows that the core of the undercurrent is not really homogeneous in oxygen con-

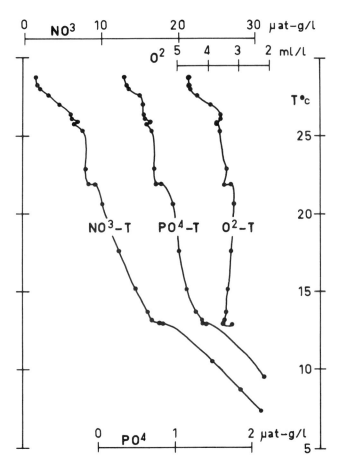

FIGURE 6 Oxygen-temperature, phosphate-temperature, and nitrate-temperature diagrams at the equator and 170°E in July 1967. (Reprinted with permission from Pickard and Rotschi, 1968.)

tent; on the contrary, there is an oxygen minimum at the upper boundary of this layer and a maximum at its lower boundary, the increase from the minimum to the maximum amounting to about 0.10 ml/l (Figure 5). It seems somewhat surprising that a mechanism of vertical mixing could create such a distribution. A further detailed study of the hydrological structure of the undercurrent has shown that in the same layer, where the oxygen content is apparently constant, the concentrations of nitrate and of phosphate increase substantially with depth. In fact, it was observed at 170° in July 1967 (Pickard and Rotschi, 1968), that in the core, where the speed was higher than 50 cm sec^{-1}, there was a chemical two-layer system. In the upper layer, a little less than 100 m thick (between 100 m and 200 m in depth) and lying between two thermal inversions, the oxygen decreased with depth (Figure 6). In the lower layer, at depths between 200 m and 300 m, oxygen increased slightly with depth. In both layers, nitrate and phosphate increased with depth, but in the deeper one, the gradient was greater.

Such a stratification and such different depth variations of the concentrations of variables that are bound to each

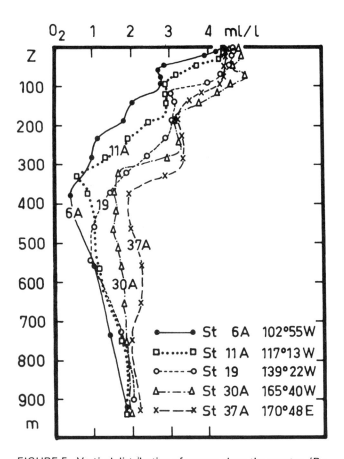

FIGURE 5 Vertical distribution of oxygen along the equator. (Reprinted with permission from Rotschi and Wauthy, 1969.)

other do not support belief in strong vertical mixing. They suggest, on the contrary, that in the western Pacific, at least, the water of the core of the Equatorial Undercurrent is stratified and that the stratification is favored by stable, quasipermanent thermal inversions.

Thus, it could well be that the apparent oxygen homogeneity of the undercurrent water is due to the formation of this current from different water sources at different depths that have approximately the same oxygen content purely by chance. This view is supported by Tsuchiya (1967) and by his analysis of oxygen distributions on the 160-cl/t isanosteric surface. Recent studies undertaken at the Centre O.R.S.T.O.M. of Nouméa have shown that the undercurrent can be formed in its upper part by Tropical and Subtropical Water of the south Pacific, crossing the equator north of New Guinea, and by water of the North Equatorial Countercurrent. As for the lower part, the same studies confirm Tsuchiya's suggestion that the Coral Sea is a likely source region.

REFERENCES

Jarrige, F. (1968) On the eastward flow in the western Pacific south of the equator. *J. Mar. Res.*, *26*(3), 286–289.

Knauss, J. A. (1960) Measurements of the Cromwell Current. *Deep-Sea Res.*, *6*(4), 265–286.

Montgomery, R. B., and E. D. Stroup (1962) Equatorial waters and currents at 150°W in July–August 1952. *Johns Hopk. Oceanogr. Stud.*, *1*. Johns Hopkins Press, Baltimore, Md., 68 pp.

Pickard, G. L., and H. Rotschi (1968) Structure hydrologique associée au courant de Cromwell dans le Pacifique occidental. *C.R. Acad. Sci. Paris, 267*, 1557–1560.

Reid, J. L. (1959) Evidence of a South Equatorial Countercurrent in the Pacific Ocean. *Nature, 184*, 209–210.

Rotschi, H, Ph. Hisard, L. Lemasson, Y. Magnier, J. Noel, and B. Piton (1967) Résultats des observations physico-chimiques de la croisière Alize du N.O. *Coriolis. Rapp. Scient. Inst. Franc. Ocean. Sci. 2*, 56 pp.

Rotschi, H. (1968) Remontée d'eau froide et convergence à l'equateur dans le Pacifique occidental. *C.R. Acad. Sci. Paris, 267*, 1459–1462.

Rotschi, H., and B. Wauthy (1969) Remarques sur le courant de Cromwell. *Cahier O.R.S.T.O.M. ser. Oceanogr., III, 2*, 27–43.

Tsuchiya, M. (1967) Distribution of salinity, oxygen content and thickness at 160 cl/t of thermosteric anomaly in the intertropical Pacific Ocean. *Stud. Trop. Oceanogr., Miami, 5*, 37–41.

III
GEOLOGY OF
THE SOUTH PACIFIC

Henry W. Menard

SCRIPPS INSTITUTION OF OCEANOGRAPHY,
LA JOLLA, CALIFORNIA

MARINE GEOLOGY

The concepts of sea-floor spreading and plate tectonics have implications and ramifications that profoundly influence thinking about all aspects of the history and development of ocean basins. The concepts permit a wide range of detailed predictions that are now being confirmed, and thus they vastly increase our certainty about some aspects of the history. On the other hand, many new uncertainties are introduced by the probability that every part of the oceanic crust has moved large distances and in various directions. This paper explores some of the confirmations and implications of the concepts, with particular reference to the South Pacific.

The motion of a plate on the surface of a sphere can be described in terms of motion along small circles relative to a pole that may be inside or outside the plate. The earth is such a sphere, and its crust is broken into such moving plates. Within a plate of the earth's crust, there is no relative horizontal motion, and no earthquakes occur other than those associated with vulcanism and vertical adjustments. Hence, the boundaries of crustal plates can be defined by the locations of lines of earthquakes and the directions of the first motions of earthquakes. Such studies reveal that enormous regions, such as half the South Pacific, are moving uniformly relative to a single pole. The entire North Pacific is moving as a single plate toward Japan and the adjacent trenches.

The boundaries of a plate consist of spreading centers, located at the edge where crust is created; trenches at the opposite edge, where crust is destroyed; and transform faults along the sides connecting these edges. When the direction of spreading is constant for some time, the transform faults are oriented along small circles parallel to the motion of a plate. The spreading centers are on radii pointing toward the pole around which the plate is moving. The record of ancient motions is fossilized in the magnetic anomalies and fracture-zone topography associated with midocean ridges. In the best surveyed regions in the north-

eastern Pacific, these fossil records show that the direction of movement of plates has changed repeatedly by small amounts, although some feedback mechanism preserves an average trend.

The movement of very large plates across enormous distances occurs at rapid rates—particularly rapid when compared with pelagic sedimentation. Thus, a meter of red clay is deposited in about a million years. During that time, most of the sea floor moves 10–50 km. On a larger but entirely realistic scale, 200 m of sediment equals 2,000–10,000-km displacement of the sea floor. More rapid deposition of calcareous oozes, at roughly 10 m per million years, occurs in the shallow water of ridge crests. The crust spreads and the sea floor subsides below the depth at which such oozes dissolve in roughly 20 million years, regardless of spreading rate. Thus, the bottom of any drilled stratigraphic section in a deep ocean basin might be expected to be 200 m of calcareous ooze with red clay above. Exceptions would be expected where the sea floor has moved under the curtain of rapid carbonate deposition along the equator. For example, the floor west of the Japan trench was initially well south of the equator, according to paleomagnetic measurements of seamounts. If the motion corresponded to that of the present, the crust moved from the direction of the East Pacific Rise. Thus, the sediment northwest of the intersection of the rise and the equator should be exceptionally thick, as it is, and should include a thick section of carbonates under red clay. On the other hand, the sediment on the crust that has spread northwestward from the west of the East Pacific Rise north of the equator should be relatively thin, which it is.

A population of abyssal benthic organisims living on the crest of the East Pacific Rise for 10 million years would be dispersed in both directions over a width of 200–1,000 km. In perhaps 100 million years, it would be spread across most of the South Pacific. If it were depth-limited, the living members would be confined to the crest, but the fossils would be spread across the South Pacific at increasing depth below the sediment.

The erosion products of the continents are deposited on the sea floor in particular environments and geographical configurations. Manganese nodules, for example, occur in deep regions where deposition is slow. As these regions are swept into island arcs, erosion products are restored to the continental crust from which they came, but not in the original proportions. Thus, sea-floor spreading, acting through geological time, may account for major regional variations in the chemical composition of the continental crust and in the composition of ores in metallogenic provinces.

Alexander P. Lisitzin
INSTITUTE OF OCEANOLOGY,
USSR ACADEMY OF SCIENCES, MOSCOW

SEDIMENTATION AND GEOCHEMICAL CONSIDERATIONS

INTRODUCTION

Until recently, the South Pacific has been one of the least thoroughly studied parts of the World Ocean from the geological point of view. Many unique sediment types accumulate here; in some areas of the southern arid zone of the Pacific, the rate of sedimentation is apparently smaller than anywhere else on our planet. Quite unusual conditions for sedimentation, hence a peculiar geochemical environment, exist here.

Bottom sediments were first sampled in the South Pacific during the English round-the-world expedition of the *Challenger* (1872–1874). The study of these and numerous other samples collected by the expedition from different areas of the World Ocean served as a basis for the development of marine geology as a science.

A new stage in marine geological investigations began with the preparations for the International Geophysical Year (IGY) and has continued to the present. Among the largest expeditions of the IGY period was the Soviet marine Antarctic expedition aboard the diesel-electric ship *Ob* (1955–1959), which occupied more than 500 geological stations in the Southern Ocean as well as the American expeditions combined in the "Deep-Freeze" operations (1954–1963).

Abundant samples of interest to marine geology were obtained in the U.S. expedition "Downwind" (1957–1958) aboard the ships *Horizon* and *Spencer F. Baird*, in the New Zealand expeditions on the *Endeavour* (since 1959), in the expeditions organized by the Lamont Geological Observatory (U.S.) on the *Vema*, and particularly during the continuing studies conducted during cruises of the U.S. research vessel *Eltanin* (1962–1967). A small number of sediment samples was obtained during Richard Byrd's expedition aboard the *Bear* (U.S.), 1939–1941, and during operation "High Jump" (U.S.) in 1946–1947.

Yet, vast areas of the Southern Ocean, even after the large-scale marine geological studies made during recent years, remain poorly studied.

Reports concerning the distribution and composition of suspended matter in the Southern Ocean are far less numerous than are those concerning bottom sediments. Despite the fact that the first determinations of suspended matter in the seas were made in 1890–1891, by Murray and Irvine using materials obtained by the *Challenger*, no investigations of this sort were carried out in later expeditions. Only during the Soviet expeditions on the *Ob* and *Vityaz* did studies of suspended matter begin to be made systematically for vast areas of the ocean. Studies of suspended matter in the Southern Ocean have recently been started as part of the Japanese Antarctic Expedition.

Unfortunately, the equipment and methods used by various expeditions differ considerably, making the results difficult to compare; in particular, C_{org} and $CaCO_3$ determinations made during Soviet and U.S. expeditions are significantly different. Methods of sampling sediment and suspended matter employed on the Soviet expeditions are described in many papers (Lisitzin and Udintsev, 1952; Sysoev, 1956; Zhivago and Lisitzin, 1957; Lisitzin and Zhivago, 1958; Lisitzin, 1956, 1959, 1960, 1961, 1962; Lisitzin and Zhivago, 1960; Lisitzin and Glazunov, 1960; Lisitzin and Barinov, 1960; etc.).

Because of the differences in methods of marine and analytical investigations, the materials obtained in the Soviet expeditions are used as the basis for the present discussion.

SEDIMENTARY SUPPLY

Of great significance for sedimentation in the South Pacific is the supply of clastic and clay sedimentary material from land as well as of the material formed as the result of the activity of organisms. These basic sources are supplemented locally by volcanogenic material. Of minor importance is the contribution of cosmogenic material.

The Supply of Sedimentary Material from Land

In the South Pacific, the most oceanic part of our planet, the ratio of drainage area to ocean surface area is minimal. Drainage areas emptying into the Pacific basin from South America and Australia make negligible contributions. Therefore, the main source of sedimentary material for the whole southern part of the ocean is Antarctica.

In Antarctica, the processes of chemical weathering are extremely weak. Predominant is the mechanical disintegration of bedrock of which the continent is composed. The leading role is played by ice, although in some places (oases, mountain ridges), wind activity is prominent.

Of great interest from the geochemical point of view is the fact that ice runoff is irregular; it is intensified in the regions with low relief beneath the ice, where a kind of ice riverbed, the so-called discharge glaciers, is formed. In contrast to the average rate of spreading of continental ice of about 20–45 m per year, the discharge glaciers move at a rate of 500–1,000 m per year and greater (Vtyurin et al., 1959; Evteev, 1964).

When moving, the glaciers cut off rock material from the surface within the ice-discharge area, crush it, and gradually carry it out of the central parts toward the ocean shores. Here the sedimentary material is partially unloaded near the terminus of the discharge glacier, giving birth to a terminal moraine on the ocean bottom. The bulk of the material, however, is carried away to the ocean by icebergs. Sedimentary material transported by glaciers is characterized by an extremely good preservation of rocks and minerals, without even traces of chemical change. According to recent data, the limit of the ice-discharge area in the Pacific goes far beyond West Antarctica and begins in the East Antarctic watershed (the Gamburtsev and Vernadsky under-ice mountains). From here, ice spreads up to the Pacific sector. This ice is supplemented by ice from other caps in West Antarctica—the Antarctic peninsula, Central, and Marie Byrd ice-caps (Kapitsa, 1968; Figure 1).

Ice discharge, according to different authors, totals for Antarctica from 550 to 780 km^3 to 1,873 km^3, with an average of 1,000–1,500 km^3 per year (Bardin and Shilnikov, 1960; Shilnikov, 1960, 1961; Maksimov, 1961; Nazarov, 1962). The average content of morainic material is about 1.6 percent by volume (Evteev, 1964), which yields a maximum discharge of terrigenous material from the continent of 16–24 km^3, or 40–60 billion (10^9) tons, per year. According to the author's observations, these figures should be reduced by at least 10 times. Annually, Antarctic glaciers cut off a layer of about 0.05 mm, which is close to the average values for river erosion of the Russian platform (0.03

mm) and of the land surface of the United States (0.04 mm per year).

According to Matveev's (1963) data, icebergs carry away to the ocean a total of about 1,900 million tons of clastic material per year. The most reliable values are 2–5 billion (10^9) tons. Of this quantity, no more than one third, i.e., 600–2,000 million tons, falls to the share of the Pacific sector. For Antarctica, the index of mechanical denudation is 134.6 tons per km^2, second in magnitude to that for Southeast Asia.

All these calculations are very rough, yet they enable one to draw the conclusion that Antarctica is the main supplier of terrigenous material to the South Pacific.

Other sources of terrigenous material are of minor importance. Australia, surrounded on the Pacific side by the Great Barrier Reef, contributes practically no sedimentary material. An essential role is played, therefore, by New Zealand and by the islands of the western portion of the ocean. The narrow, near-shore band along the South American coast belonging to the Pacific drainage area contributes coarse sedimentary material from mountains, which settles mostly in the near-shore band. This material is transported along the shores of the continent by the Peru Current and essentially does not penetrate the pelagic regions (Figure 2).

A. A. Matveev's (1963) calculations show that Antarctic ice discharge is equal to 1,184 km^3 of water. The average ice and snow mineralization, according to the same author, constitutes 12 mg per l and is due chiefly to the contribution of oceanic salts to the continent. The quantity of ice discharged to the Pacific sector does not exceed one third of the total. Whereas 15.45 million tons of salts are introduced to the ocean annually by Antarctic icebergs, only about 5 million tons per year fall to the share of the Pacific sector. The great thickness of continental ice results in the melting of a thin layer overlying the bedrock of the glacial floor, which causes a sharp increase in the mineralization and the change of salt composition. This change of salt composition is related to the rapid increase of the solubility of minerals when they are finely disintegrated (Keller et al., 1963). The annual chemical discharge owing to the under-ice melting constitutes about 1.6 million tons (Matveev, 1963).

These calculations show that the index of chemical denudation in Antarctica is many times lower than for any other continent and comprises 1.37 tons per km^2.

Ice is of importance in the further distribution of sedimentary material in the ocean. According to Kosak (1954), icebergs occupy a vast area in the Southern Ocean—62.3 million km^2, or an area almost five times as great as the area of the Arctic Ocean. In the whole of this large region, a substantial portion of the sedimentary material accumulating on the bottom has been transported by icebergs. It is natural that icebergs unload morainic material mostly near the Antarctic coasts. As a result, near the shores of the conti-

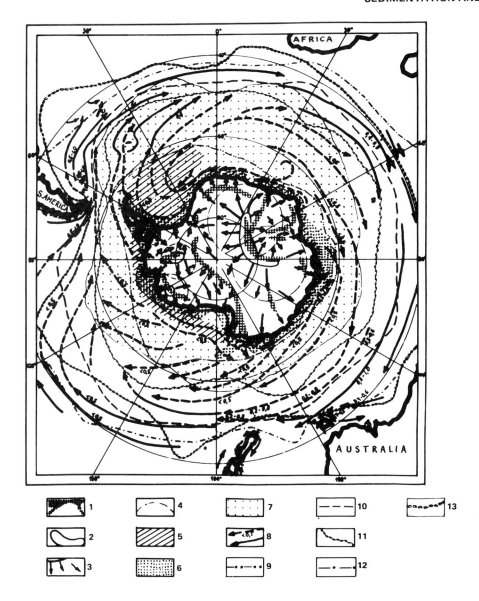

FIGURE 1 Sedimentary supply from Antarctica, the main source of terrigenous material for the South Pacific. (1) Major sub-ice ridges (ice-discharge boundaries); (2) icecaps; (3) direction of ice flow on land (paths of sedimentary transport on the continent); (4) theoretical boundaries of flow caps (after Kapitsa, 1968); (5) region of year-round sea ice; (6) region covered with icebergs in summer (more than 3/10 coverage); (7) regions where icebergs are rare in summer (1–3/10); (8) current direction and speed (in knots); (9) Antarctic Divergence; (10) Subtropical Divergence; (11) Antarctic Convergence; (12) Subtropical Convergence; (13) extreme boundary of iceberg occurrences (since the 18th century).

nent, iceberg sediments stretch as a continuous belt consisting almost entirely of land-denudation products and only in some places containing 10–30 percent biogenous siliceous material. Such sediments form a belt encircling the continent and having the width of 500–1,000 km (Lisitzin, 1968).

To the north, biogenous (siliceous and foraminiferal) sediments are found. In these sediments, iceberg material is no longer predominant, but it plays an important and peculiar role.

The rate of sedimentary supply by icebergs, as well as terrigenous supply from other continents, is far from constant. In some places it is very high, which is usually related to a wide development of discharge glaciers, while in other places it is very low. A low supply rate is most often observed near large shelf glaciers or in places where continental ice does not reach the ocean but is separated from it by a band of near-shore oases. Along with the rate of supply of rocks and other sedimentary material by icebergs, of importance also is their rate of unloading, which is determined by the thermal regime of the waters and air as well as by the paths of iceberg migration as determined by currents. Near the Antarctic coasts, icebergs and sedimentary material migrate with the East Wind Drift, and farther north with the West Wind Drift.

Petrographic study of the composition of rock material obtained from the ocean bottom at hundreds of the *Ob* oceanographic stations has shown that icebergs may transport their rock loads to a distance of 1,000–2,000 km in a straight line. In fact, the paths of icebergs are complicated curves, and the range of transportation is much greater (Lisitzin, 1958, 1961). Icebergs reach the southern extremi-

FIGURE 2 Sediment supply from the Pacific discharge area; basic zones of lithogenesis; major climatic zones of the Pacific Ocean established by the author from the relationship between the annual quantity of precipitation and evaporation, as well as thermal zonality, and upwelling areas and their associated regions of biogenous sedimentation. Boxes 1–4 show the mechanical supply of terrigenous sedimentary material (in tons per km^2 of the water discharge area), as follows: (1) 10–50, (2) 50–100, (3) 100–240, (4) >240; (5) arid regions; (6) solid discharge of the main rivers, millions of tons per year (Strakhov, 1960); (7) lines of the equal quantities of precipitation and evaporation; (8) divergence zones; (9) convergence zones (Burkov, 1963); (10) main current directions, speed below 0.5 knot; (11) main current directions, speed more than 0.5 knot; (12) trend of the bottom ridges dividing the ocean into separate basins; (13) year-round ice (December); (14) extreme boundary of ice.

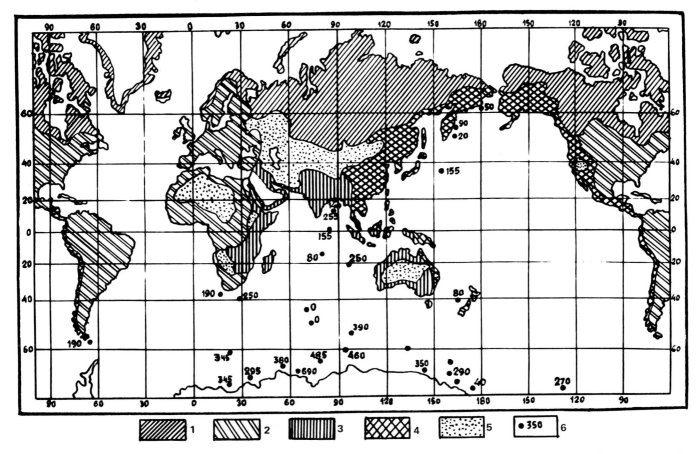

FIGURE 3   Absolute age of finely dispersed minerals of the surface-sediment layer from potassium–argon determinations. (1) Drainage area of the Arctic Ocean; (2) drainage area of the Atlantic Ocean; (3) drainage area of the Indian Ocean; (4) drainage area of the Pacific Ocean; (5) drainless areas; (6) absolute age of the upper sediment layer in millions of years (Krylov *et al.*, 1961).

ties of Australia and South America, yet the bulk of their sedimentary material is unloaded earlier, and most icebergs do not penetrate farther than the Antarctic Convergence.

River runoff in Antarctica is practically absent, though locally at the height of summer, temporary streams (streamlets and even small rivers) are formed here. The formation of terrigenous material owing to shore erosion, along with the wash-out from land, which is of great importance for other continents, is of no significance near Antarctica. This is accounted for by the fact that the Antarctic shores are composed of ice, and bedrock outcrops are uncommon. Where they are found, the shores are little changed by wave action since sea ice attenuates waves; wave abrasion of the Antarctic shores is rarely observed.

The geological structure of Antarctica leaves its imprint on the course of recent sedimentation in the Southern Ocean. The major provinces and paths of supply of sedimentary material from the continent to the ocean and the absolute age of the material in different localities have been determined on the basis of studies of the potassium–argon ratio of the finely dispersed fraction of muds from the Southern Ocean bottom and of correlations with the results

of the corresponding determinations on land (Krylov *et al.*, 1961; Figure 3). These investigations have opened up new possibilities for the mineralogical and petrographic analysis of bottom sediments. The composition and distribution of the Antarctic sedimentary material are also shown on the maps of clay minerals (Figure 4 *A–D*).

The Supply of Biogenous Sedimentary Material

Of the sediment-forming organisms, planktonic algae, especially diatoms with siliceous frustules, and foraminifera with carbonate tests are of paramount importance. The distribution of these organisms, as the studies of plankton, suspended matter, and bottom sediments proper show, is of a strictly zonal character: The northern limit in the distribution of the mass diatom accumulations is the Antarctic Convergence line. The same limit determines the replacement of biogenous siliceous sediments by carbonate sediments that stretch up to the equator, and beyond to the temperate latitudes of the Northern Hemisphere.

In the Southern Ocean, the main silica accumulation of our planet is concentrated; it is here that at least 70 percent

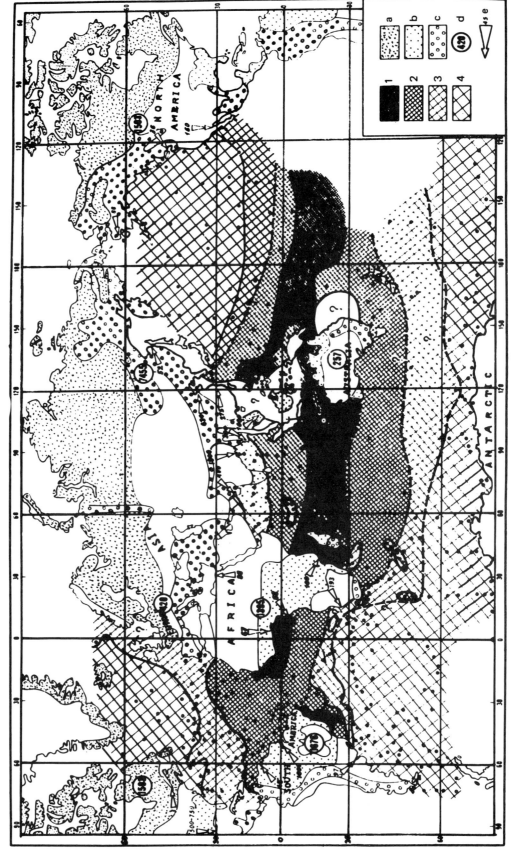

FIGURE 4 A–D Distribution of the major clay minerals in bottom sediments of the World Ocean (Rateev et al., 1966). Weathering zones on land: (a) temperate humid climate; (b) tropical humid climate; (c) tectonically active areas without the formation of crusts of weathering; (d) total solid discharge from the continents (in millions of tons per year); (e) direction and magnitude of solid discharge of the most important rivers (in millions of tons per year). Stations used indicated by filled circles.

FIGURE 4A Kaolinite. Concentration scales (in percent of kaolinite): (1) 60–40%; (2) 40–20%; (3) 20–10%; (4) <10%.

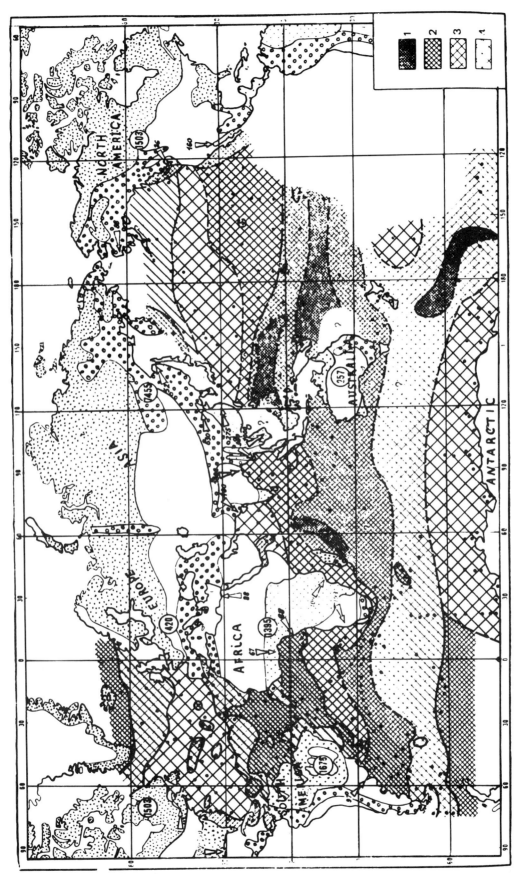

FIGURE 4B Montmorillonite. Concentration scales (in percent of montmorillonite): (1) 60–40%; (2) 40–20%; (3) 20–10%; (4) < 10%.

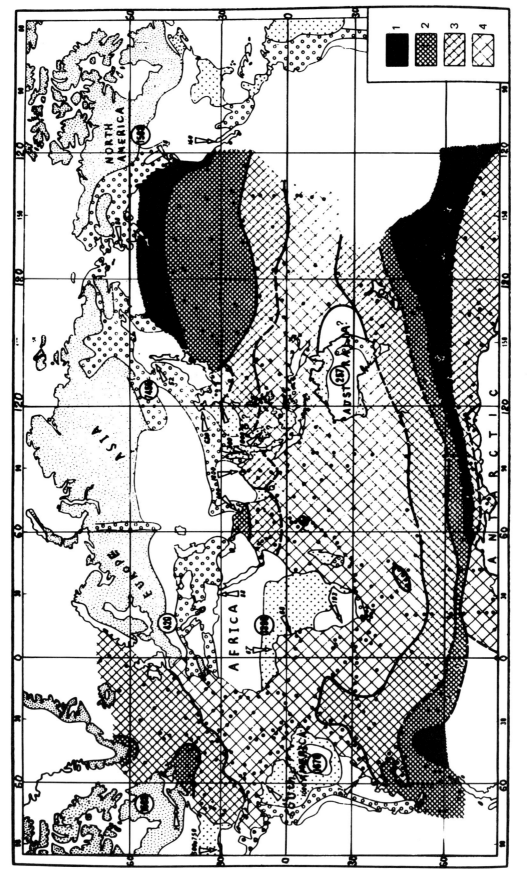

FIGURE 4C Chlorite. Concentration scales (in percent of chlorite): (1) > 30%; (2) 30–20%; (3) 20–10%; (4) > 10%.

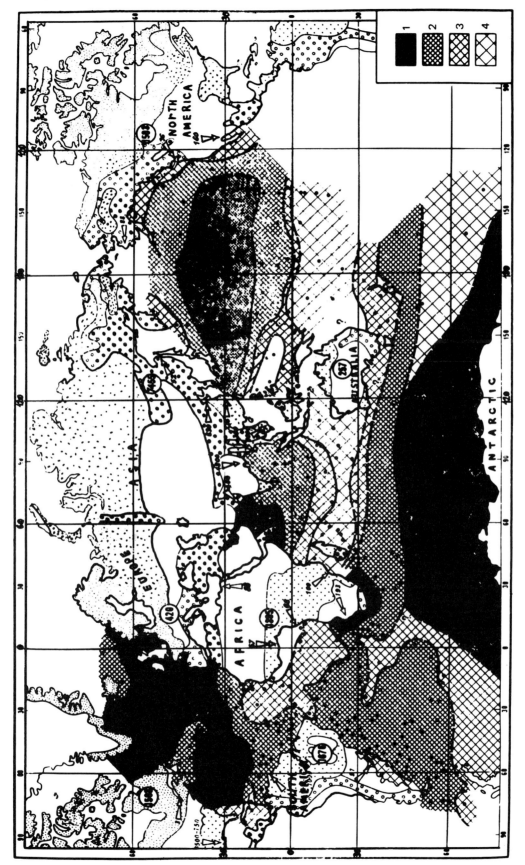

FIGURE 4D Illite (hydromica). Concentration scales (in percent of illite): (1) 80–60%; (2) 60–40%; (3) 40–20%; (4) 20%.

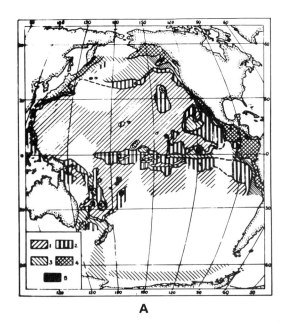

A

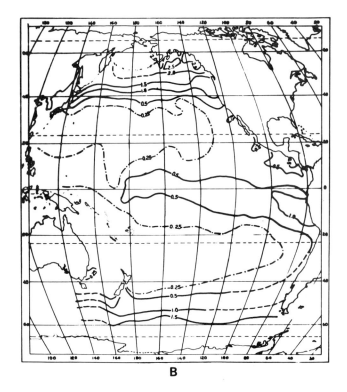

B

FIGURE 5 *A*, primary production in the Pacific Ocean (Koblentz-Mishke, 1965). *B*, the distribution of phosphates, in μg atom/l, at the surface (Bruevich, 1966, Figure 56).

Primary Production
Surface (mg of C per m^3 per day): (1) <2; (2) 2–5; (3) 5–10; (4) 10–100; (5) >100.
Water column (mg of C per m^2 per day): (1) <100; (2) 100–150; (3) 150–250; (4) 250–650; (5) >650.

of the Recent siliceous material has accumulated (Lisitzin, 1966; Lisitzin *et al.*, 1966).

As shown by geochemical studies of suspended matter (Lisitzin, 1964) and analysis of the distribution of diatoms in the suspension samples (Kozlova, 1964; Kozlova and Mukhina, 1966), diatoms are widely distributed in the water column in the Antarctic and subantarctic zones of the Southern Hemisphere. The maximum diatom concentrations are confined to the shelf zone around the continents. The number of diatoms in one cubic meter of water varies in different areas of the Southern Ocean from 0.5 million to 1.0 billion (10^9), i.e., by a factor of almost 2,000. The Antarctic Convergence is clearly defined with respect to suspended matter by a sharp decrease of the quantity of diatoms; north of the Convergence they number less than 6 million per cubic meter of water.

Another important law thus becomes evident: i.e., the Pacific surface waters contain considerably fewer diatoms than do waters of the Indian Ocean. Similar relationships have been found for the bottom sediments; the largest amounts of amorphous silica in sediments are encountered in the Indian Ocean.

If the entire annual supply of diatom remains is considered, one may notice that despite the fact that the maximum content is found in the near-shore suspended matter, the annual maximum appears to be displaced toward the pelagic parts of the ocean, and is confined to the base of the Antarctic continental slope and the periphery of the Southern Ocean floor. Thus, the average yearly maximum is displaced far northward with respect to the summer maximum of diatom plankton and is situated not in the Antarctic but in the subantarctic zone. This displacement is related to the short duration of the vegetation period on the shelf, which, during the greater part of the year, is covered with ice, which, in turn, interrupts photosynthesis. In the subantarctic zone, however, the maximum values are not high; yet the vegetation period is much longer. A prominent role in the accumulation of siliceous material in the Antarctic sediments is also played by other planktonic organisms with siliceous skeletons—radiolarians and silicoflagellates. These organisms do not dominate sediments as do diatoms in diatom oozes, but they are present constantly as admixtures in diatom and iceberg sediments (Petrushevskaya, 1966; Kozlova and Mukhina, 1966).

Foraminifers are widely distributed north of the Antarctic Convergence, where they represent the main source of sedimentary material.

Neither phytoplankton biomass nor determinations of the cell numbers in unit volume of water in suspension samples can be used to characterize the total annual biogenous supply. They are only "snapshots" of stages of the year-round process that is more fully characterized by primary production (Figures 5, 6). Attention is drawn to the close relationship, in the surface waters, between primary

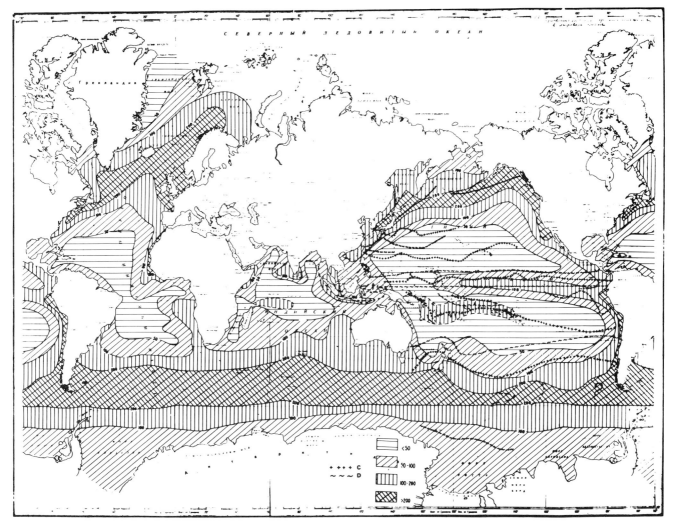

FIGURE 6   Primary production in the World Ocean (g of C per m^2 per year) (Gessner, 1959; Hela and Laevastu, 1961). D($\sim$) = divergence; C(+) = convergence (in the Pacific).

production and the content of phosphorus, the major biogenic element, as well as to the zonal distribution of primary production.

Direct determinations of primary production in the South Pacific are very scanty, and the available data usually characterize only one season. The maximum primary production values are recorded in the humid zones (equatorial and southern), and the minimum values in the southern arid (Forsberg and Joseph, 1964) and ice zones (less than 4.9 mg of C per m^3 per day). According to Koblentz-Mishke's (1965) determinations, primary production in the equatorial zone is from 150–250 to 250–650 mg of C per m^2 per day, whereas in the arid zone it is 5–10 times lower. The average value for the southern humid zone is 135 mg of C per m^2 per day, with separate values as high as 1,800 mg of C per m^2 (Klyashtorin, 1964). In the ice zone under the ice

surface, according to the same author, primary production constitutes about 9 mg of C per m^2 per day. Very high values are found within the ice zone in the ice-free water and shore leads, yet the annual production here is small because of the long polar night and the unfavorable light conditions (caused by the ice) during a short polar summer. Many authors compare the humid zone of the Southern Ocean with the productive waters in the vicinity of Greenland and in the Davis Strait. In the South Pacific, primary production is apparently somewhat lower (Saijo and Kawashima, 1963).

The bulk of primary production (more than 90 percent) is determined by diatom algae. The normal development of diatoms per unit of organic matter requires strictly definite amounts of amorphous silica, iron, manganese, and other elements and compounds. If the average for the Southern Ocean annual production is about 90 g of C per m^2 per year, its formation requires, per square meter, 136 g of amorphous silica, 3.5 g of CaCO$_3$, 26 g of iron, and 0.05 g of manganese. The sum of the biogenous components of suspended matter utilized by diatoms per square meter of the Southern Ocean surface within the 0–200-m layer is 262 g. The study of suspended matter shows that at greater depths the bulk of biogenous elements and compounds passes again into the water and joins in the geochemical cycle, and only an insignificant number of them, on having passed through the long water column, reach the ocean bottom. According to the calculations made from the studies of the distribution and composition of suspended matter and sediments from the Southern Ocean, only 1/245 of silica (of the initial amount at the surface), 1/3,750 of C$_{org}$, 1/44 of CaCO$_3$, 1/128 of iron, etc., are buried (Lisitzin, 1964) in the iceberg sediment zone.

The Supply of Volcanogenic Material

Extinct volcanoes, both above sea level and submerged, are widely distributed in the southern part of the ocean. They are particularly numerous along the South American coasts as well as near the many islands in the western portion of the ocean. The volcanic products in the form of ash, scoria, lava, and pumice fragments are often found in suspended matter and bottom sediments. Also widespread is altered volcanogenic material (palagonite, phillipsite, montmorillonite, etc.). Only rarely does volcanogenic material make up the major part of pelagic sediments, which are more often mixtures of biogenous and terrigenous materials.

The Supply of Eolian and Cosmic Material

Because the supply of clastic and clay material entering the southern part of the ocean is small, the long-range and super-long-range transportation of sedimentary material from the earth's stratospheric reservoir and the contribution of material from space become relatively more important. In accordance with the general atmospheric circulation, the bulk of sedimentary material that penetrates the stratosphere in any part of the planet precipitates only in definite localities—the global precipitation belts. These belts were established through study of radioactive fallout resulting from nuclear explosions (Figure 7). In the Southern Hemi-

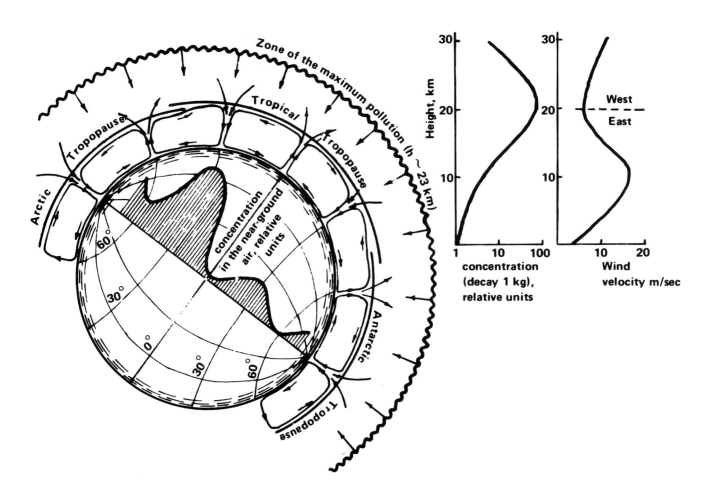

FIGURE 7 Atmospheric circulation and global precipitation of active aerosols from the atmosphere (Lavrenchik, 1965).

sphere, these belts coincide with the southern arid zone as well as with the regions adjacent to the South Pole. It is in these areas that the concentrations of the material transported in the lower atmospheric layers (particularly by trade winds) are formed. In the central parts of Antarctica, appreciable amounts of cosmic material, including magnetic materials, fall out. These amounts are related to the structure of the earth's magnetic field and to the existence of "windows" in the near-polar regions. In particular, this relationship can be seen from the distribution of nickel and other elements in firn (Picciotto, 1967).

In addition to cosmic material, eolian and long-range volcanic material is precipitated in these belts. Volcanogenic material (ash) that penetrated the stratospheric reservoir (higher than 11 km) falls to the earth's surface not in places of its eruption but in the latitudinal precipitation zones, sometimes thousands of kilometers distant from the volcanic focus. Particularly great precipitation of volcanogenic material may be expected in the southern arid belt, as confirmed by the mineralogical studies of bottom sediments. Many authors relate this to the specific development of volcanism in this area of the ocean. In fact, this is the zone of the global precipitation of volcanogenic (as well as of eolian and cosmogenic) material; it is especially evident here due to the weakening of agents dominant in other places (the supply of terrigenous and biogenous material). The long-range transport of eolian material is clearly seen, especially from the distribution of quartz (Figure 8).

A complex combination of different factors in space and time results in the deposition on the bottom of the South Pacific of rather peculiar sediments. Among them, the most widely distributed are terrigenous and biogenous (carbonate and siliceous) sediments (Figure 9).

THE DISTRIBUTION OF SEDIMENTARY MATERIAL (SUSPENSION) IN THE WATER COLUMN

To understand hydrochemistry and geochemistry it is important to know the amount and qualitative composition of suspended matter from various depths of the ocean. Suspended matter is one of the most important stages in the geochemical history of the majority of elements and compounds.

Data concerning the amount of suspended matter were obtained during the Soviet Antarctic expeditions and have been published in the Atlas of Antarctica (1966).

The quantity of suspended particles in the surface waters of the South Pacific (Figure 10) varies over a wide range, from less than 0.25 g per m³ to 2-4 g per m³ (Lisitzin, 1961, 1962, 1964, in press; Lisitzin and Bogdanov, 1968; Lisitzin, in press). The maximum concentration is found in the regions confined to the shelf. High values are also observed in the southern part of the ocean up to the latitude

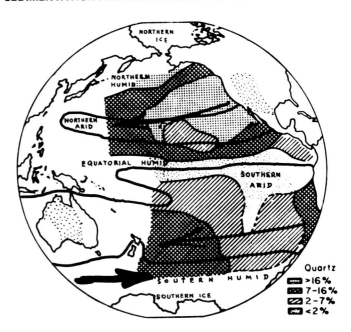

FIGURE 8 Quartz distribution in bottom sediments of the Pacific (in percent of carbonate-free and silicate-free material). The major deserts on land and the boundaries of climatic zones are shown (Goldberg and Rex, 1963).

of New Zealand. Farther north, they decrease sharply and show a new increase only near the equator. Suspension concentrations are associated with both terrigenous and biogenous (mainly diatom) supply. Especially high concentrations are observed in those areas where the abundant terrigenous supply coincides with phytoplankton "bloom."

The pattern of quantitative distribution of suspended matter along the series of meridional sections running through the southern part of the ocean is complex (Figure 11 A-D). This complexity is accounted for by both the irregular sedimentary supply from land and a complex distribution of planktonic organisms, as well as by the effect of hydrological factors (pycnocline, upwelling, and sinking) and, at depth, also by the dissolution of some particles. It can be seen from the sections that the most complex is the distribution of suspended matter in the upper 200 m (active layer), where the development of phytoplankton is possible.

At greater depths, three regions with high suspended-matter concentrations through the whole of the water column can be seen on the meridional section—southern, equatorial, and northern. Between them are waters with a minimum content of sedimentary material, corresponding to the earth's arid zones. Other sections of the southern part of the ocean and near the equator testify to the complex structure of these regions of enrichment.

One may notice that the amount of suspended matter usually decreases somewhat with depth. If the bottom line is displaced on the sections, it becomes apparent that in all the zones, the amount of sedimentary material that would

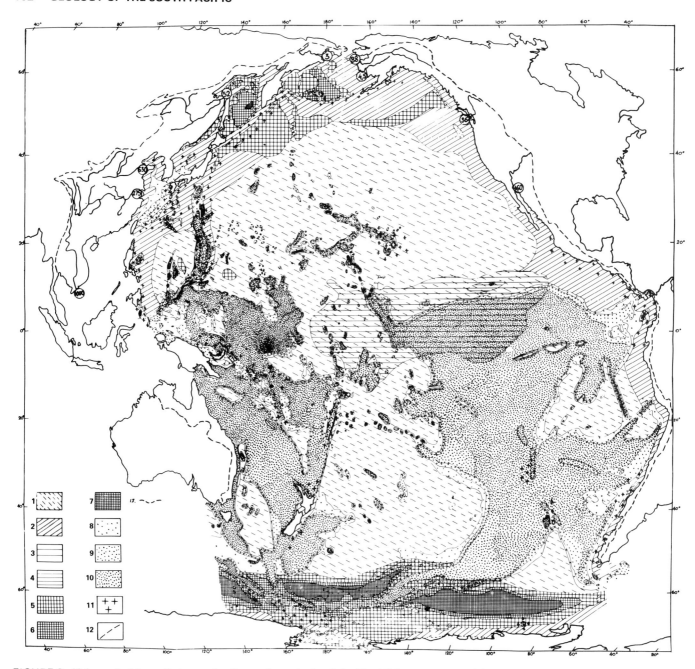

FIGURE 9 Major material-genetic types of sediments from the South Pacific. (1) Pelagic red clay; (2) terrigenous beyond the iceberg zone and iceberg (off the Antarctic coasts); (3) low-silica radiolarians (5–10% amorphous silica); (4) low-silica radiolarian—diatoms of the equatorial region (10–30% of amorphous silica); (5) low-silica diatoms (10–30% amorphous silica); (6) siliceous diatoms (30–50%); (7) high-silica diatoms (>50% amorphous silica); (8) low-carbonate foraminiferal (10–30% $CaCO_3$); (9) carbonate (30–50% $CaCO_3$); (10) high-carbonate (>50% $CaCO_3$); (11) volcanogenic material; (12) volcanogenic material; (13) limits of drainage into the Pacific.

reach the bottom at a depth of 1,000 m is considerably greater than that at 5,000 m. This phenomenon is called vertical zonality. We shall see below that depth influences not only the quantity of sedimentary material but also its qualitative composition (the dissolution of $CaCO_3$, C_{org}, P, etc.).

The regions with the maximum content of sedimentary material in the water should, in general, correspond to the zones of highest rates of sedimentation. And conversely, we have a right to expect the minimum rates of sedimentation in the arid zones (in particular, at great depths).

According to the author's calculations (Lisitzin, 1964), the amount of suspended matter in the Pacific Ocean and adjacent seas is 724 billion (10^9) tons, of which less than

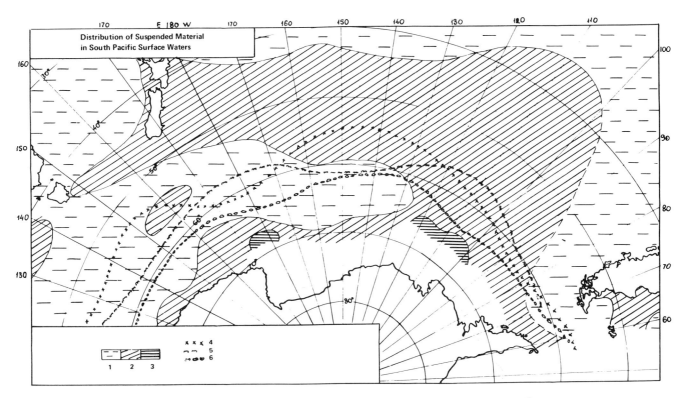

FIGURE 10 Quantitative distribution of suspended material in the surface waters of the South Pacific (in g/m^3): (1) <1; (2) 1–2; (3) 2–4; (4) Antarctic Convergence; (5) iceberg limit; (6) sea-ice limit.

one half is in the southern part. Nearly 6 billion tons of suspended matter are contributed annually to the Pacific from the Indian Ocean, and about 4.5 billion tons are carried away through the Drake Passage. Thus, of the supply of suspended material from the Indian Ocean, about 1.5 billion tons are deposited in the South Pacific each year, exceeding the sedimentary supply from land (Lisitzin, 1964).

Many important details can be obtained from a number of published papers (Lisitzin, 1956, 1959, 1960, 1961, 1966; Lisitzin and Bogdanov, in press; etc.).

An essential conclusion to be drawn from these papers is that there are climatic and vertical zonalities in the quantitative distribution of suspended material, which suggests that the accumulation of sedimentary material on the bottom should, to some extent, reflect these features.

CALCIUM CARBONATE

Calcium carbonate is the major biogenous component of bottom sediment, of which it often constitutes 95–98 percent.

Determinations of $CaCO_3$ in suspended matter and in sediment were conducted by the classical method of Knopp-Frezenius as well as in HCl extracts; the mineralogical composition was studied under a microscope with the application of staining technique and by x-ray diffraction analysis.

The genesis of carbonate sediments was studied by micropaleontological methods, with a particularly thorough investigation of planktonic (Belyaeva, 1964) and benthic (Saidova, 1961) foraminifers and coccolithophorids.

Percentage concentration of $CaCO_3$ in the surface suspended matter is very low. Even in the areas of a wide distribution of purely carbonate sediments on the bottom north of the Antarctic Divergence, suspension seldom contains more than 8–10 percent of $CaCO_3$ (Figure 12) because the bulk of planktonic foraminifers, which are the principal $CaCO_3$ suppliers to the oceanic sediment, do not inhabit the ocean surface; instead they live at depths of 100–500 m. Samples from the surface are impoverished in foraminifers. Thus, suspended matter in the surface waters contains diluting material in far greater amounts than $CaCO_3$; only the $CaCO_3$ reaches the deep-water sediments because the diluting material is practically completely dissolved in the water column.

To eliminate the diluting effect of silica, organic matter, and terrigenous material, $CaCO_3$ content of suspension is calculated in units of water volume (in $\mu g/l$; see Figure 13). An unexpected law is thus arrived at: The highest $CaCO_3$ concentrations at the surface appear to be related not to the waters of the arid zones, where the most pure carbonate sediments accumulate on the bottom, but to the periphery

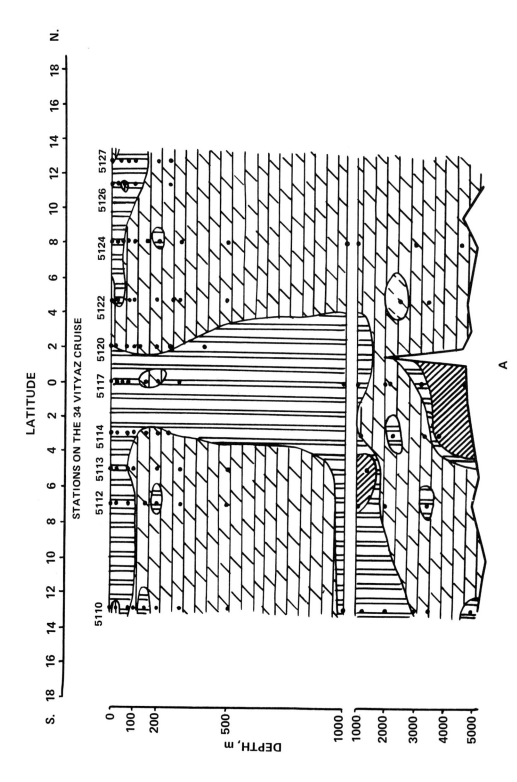

FIGURE 11 *A–D* Quantitative distribution of suspended material in the vertical direction along three meridional sections through the equatorial Pacific (membrane ultrafiltration method). Suspension concentration in g/m³: (1) <0.1; (2) 0.1–0.25; (3) 0.25–0.5; (4) 0.5–1.0; (5) 1.0–2.5; (6) >2.5.

FIGURE 11A Quantitative distribution of suspended material along 154°W.

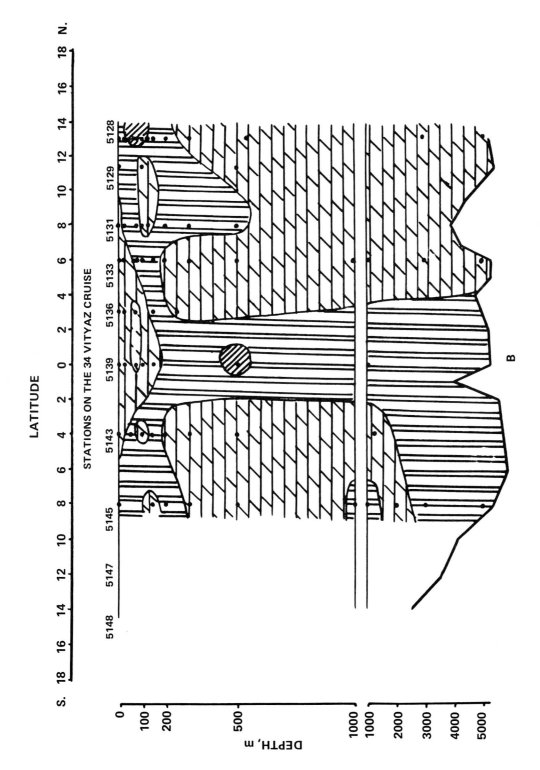

FIGURE 11B Quantitative distribution of suspended material along 176°W.

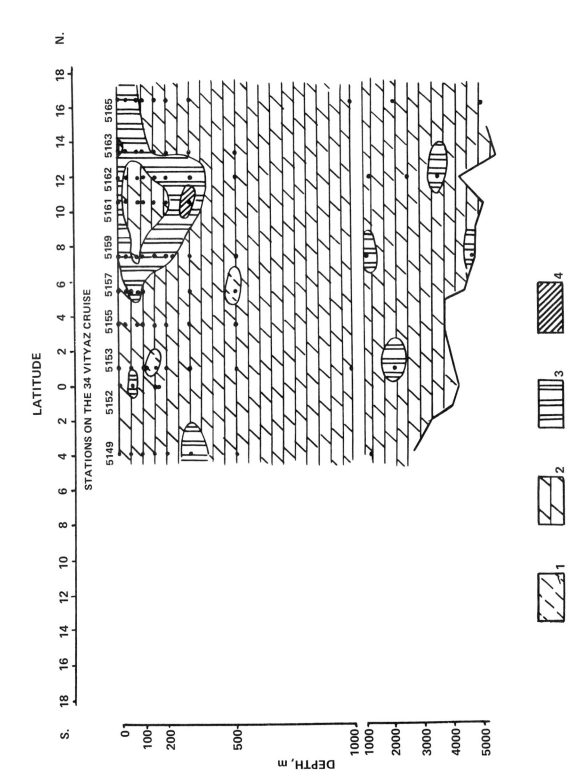

FIGURE 11C Quantitative distribution of suspended material along 160–162°E.

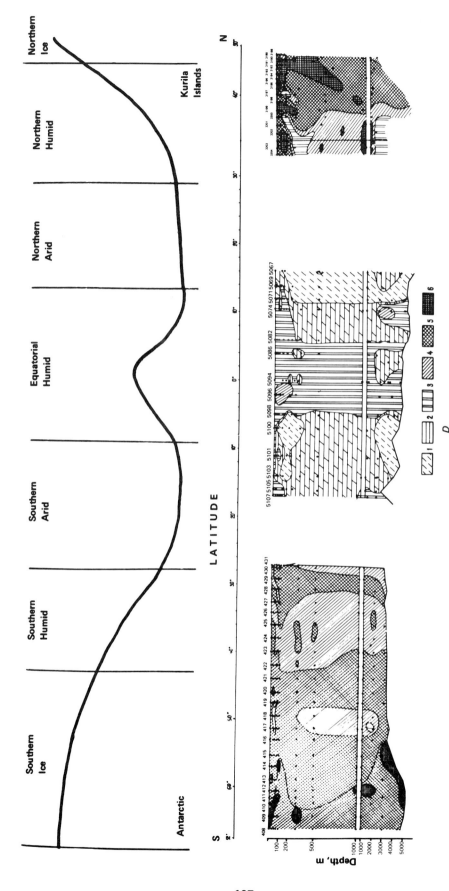

FIGURE 11D Quantitative distribution of suspended material in the transoceanic section from the Antarctic to the Aleutian Islands.

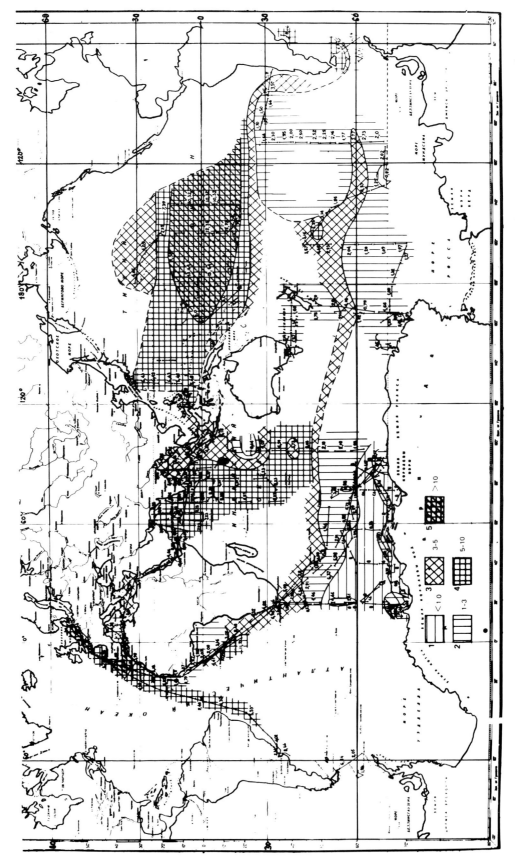

FIGURE 12 Surface distribution pattern of CaCO₃ in suspended matter (0–5-m layer) as a percentage of dry suspension: (1) < 1; (2) 1–3; (3) 3–5; (4) 5–10; (5) > 10.

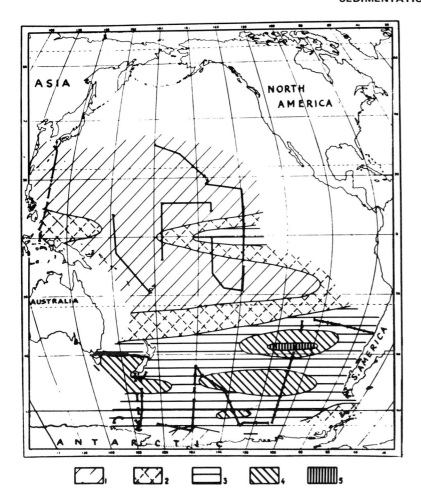

FIGURE 13 Surface distribution pattern of CaCO$_3$ in suspended matter (0–5-m layer), in μg/l of water from which suspension was recovered: (1) $<$5; (2) 5–10; (3) 10–50; (4) 50–100; (5) $>$100.

of the humid zones, where siliceous-carbonate sediments are deposited on the bottom. The southern belt of carbonate suspension is especially strongly pronounced in the Indian Ocean, where it lies within the 40°–60°S band. Such a distribution of carbonates in suspended matter is determined mainly by trophic relations (the higher the primary production of diatoms, which are the basic producers of organic matter, the higher the production of foraminiferal zooplankton consuming this organic matter). Of great importance for a number of foraminiferal species is temperature. Many planktonic species do not reproduce at water temperature below +10°C. The coccolitho-foraminiferal suspension zone can be traced from 50°N to 50°S, and generally coincides with +10°C isotherms of the surface waters.

The distribution of coccolithophorids, which represent the biogenous component of second importance, is also associated with climatic zonality. They are contained in 5–10 times greater amounts in suspension in the equatorial zone than in the neighboring arid zones; their abundance also increases in the vicinity of the temperate humid zones.

The microscopic and mineralogical analysis of carbonate material from the Pacific suspended matter shows that this is an exclusively biogenous material. Chemogenic CaCO$_3$ is not encountered in suspension despite the fact that the Pacific waters are locally 2–3 times oversaturated in carbonate. Thus, the geochemistry of this component in suspension, like that of amorphous silica, is determined by biogenous factors.

Near the shores and islands, planktonogenic material is supplemented by benthogenic material: coral, algae, shells and shell fragments, and local bryozoans.

The fate of carbonate material in the ocean depths turns out to be related to its mineralogical composition, size, and a number of other factors. Calcite tests of planktonic foraminifers are of 0.05–0.25 mm; they settle at a rate of 0.15–2.0 cm sec^{-1} and reach depths of 5 km in several days. Their study in suspended matter indicates that essentially no dissolution is observed in the water column during this time, and well-preserved tests reach the surface layer of sediments. They lie at the sediment surface during many hundreds, or thousands, of years, and therefore, the basic dissolution processes of CaCO$_3$ of foraminifers take place not in suspension but in the uppermost sediment layer.

Quite different is the fate of coccolithophorids, which

are much smaller than foraminifers (coccolithophorids usually average 0.02–0.005 mm). Their settling velocity is very low; many years, even tens of years, are required for them to reach the bottom. Microscopic study of suspension material has shown that unbroken coccolithophorids are found in the Pacific from the surface down to 200-m depths, while deeper, only fragments of these algae and, in particular, finely dispersed carbonate, are encountered. The content of the latter increases with depth simultaneously with the decrease in the amount of remains. Thus, the bulk of coccolithophorids are decomposed in the water column, and only their finely dispersed remains reach the deep-sea sediments.

Quantitatively, foraminifers usually predominate over coccolithophorids; there is thus every reason to expect that the carbonate belts established on the basis of suspended material will be reflected on the bottom as carbonate sediment belts, as observed for silica.

When examining the map of $CaCO_3$ distribution in the sediments of the Pacific, compiled on the basis of all the available data (Lisitzin and Petelin, 1967), one may note that these belts are not found in the bottom sediments (Figures 14 A and B). Distribution of $CaCO_3$ in sediments is much more complicated than that in suspension. The concentration of this component in sediments is evidently influenced by new factors that we did not come across when studying suspension materials. Locally, the carbonate suspension zones correspond on the bottom to carbonate-free sediments.

As in the case of amorphous SiO_2, a direct reflection of $CaCO_3$ content in sediments cannot be expected because concentration is affected by dilution by terrigenous and other types of sedimentary material, the impossibility of deposition on submarine rises, etc. Along with these factors, common for all biogenous components, $CaCO_3$ distribution in bottom sediments shows an especially pronounced reflection of one more factor—depth effect, or vertical zonality.

As far back as the *Challenger* expedition, it was observed that a certain relationship existed between $CaCO_3$ distribution in bottom sediments and depth. Carbonate accumulation is impossible at depths known as compensating or critical depths. This relationship was discussed by many authors (Murray and Renard, 1891; Pia, 1933; Revelle, 1944; Lisitzin and Zhivago, 1958; Lisitzin, 1960, 1961; Bezrukov, 1961; Bezrukov, Lisitzin *et al.*, 1961; Lisitzin and Petelin, 1967; Lisitzin, 1965).

Such a relationship has been interpreted in various ways; mention is made of the effect of water temperature, free CO_2 and pH content of the water (Park, 1966), C_{org} content of bottom sediments (Wiseman, 1954), the activity of mud-eaters, and underwater volcanicity (Petersson, 1953; Figure 15).

When studying in detail the depth dependence of $CaCO_3$ distribution, attention is drawn to the fact that this relationship manifests itself differently in different climatic zones. Within one ocean, critical depth may vary by 1,000–2,500 m (critical depth is the depth at which $CaCO_3$ content decreases to 10 percent and below which carbonate sediments are replaced with carbonate-free sediments).

Hydrochemists pointed out long ago that there was a certain dependence of carbonate accumulation upon the distribution of the elements of the carbonate system in the water column and in the bottom layer.

On the meridional section through the Pacific, pH values in the bottom layer are practically constant (7.8–7.9); zonal features can be traced only in the surface waters down to about 1,000-m depths (Ivanenko, 1966; Figure 16). Also, specific alkalinity is practically constant in the bottom layer (0.128–0.130), as is temperature ($+1°$ to $+2°$C). At maximum depths, the temperature difference between separate basins of the bottom does not exceed 1°C. C_{org} content of suspension materials in the deep-water layers as well as in the surface-sediment layer does not change sharply either (the deep-water pelagic sediments of the Pacific usually contain less than 0.25 percent of C_{org}).

Of great significance is the increase with depth of the hydrostatic pressure and of the value of the second constant of dissociation of carbonic acid, which results in $CaCO_3$ dissolution.

Actually, all the elements of the carbonate system change gradually with depth and cannot account for such a sharp increase of $CaCO_3$ solubility or for the appreciable changes of the critical depth. We may say that $CaCO_3$ dissolution in sediments between about 45°N and S begins everywhere at a definite depth of 3,500–3,800 m, which is called the depth of the beginning of dissolution. This depth has recently been determined experimentally (Peterson, 1966; Park, 1966). In the temperate and cold waters, the depth at which dissolution begins decreases with increase in latitude, and reaches the surface at 60°N and S. The depth of $CaCO_3$ penetration (the "slip" of carbonate material) of the dissolution zone changes depending on a number of factors, first of all, on the rates of $CaCO_3$ supply to sediments and on its dilution by carbonate-free material. The critical depth is determined, therefore, by a combination of the dissolution depth, which depends on the elements of the carbonate system, and the quantitative and qualitative peculiarities of carbonate sedimentary material. One may expect that the maximum $CaCO_3$ penetration to the oceanic depths will occur in those places where its sedimentation rate is the highest and where, at the same time, its dilution with carbonate-free material is lower. These factors, as we have already seen from the distribution and composition of suspended matter, depend on latitudinal zonality.

Let us consider the distribution of the critical depth of carbonate accumulation in the Pacific by major climatic zones (Figures 17, 18, 19). Within each latitudinal zone (20° of latitude) considerable scatter in $CaCO_3$ content is observed over the given range of depths. It can be noted, however, that the maximum amounts (connected by curves) decrease rather regularly. Beginning with a depth of about 4,500 m, a sharp decrease of $CaCO_3$ content in sediments takes place due to dissolution. Examination of a series of the connecting curves for different latitudinal zones of the oceans shows that the zonal relations of $CaCO_3$ are of a general character.

The maximum values of the critical depth in the Pacific and other oceans are found in the equatorial humid zone, where the high rate of carbonate accumulation coincides with an extremely low dilution.

One can see from the meridional section through the ocean that the critical depth rises gradually from the high latitudes toward the equator; this is also typical of other oceans. Within the equatorial zone, areas corresponding to 20 degrees of longitude have been distinguished, making it possible to study the change of critical depth when moving from the Asian coasts to America. It has turned out that the critical depth changes here regularly with longitude; the maximum values are found in the central part of the ocean (5,100 m), whereas, as the coast is approached and the role of diluting material increases, the critical depth decreases (Figure 20).

In space, the critical depths form a surface resembling that of a bowl, the bottom of which becomes deep in the region corresponding to the equator. Such a form of the surfaces of the critical depths is the result of a long-term dynamic equilibrium between the processes of $CaCO_3$ supply by planktonic organisms and its dissolution (and dilution) on the bottom.

The decrease of $CaCO_3$ content and the increase of the critical depth landward is one of the manifestations of circumcontinental zonality. The dynamic competition of terrigenous and carbonate material results in the shift of carbonate sediments from the land toward the pelagic parts of the ocean. This shift is greater in the humid zones, where more terrigenous material is supplied from land; it is minimal in the arid zones.

The critical depths prove to be sharply different for two carbonate materials—calcite and aragonite. Due to the lower resistance of aragonite against dissolution, the critical depth of its penetration to the sediments does not exceed 2,000 m, below which the realm of calcite begins. The critical depth of aragonite may be considered as one of the most reliable indicators of sedimentation depths.

Thus, the distribution of the basic genetic and mineralogical types of carbonates in the bottom sediments of the Pacific is determined by a combination of climatic and vertical zonalities. In the warm waters down to a depth of 2,000 m, aragonite is predominant, while in the waters of the temperate and cold zones, as well as at depths of more than 2,000 m, calcite dominates. Only rarely is this general law violated. In arid zones, the composition of carbonate material is usually characterized by the presence of separate dolomite grains associated with eolian transport from deserts.

The most representative measure of the real carbonate accumulation on the ocean bottom is the absolute quantity.

Attention is drawn to the fact that the maximum rates of $CaCO_3$ supply to the pelagic sediments are observed not in the arid zones, where the most pure carbonate sediments are found, but at their boundaries with the humid zones. The carbonate sedimentation maxima in the oceans coincide, therefore, with the maxima of silica accumulation but are somewhat displaced toward the equator. As for silica accumulation, the meridional section through the ocean shows three such maxima, all of them related to the humid zones and to regions of upwelling.

The finely dispersed carbonate material widely distributed in the deep-sea sediments was interpreted by many investigators as resulting from chemogenic sedimentation or the activity of bacteria. The study of suspended matter makes it possible not only to speak with confidence about the origin of this material from foraminiferal tests and from coccolithophorids but also to determine at what depths and in what amounts it usually appears. To the process of decay of the tests into their constituent crystalites is related the "pelletization" of the deep-sea sediments, as compared with suspension (Lisitzin, Murdmaa, Petelin, Skornyakova, 1966). One should consider that, basically, the geochemistry of carbonate material like that of $SiO_{2\ amorph}$ is biogeochemistry.

GEOCHEMISTRY OF SILICA

Over many years, investigations of silica in suspended matter and in bottom sediments of the seas and oceans have shown its biogenous origin. Hypotheses about volcanogenic, chemogenic, or sorptive deposition of silica, which are popular among many geochemists, have not been confirmed by observations of sediments and suspended matter. Quantitative determinations of silica in suspended material and sediments have been made chiefly by the chemical method. (Methods for the Study of Sedimentary Rocks, Strakhov, 1957); its qualitative composition has been studied by the optical and electron microscope as well as by x-ray diffraction methods.

Most organisms utilizing silica for the construction of their skeletons inhabit the surface (0–200 m) waters of the ocean. Below this depth only dead frustules and rare radiolarians are encountered. In ocean waters, dissolved silica usually predominates; in Antarctica it is found in amounts 3–4 times greater than in suspended form. In other places, the excess is 5–50-fold. Thus, the organisms can convert to suspended matter only an insignificant part of silica (Figure 21). The amount of amorphous silica in the suspended

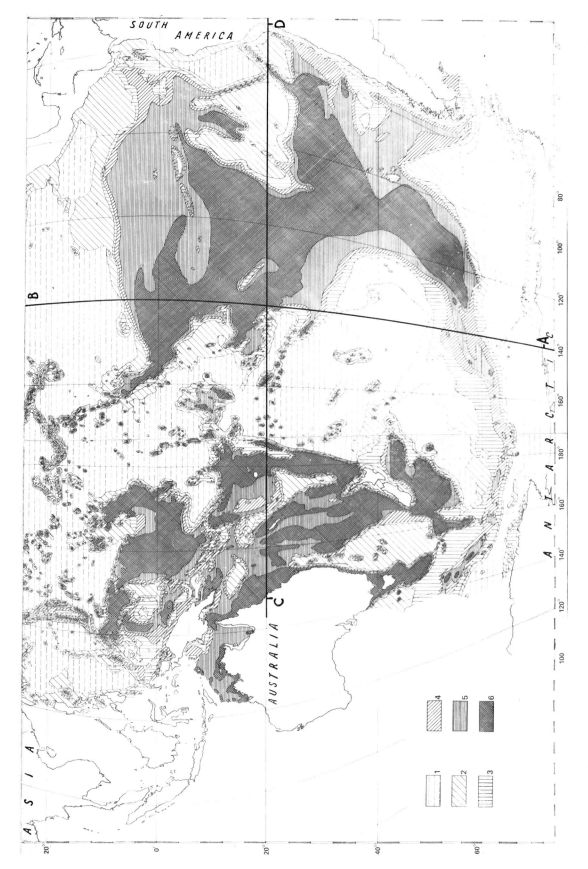

FIGURE 14 *A*, CaCO₃ distribution in the surface sediment layer of the South Pacific (in percent of dry sediments): (1) <1; (2) 1-10; (3) 10-30; (4) 30-50; (5) 50-70; (6) >70. Lines *AB* and *CD* show the location of sections.

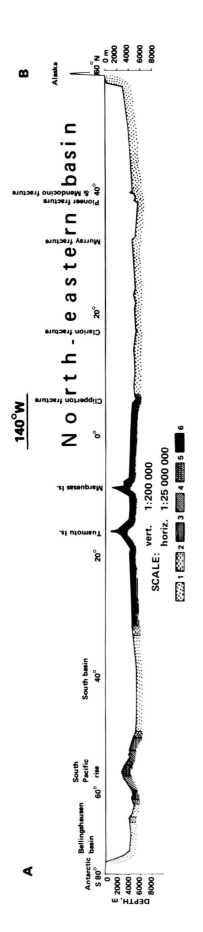

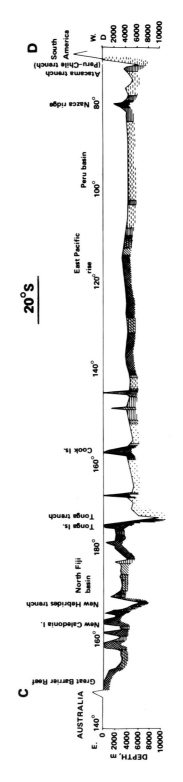

Figure 14 *B*, CaCo₃ quantitative distribution in bottom sediments. *AB* = meridional section along 140°W; CD = section along 20°S. Notations are the same as in *A*.

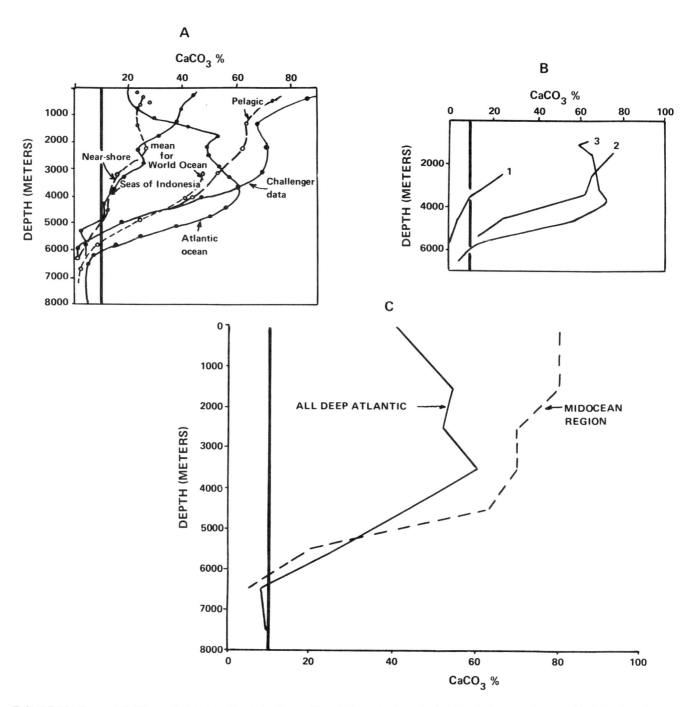

FIGURE 15 Types of $CaCO_3$ vertical distribution in bottom sediments from the data obtained by different authors: *A*, Neeb (1943). The Indonesian seas from the materials of the "Snellius" expedition. For the Atlantic Ocean, Pia's data (1933), the average for the World Ocean (Trask, 1937), for the near-shore (closer than 800 km from the shore) and pelagic (farther than 800 km from the shore) parts separately: *B*, 1. for the Pacific Ocean from 10° to 50°N (Revelle, 1944); 2. for the Pacific Ocean from 10°N to 50°S (Revelle, 1944); 3. for the Atlantic Ocean (Pia, 1933). *C*, for the Atlantic Ocean as a whole, for the midocean region separately (Turekian, 1964).

matter of the surface waters of the South Pacific ranges from less than 1 percent to 30 percent or more (Figure 22). One may notice close relationships between the regions of high concentrations of dissolved silica, of phytoplankton (diatoms), of suspended silica, and of the quantitative distribution of diatom frustules in the surface suspended matter (Figures 23, 24, and 25).

The zones in which there is upwelling of deep water to the ocean surface are regions of high content of dissolved silica. It is here that phytoplankton is particularly abundant and that the utilization of dissolved silica for the construction of frustules proceeds on the largest scale.

North of the Antarctic Convergence, silica accumulation is extremely slow, and only in the equatorial zone, due to the upwelling (Figure 26), is a new increase of silica content of suspended matter (diatoms and radiolarians) observed.

Quantitative microscopic analysis shows that diatoms are of basic importance in recent silica accumulation. They are followed in importance by radiolarians, which are encountered in sediments from the Arctic to Antarctica and are most widely developed in the equatorial zone. The third place is occupied by silicoflagellates and siliceous sponges. The relationships between these organisms in suspended matter of the Pacific vary with latitude. In the Antarctic belt of silica accumulation in suspended matter, diatoms constitute more than 99 percent; radiolarians, 0.1–0.01 percent; and silicoflagellates, less than 0.1 percent. In the equatorial belt, the relationships are different: Diatoms make up 1–36 percent of silica; radiolarians, 60–90 percent; and silicoflagellates, 0.1–2 percent.

Diatoms are also the main producer of C_{org} in suspended matter. Therefore, the causes of latitudinal zonality in silica distribution in suspension are the same as for C_{org}. Primary production and silica production are determined mostly by the supply of nutrient elements necessary for the development of phytoplankton (global and local divergences). One can see a coincidence of the regions of high primary production, high silica content of the water and suspended matter, and silica accumulation belts in bottom sediments.

Based on the close relationship between C_{org} and silica content of diatom frustules, the author has found that the greatest silica amounts are utilized during a year to form the frustules in the South Pacific; the rate of silica accumulation in suspended matter of the Pacific equatorial zone is 5–10 times lower (Lisitzin, 1968).

After the death of planktonic organisms, diatom frustules settle to the bottom at a rate equivalent to that of quartz particles 1–5 μm in diameter. Large and heavy frustules of oceanic diatoms reach the bottom in 30–100 days; their finer fragments settle to the bottom during many tens of years.

Below 200 m, practically all diatom frustules are dead, whereas the bulk of radiolarians are alive when they reach the near-bottom layer of the ocean. Figure 25 compares the

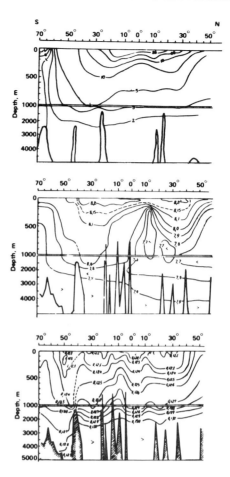

FIGURE 16   Water temperature, pH, and specific alkalinity in the meridional section (170°–160°W) through the Pacific Ocean (Institute of Oceanology, 1966).

quantitative distribution of frustules with the distribution of dissolved silica. The sections show that diatom frustules decrease in number in all zones when settling to the ocean bottom, the near-bottom layer being attained by 1/10–1/100 of the frustules at the surface. Their specific composition also changes because of the dissolution of the finest forms. The best preserved are coarsely silicified forms of the oceanic complex. For silicoflagellates, only a twofold decrease of the number of skeletons is observed, while for radiolarians no decrease occurs.

Despite considerable losses through dissolution, one may see that the zones of high silica concentration established at the surface are projected in the deep-water layers of the ocean. They are also reflected in the surface layer of bottom sediments (Figure 26).

Two belts of siliceous sediments (silica content more than 30 percent) are clearly defined in the South Pacific sediments: a southern belt and an equatorial belt. Both are reflections of the siliceous belts of suspended matter.

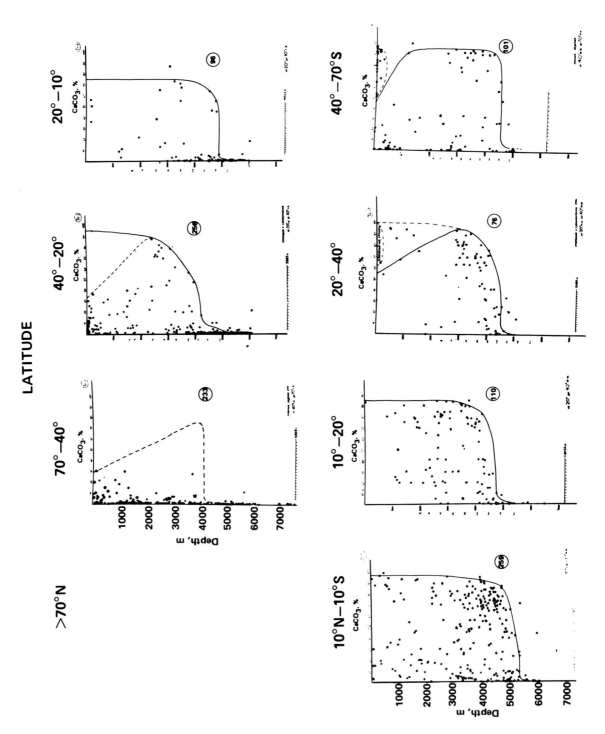

FIGURE 17 CaCO₃ distribution by depth in the surface sediment layer of the Pacific as dependent on latitude. Figures in circles show the number of analyses within each latitudinal zone.

116

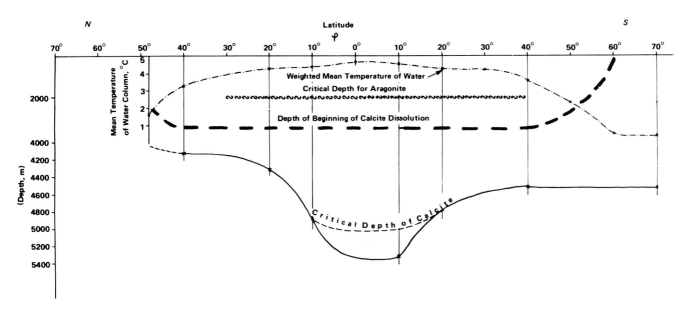

FIGURE 18 Weighted mean temperature for the water column from the surface to the bottom as dependent on latitude in the meridional section through the Pacific Ocean, the location of depth from which dissolution begins and of the critical depths for calcite and aragonite.

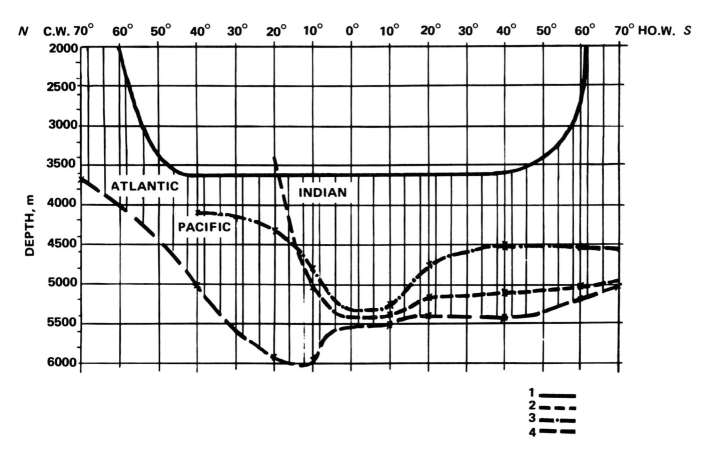

FIGURE 19 Critical depth of CaCO₃ accumulation by the latitudinal zones of the ocean. (1) The location of depth at which CaCO₃ dissolution begins; (2) critical depth for the major zones of the Indian Ocean; (3) the same for the Pacific Ocean; (4) the same for the Atlantic Ocean.

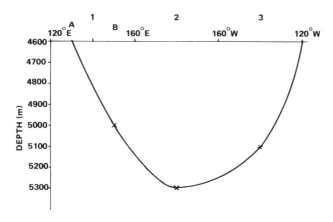

FIGURE 20   East–west variation in $CaCO_3$ (calcite) critical depth in the Pacific bottom sediments along the section in the equatorial zone within $10°N–10°S$.

The southern belt is characterized by the highest silica concentrations and the greatest width. It is here, in the Antarctic region, that more than three fourths of the World Ocean silica accumulates (Lisitzin, 1966). The width of the southern siliceous sediment belt of the Pacific is 1,000–1,200 km; the maximum amount of SiO_2 $_{amorph}$ constitutes 58 percent (in the Indian sector, 72 percent).

The equatorial zone of low-siliceous and siliceous sediments is closely related to depth. The highest silica concentrations are found at depths greater than the critical depth, below which no dilution of $CaCO_3$ is observed. When expressing silica on a carbonate-free basis, the equatorial belt is most strongly pronounced (Lisitzin, 1966). In its western part, high concentrations are found in separate bottom depressions. The map of SiO_2 $_{amorph}$ distribution also shows that the siliceous-sediment belts are related to upwellings (divergences). In the equatorial zone, two divergences are combined, yet their influence is far smaller than that of the Antarctic Divergence. The conclusion about a considerable weakening of the Antarctic Divergence in the Pacific eastward of New Zealand (Ivanov, Neyman, 1965) is confirmed by the data on the distribution of dissolved silica, phosphorus, suspended silica, and silica in bottom sediments.

A thorough microscopic study of silica in bottom sediments shows that, as for suspended matter, this silica is exclusively biogenous; no other silica in any appreciable amounts is found in sediments.

The pattern of the quantitative distribution of diatom frustules, radiolarian skeletons, and silicoflagellates in the surface layer of the Pacific sediments coincides with the map of the quantitative distribution of silica (Lisitzin, in press). Within the major silica accumulation belts, a simultaneous increase of the number of frustules of all the three planktonic organisms with siliceous skeletons is observed.

The actual intensity of silica accumulation may not in some places coincide with the percentage of SiO_2 $_{amorph}$ in bottom sediments. This is because of the supply to the sediments, along with silica, of the diluting material (terrigenous, $CaCO_3$). To elucidate the real picture of the rates of sedimentation, it is necessary to make use of the absolute masses. These have been computed by the present author for the Indian Ocean as well as for the Bering and Okhotsk seas and for the North Pacific (Lisitzin, 1966). Generalization of all information concerning absolute mass determinations (Holmes, 1965, etc.), stratigraphic investigations (Hays, 1967; Hays and Opdike, 1967; Opdike et al., 1966; Donahue, 1967; Blair, 1965; etc.) (especially diatom and radiolarian analysis), and paleomagnetic determinations (Goodell and Watkins, in press; Watkins and Goodell, 1967; Mather, 1966; Koster, 1966) allows one to speak, as a first approximation, about the absolute masses of silica accumulation in the South Pacific.

Of particularly great significance are the data obtained by the American Antarctic expeditions on the *Eltanin*, processed under the guidance of G. Goodell. The pattern of the distribution of sediment absolute masses is shown in Figures 27 and 28. It is seen from these figures that the distribution pattern has a zonal character: The highest rates of sedimentation are in the humid zones and the lowest ones are in the arid zone. The submarine ridges are characterized by a very mixed distribution of sedimentation rates and absolute masses, therefore, such localities are excluded from our further consideration. Sedimentation on the shelf is to a considerable extent associated with local conditions and is only a feeble reflection of the general zonal regularities. Attention is drawn also to the fact that the regions of high sedimentation rates are related to the areas with high suspended-matter content of the waters.

Proceeding from the scheme of the absolute masses and quantitative distribution of silica in sediment cores, the absolute masses of silica can be established for the Holocene. The rate of silica accumulation per 1,000 years in the Indian sector is 0.1–0.5 to 2.9 g per cm^2 for the lower part of the continental slope. In the arid zone, the values are below 0.001 g per cm^2 per 1,000 years. In the Bering Sea and the North Pacific, the values are 2–4 and 0.1–0.01 g per cm^2 per 1,000 years, correspondingly (Lisitzin, 1966). In the equatorial belt, the most common values are about 0.01, with an increase to several grams and even to 10 g per cm^2 per 1,000 years in the bottom depressions where Ethmodiscus oozes accumulate. High values are also noted in the Gulf of California where diatom sediments are associated with the local divergence (on the average, 50 g per cm^2 per 1,000 years, Calvert, 1966).

The major part of silica utilized by organisms to construct their frustules is dissolved in the water column. In this way, the major part of biogenous silica enters the exchangeable (dissolved) part of the siliceous balance of the ocean. The amount of silica deposited in bottom sediments is approximately equal to its supply from land (452×10^6 tons for all the oceans and seas). Recent silica accumulation is a biogenous process occurring on a global scale.

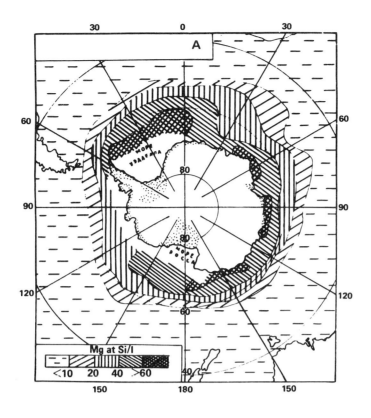

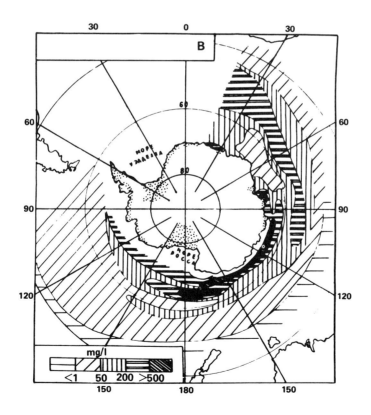

FIGURE 21   *A*, dissolved silicic acid distribution (in μg-at Si/l) at the surface in Antarctica. *B*, phytoplankton biomass distribution in the 0–100-m layer (in mg/l).

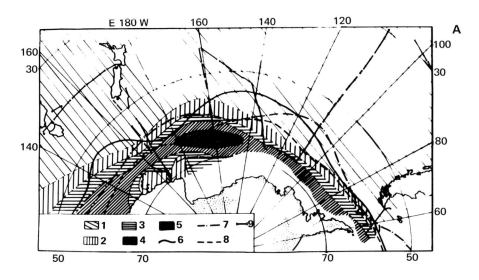

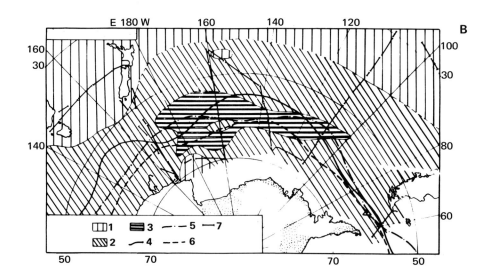

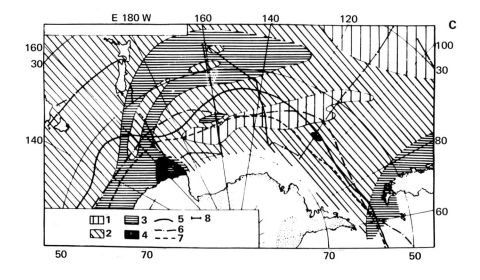

FIGURE 22 The distribution of suspended amorphous silica, phosphorus, and organic carbon (in percent of dry suspension). *A*, amorphous silica distribution: (1) <1; (2) 1–5; (3) 5–10; (4) 10–30; (5) >30; (6) Antarctic Convergence; (7) limit of the greatest iceberg distribution; (8) limit of the greatest ice distribution; (9) sampling localities. *B*, phosphorus distribution: (1) 0.1–0.3; (2) 0.3–0.6; (3) 0.6–1.0; (4) Antarctic Convergence; (5) extreme boundary in the distribution of icebergs; (6) extreme boundary in ice distribution; (7) sampling localities. *C*, organic carbon distribution: (1) <2; (2) 2–5; (3) 5–10; (4) 10–20; (5) Antarctic Convergence; (6) extreme boundary in the distribution of icebergs; (7) extreme boundary in ice distribution; (8) sampling localities.

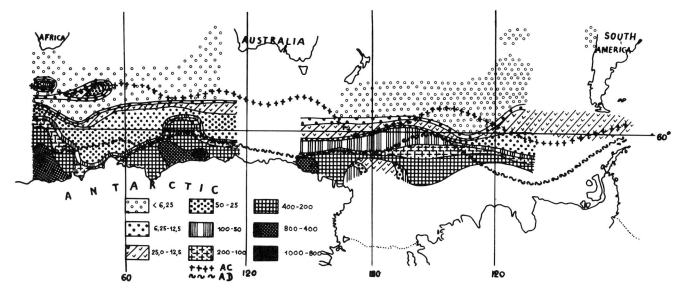

FIGURE 23 Diatoms in suspended matter from the surface waters of the South Pacific and Indian oceans (million of frustules per m^3 of water, Kozlova, 1964).

GEOCHEMISTRY OF IRON, MANGANESE, ORGANIC MATTER, AND MINOR ELEMENTS

In this brief report, it is impossible to dwell on problems of the geochemistry of other elements and compounds in suspended matter and bottom sediments of the South Pacific. A number of them have been studied, however, and some of the data have been published.

The geochemical features of iron and manganese from the Pacific sediments are dealt with in many papers (Skornakova, 1965; Strakhov, 1963). The geochemistry of organic matter in suspended material and bottom sediments is discussed in papers by Bogdanov et al. (1970), and Bogdanov and Lisitzin (1968), the distribution of bitumens in suspended matter downward from the surface and along the vertical sections, is discussed in papers by Bogdanov (1965), Bogdanov and Ovchinnikova (1967), Bogdanov (1966), Lisitzin et al. (1967), and Lisitzin and Bogdanov (1968). Migration and geochemistry of the rare-earth elements from the Southern Ocean is dealt with in a paper by Balashov and Lisitzin (1968), and minor elements in seawater are discussed in a series of papers by Schutz and Turekian (1965), and Turekian et al. (1967). The problems of the geochemistry of radium in suspended matter and bottom sediments as well as of the geochemistry of thorium, ionium, protactinium, and uranium are considered in papers by Starik et al. (1962), Kuznetsov et al. (1964), and others.

The above-mentioned references show that in recent years geochemical study of the South Pacific has advanced considerably.

CLIMATIC ZONALITY, TYPES OF SEDIMENTOGENESIS, AND GEOCHEMICAL ZONES

Sedimentation in the South Pacific, like that in other oceans, is closely related to climatic and vertical zonalities (Lisitzin, in press). The latitudinal zonality of natural processes is associated with the uneven heating of the earth's surface. Thermal heterogeneity causes the motion of air and water masses, and hence the general circulation of the atmosphere and hydrosphere. It is as though water were pumped out of the arid zones of the earth and precipitated in the greatest amounts in three latitudinal humid zones.

A combination of heat and moisture results in the formation of various crusts of weathering of the rocks of the land. The highest rates of chemical weathering are observed in the equatorial humid zone, where high temperature is accompanied by an abundance of water. It is here that the greatest chemical alteration of the initial minerals of rocks takes place (Strakhov, 1960).

The distribution of soils, vegetation cover, and water-discharge basins is also zonal in character. In the humid zones, with their abundance of water, sedimentary material is discharged by rivers, whereas in the arid zones eolian transport prevails. Climatic zonality is apparent in both the quantity and the qualitative composition of terrigenous sedimentary material contributed to the South Pacific.

No-less-distinct relationships are found for biogenous material; they are also reflected both in quantitative and in qualitative composition. The relation to temperature and humidity is indirect here, though temperature barriers exist

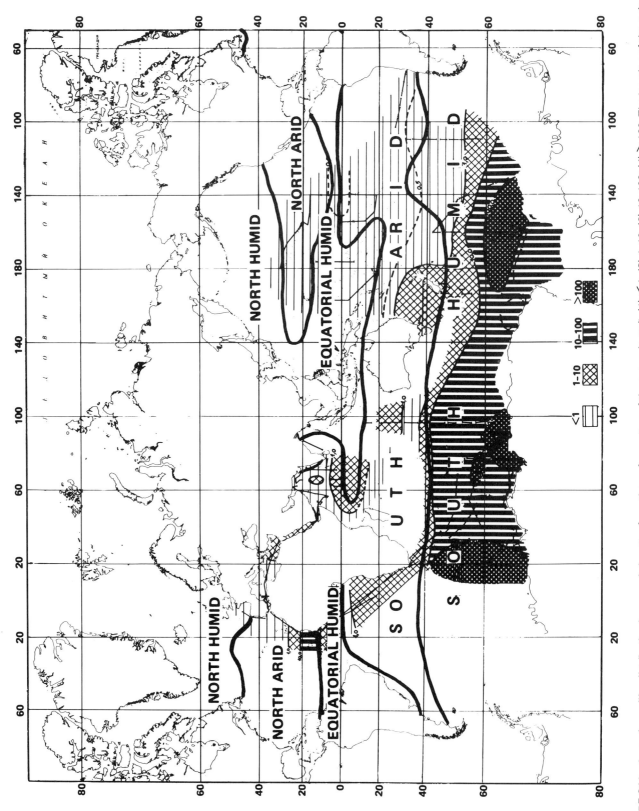

FIGURE 24A   Amorphous silica distribution in suspended matter from the surface waters of the ocean, in μg/l. (1) <1; (2) 1–10; (3) 10–100; (4) >100. The location of the major climatic zones is shown. (Lisitzin et al., 1966, with changes.)

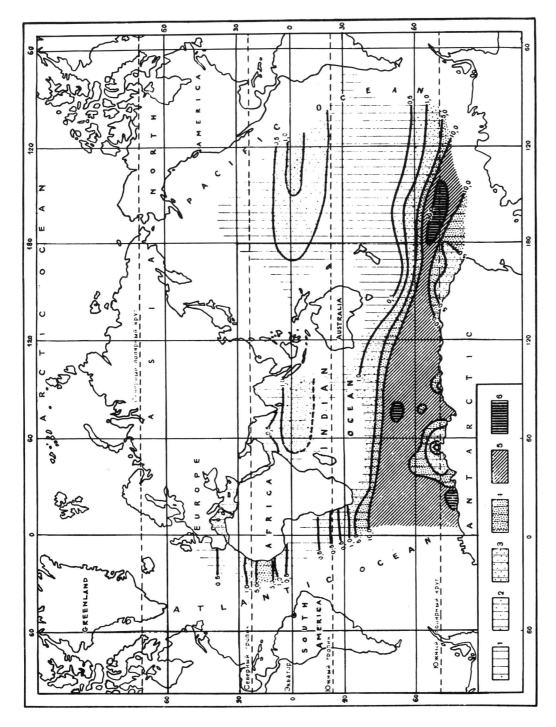

FIGURE 24B Amorphous silica distribution in suspended matter from the surface of the ocean, in percent of dry suspension. (1) <0.5; (2) 0.5–1.0; (3) 1–5; (4) 5–10; (5) 10–30; (6) >30.

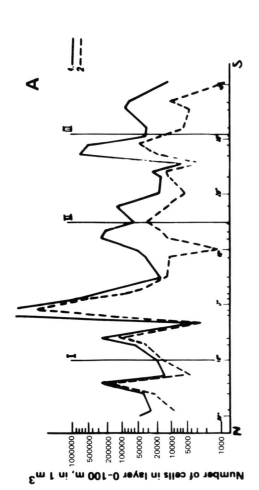

A

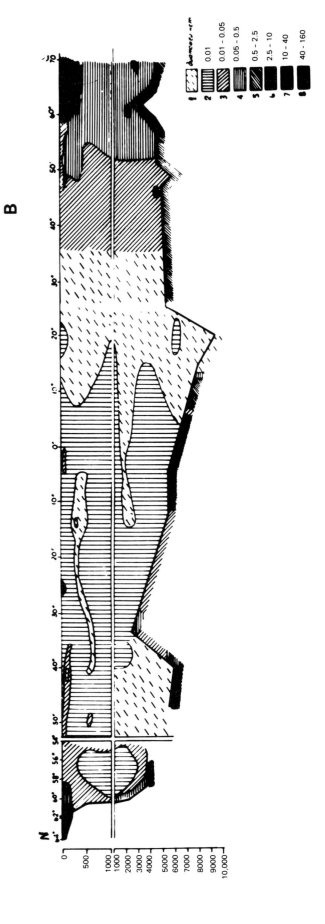

B

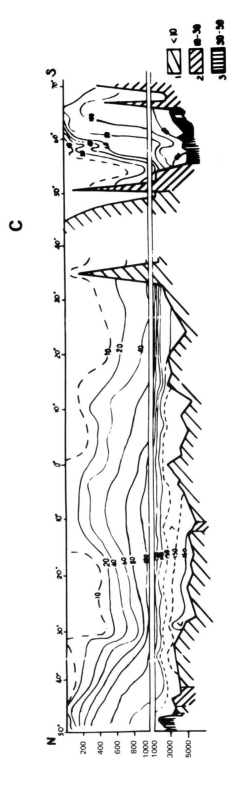

FIGURE 25 Phytoplankton and diatom numbers in suspended matter and in sediments (in millions per m³ and millions per g) and dissolved silica distribution in a combined section through the Pacific Ocean. A, phytoplankton numbers in the 0–100-m layer in the section along 174°W (Semina, 1963): (1) total number of cells; (2) diatom numbers. B, diatom numbers in suspended matter and in bottom sediments: (1) no diatoms; (2) < 0.01; (3) 0.01–0.05; (4) 0.05–0.5; (5) 0.5–2.5; (6) 2.5–10; (7) 10–40; (8) 40–160 (Kozlova and Mukhina, 1966). C, silicic acid distribution in μg at./l (Institute of Oceanology, 1966).

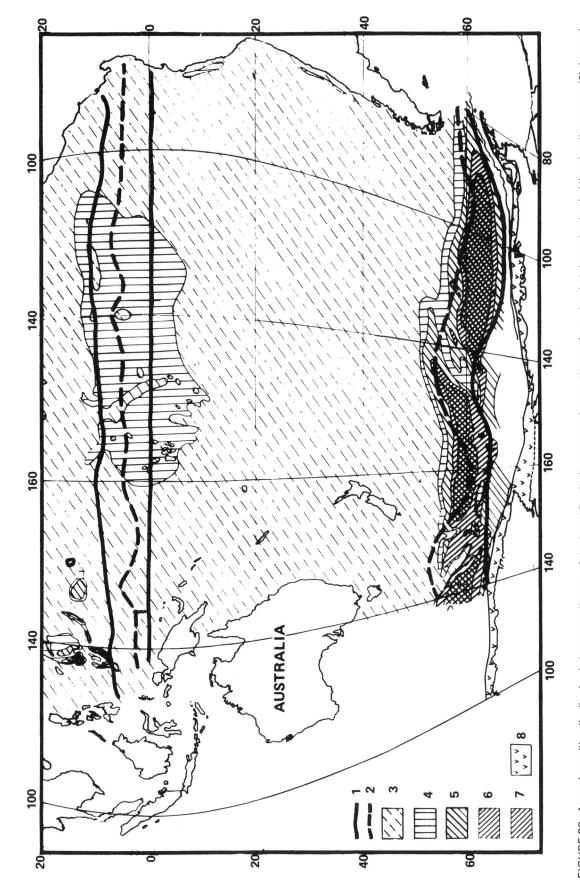

FIGURE 26 Amorphous silica distribution in bottom sediments of the South Pacific (in percent of dry sediments from chemical determinations): (1) major divergences; (2) Antarctic and North Tropical convergences; (3) < 1; (4) 1–10; (5) 10–30; (6) 30–50; (7) 50–70; (8) terrigenous iceberg sediments with the local regions of spicule sediments, regions of year-round ice occurrence.

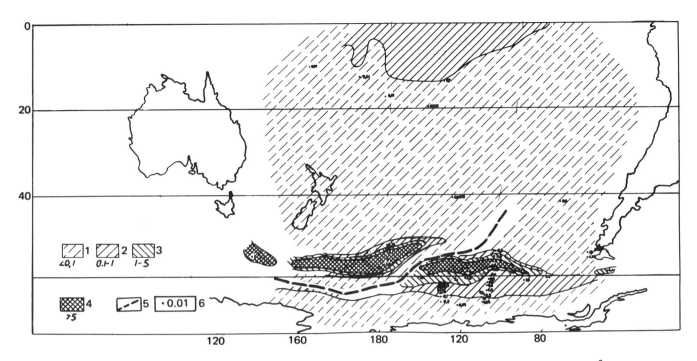

FIGURE 27   (1–4) Absolute masses of amorphous silica accumulation in bottom sediments of the South Pacific (in g per cm^2 per 1,000 years); (5) trend of the bottom ridge; (6) sampling localities and absolute masses.

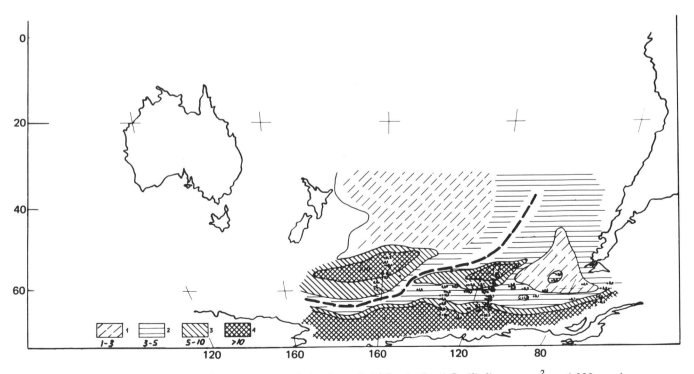

FIGURE 28   Absolute masses of sediment accumulation (as a whole) for the South Pacific (in g per cm^2 per 1,000 years).

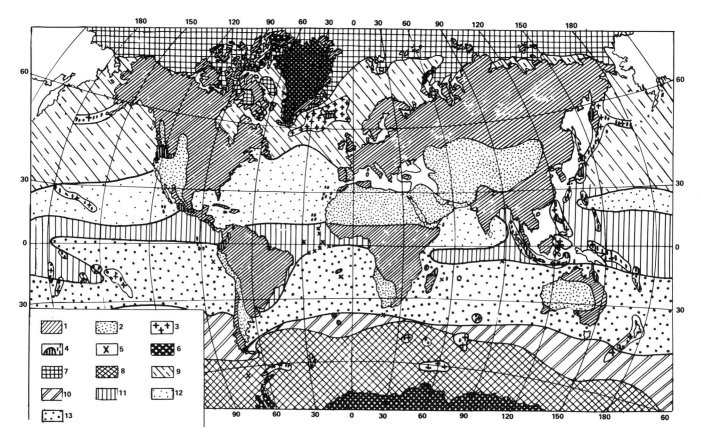

FIGURE 29 Major types of lithogenesis for the South Pacific. On land (according to Strakhov, 1960): (1) regions of humid lithogenesis of the continents with a developed crust of weathering; the maximum development of the crust of weathering is observed in the tropical and equatorial regions. Thickness decrease and complete disappearance of the crust in tectonically active regions; (2) regions of arid lithogenesis of land; (3) regions of effusive sedimentary lithogenesis on land and in the adjacent parts of the oceans; (4) volcanoes and volcanic regions; (5) single volcanoes and eruptions; (6) regions of ice lithogenesis. In the ocean (according to the author): (7) region of ice lithogenesis of the Northern Hemisphere; (8) region of ice lithogenesis of the Southern Hemisphere; (9) northern region of humid temperate lithogenesis; (10) southern region of humid temperate lithogenesis; (11) equatorial region of humid lithogenesis; (12) northern region of arid lithogenesis; (13) southern region of arid lithogenesis.

for some organisms. Of paramount importance is the supply to the ocean surface, to the photic zone, of nutrient salts from the ocean depths. The upwelling zones (divergences) are associated with cyclonic gyres included in the system of the general circulation of the hydrosphere and atmosphere.

In particular, two of the largest upwelling zones exist in the South Pacific: the equatorial and Antarctic divergences. Between them, there is an anticyclonic gyre in which sinking occurs; evaporation sharply predominates over atmospheric precipitation. Oceanic deserts, which are a continuation of the land's deserts, are widely developed here. The ascent of deep waters to the ocean surface is inhibited by the halocline. Beyond the arid zone, the halocline is replaced by the thermocline, which in the temperature latitudes admits water mixing in spring and autumn.

Thus, temperature conditions and the quantities of atmospheric precipitation and evaporation above the ocean and above the land appear to distinguish the major climatic zones (Figures 2 and 29), in which sedimentogenetic and geochemical processes have different courses.

The quantitative and qualitative differences of sedimentary material in these climatic zones are so essential and natural that several types of lithogenesis can be distinguished for the South Pacific: equatorial humid, southern arid, southern humid, and ice (Figure 29).

In accordance with the peculiarities of the atmospheric and water circulation of the South Pacific, the equatorial humid zone stretches in the eastern part of the ocean almost along the equator, while to the west its boundary reaches only to 20°S. In this area, the maximum rates of sedimentary discharge from land (up to 240 tons per square kilometer) are observed, as well as the deepest transformations of the initial rocks, with the formation of lateritic crusts of weathering as a result. The most characteristic feature is the general east–west replacement of sedimentary material, which leads to the greater role of the terrigenous factor in the eastern part of the ocean. The equatorial portion of the Pacific drainage basin is extremely narrow and mountainous.

As compared with other oceans, a negligible amount of sedimentary material is supplied to the equatorial Pacific from land; therefore, along with terrigenous sediments, biogenous sediments are also widely developed here. Because of the high humidity, the role of the eolian transport is insignificant.

The southern arid zone is, as it were, a continuation of the Australian and South American deserts. The supply of terrigenous material from land is at a minimum here, as is the contribution of biogenous material, since deep waters are cut off from the surface waters by the halocline. The insignificant supply of terrigenous and biogenous material results in a stronger influence for volcanogenic sedimentation and for the contribution of cosmic material as well as of long-distance eolian material. The relative importance of the role of these factors here increases, not because of their greater activity, but because in other places they are literally choked out by the supply of great masses of biogenous and terrigenous material. It is here, in the southern arid zone, that the pole of minimum sedimentation rates of our planet is situated. Only near the western coast of South America do local divergences occur to which sediment patches with high sedimentation rates correspond.

The southern humid zone in the Pacific Ocean is very distant from land, and the main sediment type within it is biogenous (carbonate and siliceous) sediment. Only in the vicinity of the South American coasts is biogenous sediment replaced by terrigenous material.

Finally, in the region where water is found in the solid state practically year round, one further type of lithogenesis, that of ice, is distinguished. This type is basic to the contribution of sedimentary material to the South Pacific.

The zonality of sedimentogenesis also predetermines the geochemical zonality. The sedimentogenic zones considered above should be supplemented for the arid and equatorial humid zones with the subzone of abyssal red clay. The following are characteristics of the geochemical zones of the South Pacific:

Ice Zone A large supply of disintegrated material from the land as well as clay minerals of the hydromica-chlorite complex; a minor role for biogenous material—i.e., distinct predominance of the terrigenous process and accumulation of clastic and clay (hydromica-chlorite) sediments.

Temperate Zone A weaker supply of terrigenous material owing to greater distance from the Antarctic coasts and the absence of drainage basins on land; a sharp intensification of the biogenous process owing to the supply of nutrients in the divergence region, siliceous organisms in the southern part of the zone, and carbonate organisms in the northern part. Because of the high humidity, eolian material is of no importance; pyroclastic material and lava are deposited near volcanic focuses. Accumulation of biogenous (siliceous and carbonate) and clastic and clay (hydromica-chlorite-montmorillonite) sediments is characteristic.

Arid Zone Along the coasts are desert regions having no discharge to the ocean, and in many places blocked by coral reefs. The supply of terrigenous material from rivers is insignificant. The leading role is played by the contribution of local and long-distance eolian material, including global precipitation of these and pyroclastic and fine cosmogenic material from the stratospheric reservoir. The biogenous process is very weak because of the predominance of sinking (convergence); carbonate sedimentation is extremely slow. Below the critical depth, there is deposition of pelagic red clay (hydromica-montmorillonite) enriched in eolian and pyroclastic material of both local and long-range transportation. There is also a sharp change of the primary minerals, and an authigenic mineral formation associated with diagenesis.

Equatorial Humid Zone The supply of terrigenous material is stronger here than in the arid zone and is concentrated in a narrow band crossing the ocean in the latitudinal direction. Among the clay minerals, kaolinite and montmorillonite prevail; gibbsite and illite are often encountered. The divergence along the equator results in an increase in primary production and in the development of carbonate and siliceous plankton; hence, carbonate and carbonate-siliceous sediments develop on the bottom. Below the critical depth, pelagic red clay enriched in silica is found; in bottom depressions Ethmodiscus siliceous oozes develop. Because of the high humidity, eolian material is of no significance. The deposition of pyroclastic material proceeds near volcanic focuses.

Thus, along with some features that are similar, either more distinct or hardly visible differences can be distinguished among the geochemical zones and subzones. The conditions of existence in each zone, the relationships between the volumes of biogenous, terrigenous, and eolian materials, predetermine the history of the major elements and compounds in these zones. We could recognize this history from the geochemistry of the two most important components (calcareous and siliceous), suspended matter and bottom sediments, alone. Zonality is no less important for other compounds and elements of suspended matter and bottom sediments. Their study with respect to zonality should be carried out in the near future.

REFERENCES

Arrhenius, G., and E. Bonatti (1965) Neptunism and vulcanism in the sea. *Prog. Oceanogr. 3*, 7–22.

Atlas Antarktiki (1966), vol. 1. Glavnoe upravleniie geodesii i kartografii MG SSSR.

*Balashev, Yu. A., and A. P. Lisitzin (1968) Migration of rare earth elements in the ocean. *Okeanol. Issled., 18*, 213–283.

Bardin, V. I., and V. I. Shilnikov (1960) Productivity of the coast of east Antarctica. *Inform. Biull. Sov. Antarkt. Exped., 23*, 100–103 (English ed.)

Belyaeva, N. V. (1966) Climatic and vertical zonation in the distribution of planktonic foraminifera in Pacific Ocean sediments, pp. 35–36 in: *Second International Oceanographic Congress* (abstr. of papers), Nauka, Moscow.

Bezrukov, P. L., A. P. Lisitzin, V. P. Petelin, and N. S. Skornyakova (1961) Map of bottom sediments of the global ocean, pp. 73–85 in: *Recent sediments of seas and oceans*. Izd. Akad. Nauk, Moscow.

Bezrukov, P. L., A. P. Lisitzin, E. A. Romankevich, and N. S. Skornyakova (1961) Recent sediment formation in the north Pacific Ocean, pp. 98–123 in: *Recent sediments of seas and oceans*. Izd. Akad. Nauk, Moscow.

Blair, D. G. (1965) The distribution of planktonic foraminifera in deep sea cores from the Southern Ocean Antarctica, M. S. Thesis, Florida State University, Talahassee, Florida.

Bogdanov, Yu. A. (1965) Suspended organic matter in the Pacific. *Okeanologiia, 5*, 286–297.

*Bogdanov, Yu. A., and A. P. Lisitzin (1968) Suspended organic matter in Pacific Ocean waters. *Okeanol. Issled., 18*, 75–156.

*Bogdanov, Yu. A., A. P. Lisitzin, and E. A. Romankevich (1970) Organic matter in suspension and in bottom deposits of the global ocean in: *Geochemistry of Organic Matter*, N. B. Vassoevich, editor, Nauka, Moscow.

Bogdanov, Yu. A., and L. I. Ovchinnikova (1967) On a method of determining bituminous substance in suspended matter. *Okeanologiia, 5*(2), 366–371.

Bruevich, S. V., editor (1966) *Chemistry of the Pacific Ocean*. Akademiya Nauk, Inst. Okeanol., Moscow.

Burkov, V. A. (1963) Water circulation in the northern Pacific. *Okeanologiia, 3*(5), 761–776.

Calvert, S. E. (1966) Origin of diatom-rich, varved sediments from the Gulf of California. *J. Geol., 74*(5), 546–565.

Donahue, J. G. (1967) Diatoms as indicators of Pleistocene climatic fluctuations in the Pacific sector of the Southern Ocean. *Prog. Oceanogr. 4*, 133–140.

Edwards, D. S. (1968) The detrital mineralogy of surface sediments of the ocean floor in the area of the Antarctic Peninsula, Antarctica. M. S. Thesis, Florida State University, Talahassee, Florida.

*Evteev, S. A. (1964) Geological work of glaciers of east Antarctic *Glaciology*, IX Section of IGY Program N12, Nauka, Moscow, 119 pp.

Forsberg, E., and J. Joseph (1964) Phytoplankton production in the southeast Pacific Ocean. *Okeanologiia, 4*(4), 687–689.

Gessner, F. (1959) *Hydrobotanik II. Stoffhaushalt*. VEB Deutscher Verlag der Wissenschaften, Berlin, 701 pp.

Goodell, H. G., and N. D. Watkins (1968) The paleomagnetic stratigraphy of the Southern Ocean, 20° West to 160° East longitude. *Deep-Sea Res., 15*, 89–112.

Goodell, H. G., N. D. Watkins, T. T. Mather, and S. Koster (in press). The Antarctic glacial history in sediments of the Southern Ocean. *J. Geol.*

Griffin, J. J., and E. D. Goldberg (1963) Clay mineral distribution in the Pacific Ocean, pp. 728–741 in: *The Sea, vol. 3*, M. N. Hill, editor, Wiley-Interscience, New York.

Hays, J. D. (1967) Quaternary sediments of the Antarctic Ocean. *Progr. Oceanogr., 4*, 117–131.

Hays, J. D., and N. D. Opdyke (1967) Antarctic radiolaria, magnetic reversals and climatic change. *Science, 158*(3804), 1001–1011.

Hela, I., and T. Laevastu (1961) *Fisheries Hydrography*. Fishing News (Books) Ltd.

Holmes, C. W. (1965) Rates of sedimentation in the Drake Passage. PhD Thesis, Florida State University, Talahassee, Florida.

Ivanov, Yu. A., and V. G. Neiman (1965) Frontal zones of the Southern Ocean of the Antarctic. *Antarktika*, Dokl. Kom. *1964*, 98–109.

*Kapitsa, A. P. (1968) Sub-glacial relief of Antarctica in: *Results of the Investigations on the International Geophysical Project*. Nauka, Moscow, 98 pp.

Keller, W. D., W. D. Balgoro, and A. L. Reesman (1963) Dissolved products of artificially pulverized silicate minerals and rocks. Part I. *J. Sediment Petrol., 33*(1), 191–204.

Klyashtorin, L. B. (1964) Primary production investigations in the Antarctic. *Okeanologiia, 4*(3), 458–461.

Koblentz-Mishke, O. I. (1965) The value of primary production in the Pacific Ocean. *Okeanologiia, 5*(2), 325–337.

Kosak, H. P. (1954) *Antarktis*, Heidelberg.

Koster, S. (1966) Recent sediments and sedimentary history across the Pacific-Antarctic Ridge. M. S. Thesis, Florida State University, Talahassee, Florida.

Kozlova, O. G. (1964) *Diatom Algae in the Indian and Pacific Sectors of the Antarctic*. Nauka, Moscow. 168 pp.

Kozlova, O. G., and V. V. Mukhina (1966) Diatoms and silicoflagellates in suspension and bottom deposits of the Pacific Ocean, pp. 192–218 in: *Geochemistry of Silica*. Nauka, Moscow.

Krylov, A. Ya., A. P. Lisitzin, and Yu. I. Silin (1961) Significance of the argon-potassium ratio in oceanic muds. *Izv. Akad. Nauk SSSR* (geol.), *3*, 87–100.

Kuznetsov, Yu. v., V. K. Legin, A. P. Lisitzin, and A. N. Simonyak (1964) Radioactivity of oceanic suspended matter. Report I. Thorium isotopes in oceanic suspended matter. *Radiokhimiya, 6*(2), 242–254.

*Lavrenchik, V. N. (1965) Global fallout of nuclear explosion products. Atomizdat, Moscow, 170 pp.

Lisitzin, A. P. (1956) A method of studying suspension with geological objectives. *Trud. Inst. Okeanol., 19*, 204–231.

Lisitzin, A. P. (1958) On types of marine sediments related to the action of ice. *Dokl. Akad. Nauk SSSR, 118*(2), 373–376.

Lisitzin, A. P. (1959) New data on the distribution and composition of suspended matter in seas and oceans in connection with problems of geology. *Dokl. Akad. Nauk SSSR, 126*(4), 863–866.

Lisitzin, A. P. (1960) Sediment formation in the southern Pacific and Indian Oceans, pp. 69–87 in: *Marine Geology*. Nauka, Moscow.

Lisitzin, A. P. (1960) Bottom sediments of the eastern Antarctic and the southern Indian Ocean. *Deep-Sea Res., 7*, 89–99.

Lisitzin, A. P. (1961) Distribution and composition of suspended material in seas and oceans, pp. 175–231 in: *Recent sediments of seas and oceans*. Izd. Akad. Nauk, Moscow.

Lisitzin, A. P. (1961) Regularities of ice dispersion of coarsely fragmental material, pp. 232–284 in: *Recent sediments of seas and oceans*. Izd. Akad. Nauk, Moscow.

Lisitzin, A. P. (1961) Processes of recent sediment formation in the southern and central Indian Ocean, pp. 124–174 in: *Recent sediments of seas and oceans*. Izd. Akad. Nauk, Moscow.

Lisitzin, A. P. (1962) Use of submersed and deck pumps for collecting deep, large-volume water samples. *Trud. Inst. Okeanol., 55*, 137–168.

Lisitzin, A. P. (1962) Suspended matter in the ocean. *Trud. Okeanogr. Kom. 10*(3), 9–37.

*Unable to check.

*Lisitzin, A. P. (1964) Distribution and chemical composition of suspended matter in Indian Ocean waters. Okeanologiya X razdel programmy MGG No. 10, 164 pp.

Lisitzin, A. P. (1966) *Processes of sediment formation in the Bering Sea.* Nauka, Moscow, 574 pp.

Lisitzin, A. P. (1966) Basic regularities of distribution of recent silica sediments and their connection with climatic zonation, pp. 321–370 in: *Geochemistry of silica.* Nauka, Moscow.

Lisitzin, A. P. (1969) Bottom sediments and suspended matter. *Atlas Antarktiki,* vol. 2, Glavnoe upravleniie geodesii i kartografii MG SSSR.

Lisitzin, A. P. (in press) Distribution of silica microfossils in suspension and in bottom sediments in: *Micropaleontology of marine bottom sediments.* Cambridge Univ. Press.

Lisitzin, A. P. (in press) Distribution of carbonate microfossils in suspended matter and in bottom sediments. In: *Micropaleontology of marine bottom sediments.* Cambridge Univ. Press.

Lisitzin, A. P. (in press) *Sediment formation in oceans.*

Lisitzin, A. P., and L. P. Barinov (1960) New large-diameter core sampler "Antarctica." *Trud. Inst. Okeanol., 44,* 123–133.

Lisitzin, A. P., Yu. I. Belyaev, Yu. A. Bogdanov, and A. N. Bogoyavlenskii (1966) Regularity of the distribution and form of silica suspended in waters of the global ocean, pp. 37–89 in: *Geochemistry of silica.* Nauka, Moscow.

Lisitzin, A. P., Yu. A. Bogdanov, and L. I. Ovchinnikova (1967) Some results of studying bituminous matter in water suspension of the Pacific Ocean. *Okeanologiia, 7*(1), 120–129.

*Lisitzin, A. P., and Yu. A. Bogdanov (1968) Suspended amorphous silica in Pacific Ocean waters. *Okeanol. Issled, 18,* 5–42.

Lisitzin, A. P., and Yu. A. Bogdanov (in press) Suspended matter in Pacific Ocean waters. In: *Sedimentation in the Pacific Ocean.* Akademiya Nauk, Inst. Okeanol., Moscow.

Lisitzin, A. P., and V. A. Glazunov (1960) Construction and test of operation of a 200-liter bathometer. *Trud. Inst. Okeanol., 44,* 112–122.

Lisitzin, A. P., I. O. Murdmaa, V. P. Petelin, and N. S. Skornyakova (1966) Granulometric composition of deep water Pacific Ocean sediments. *Litologiya i Poleznye Iskopaemye,* No. 2, 5–26.

*Lisitzin, A. P., and V. P. Petelin (1967) Peculiarities of the distribution and modification of CaCO₃ in Pacific Ocean bottom deposits. *Litologiya i Poleznye Iskopaemye,* No. 5, 50–65.

*Lisitzin, A. P., and G. B. Udintsev (1952) A new bottom sampler "Okean-50". *Met. i Gidrol.,* No. 8.

Lisitzin, A. P., and A. V. Zhivago (1958) Contemporary methods of studying geomorphology and bottom deposits of the Antarctic Sea. *Izv. Akad. Nauk SSSR (geogr),* No. 6, 88–97.

Lisitzin, A. P., and A. V. Zhivago (1959) Marine geological work of the Soviet Antarctic Expedition 1955–57. *Deep-Sea Res., 6,* 77–87.

Maksimov, N. V. (1961) The influence of glacial discharge of Antarctica on the hydrological regime of the Southern Ocean. *Inform. Biull. Sov. Antarkt. Exped.,* No. 25, 36–38.

Mather, T. T. (1966) The deep-sea sediments of the Drake Passage and Scotia Sea. M.S. Thesis, Florida State University, Talahassee, Florida.

Matveev, A. A. (1963) On a study of chemical discharge of Antarctica. *Antarktika,* Dokl. Kom. *1962,* 109–121.

Murray, J., and L. R. Irvine (1891) On silica and the siliceous remains of organisms in modern seas. *Proc. Roy. Soc. Edinb.(b), 18,* 229–250.

Murray, J., and A. F. Renard (1891) Report on deep-sea deposits based on the specimens collected during the voyage of H.M.S. *Challenger* in the years 1872 to 1876. *Challenger Rep.,* 525 pp.

*Unable to check.

Nazarov, V. C. (1962) Ice of Antarctic waters. Okeanologiya X razdel programmy MGG No. 6, 80 pp.

Neeb, G. A. (1943) The composition and distribution of the samples. *Snellius – Exped. 5,* part 3, sect. 2, 265 pp.

Opdyke, N. D., B. Glass, J. D. Hays, and J. Foster (1966) Paleomagnetic study of Antarctic deep-sea cores. *Science, 154*(3747), 349–357.

Park, K. (1966) Deep-sea pH. *Science, 154*(3756), 1540–1541.

*Petrushevskaya, M. G. (1966) Radiolaria in plankton and in bottom deposits. In *Geochemistry of silica.* Nauka, Moscow, 219–246

Pettersson, H. (1953) The Swedish deep-sea expedition 1947–48. *Deep-Sea Res., 1,* 17–24.

Peterson, M. N. A. (1966) Calcite: Rates of dissolution in a vertical profile in the central Pacific. *Science, 154*(3756), 1542–1544.

Peterson, M. N. A., and J. J. Griffin (1964) Volcanism and clay minerals in the southeastern Pacific. *J. Mar. Res., 22,* 13–21.

Pia, J. (1933) *Die rezenten Kalksteine.* Akad. Verlagsges, Leipzig, 420 pp.

Picciotto, E. E. (1967) Geochemical investigations of snow and firn samples from east Antarctica. *Antarctic J., U.S., 2*(6).

*Rateev, M. A., Z. N. Gorbunova, A. P. Lisitzin, and G. I. Nosov (1966) Climatic zonation of distribution of clay minerals in the global ocean sediments. *Litologiya i Poleznye Iskopaemye,* No. 3, 3–22.

*Rateev, M. A., Z. N. Gorbunova, A. P. Lisitzin, and G. I. Nosov (1968) Climatic zonation of distribution of clay minerals in the global ocean sediments. *Okeanol. Issled., 18,* 283–310.

Revelle, R. (1944) Marine bottom samples collected in the Pacific Ocean by the *Carnegie* on its seventh cruise. Carnegie Inst. Wash. Pub. 556, Oceanogr. 2(1), 133 pp.

Revelle, R., and R. Fairbridge (1957) Carbonates and carbon dioxide. *Mem. Geol. Soc. Amer., 67*(1), 239–295.

Rex, R. W., and E. D. Goldberg (1958) Quartz contents of pelagic sediments of the Pacific Ocean. *Tellus, 10,* 153–159.

*Saidova, Kh. M. (1961) Ecology of foraminifera and paleogeography of far eastern seas of USSR and the northwest Pacific Ocean. Izd. Akad. Nauk, Moscow, 170 pp.

Saijo, Y., and T. Kawashima (1964) Primary production in the Antarctic Ocean. *J. Oceanogr. Soc. Japan, 19*(4), 22–28.

Schutz, D. F., and K. K. Turekian (1965) The distribution of cobalt, nickel and silver in ocean water profiles around Pacific Antarctica. *J. Geophys. Res., 70*(22), 5519–5528.

Semina, H. J. (1963) Phytoplankton of the central Pacific along 174°W Section II. The number of phytoplankton cells. *Trud. Inst. Okeanol., 71,* 5–21.

Shilnikov, V. I. (1960) Volume and quantity of icebergs in the Antarctic. *Inform. Biull. Sov. Antarkt. Exped., 21,* 34–37.

*Skornyakova, N. S. (1964) Dispersion of iron and manganese in Pacific Ocean sediments. *Litologiya i Poleznye Iskopaemye,* No. 5.

Skornyakova, N. S., and V. P. Petelin (1967) Sediments of the central region of the south Pacific Ocean. *Okeanologiia, 7*(6), 1005–1019.

Starik, I. E., A. P. Lisitzin, and Yu. V. Kuznetsov (1962) On the mechanism of the removal of radium from sea water and its accumulation in bottom deposits of seas and oceans. *Antarktika,* Dokl. Kom., *1961,* 70–135.

Strakhov, N. M., ed. (1957) Methods for the study of sedimentary rocks, vols. 1 and 2, Moscow.

*Strakhov, N. M. (1960–1962), Principles of the theory of lithogenesis, Vol. 1 (1960), 212 pp; Vol. 2 (1960), 574 pp; Vol. 3 (1962), 550 pp. Izd. Akad. Nauk, Moscow.

*Strakhov, N. M. (1963) Types of lithogenesis and their evolution in Earth's history. Gosgeoltekhizdat, Moscow, 535 pp.

*Strakhov, N. M., N. G. Brodskaia, L. M. Krjazeva, A. N. Razjivina, M. A. Rateev, D. G. Sapognivev, and E. S. Shishiva (1954) *Formation of sediments in recent water bodies.* Izd. Akad. Nauk, Moscow, 791 pp.

Sysoev, N. N. (1956) Rational construction of percussion bottom corer. *Trud. Inst. Okeanol.,* 19, 238–239.

Trask, P. D. (1937) Relation of salinity to the calcium carbonate content of marine sediments. *U.S. Geol. Survey Prof. Paper* No. 186-N, 273–299.

Turekian, K. K. (1964) The geochemistry of the Atlantic Ocean basin. *Trans. N.Y. Acad. Sci.,* Ser. II, 26(3), 312–330.

Turekian, K. K., D. F. Schutz, P. Bower, and D. G. Johnson (1967) Alkalinity and strontium profiles in Antarctic waters. *Antarctic J.U.S.,* 2(5), 186–188.

*Vtyurin, B. N., L. D. Dolgushin, A. P. Kapitsa, and Yu. M. Model (1959) Recent glacial cover of eastern Antarctica and its dy-

namics. In *Scientific results of the First Continental Expedition 1956-57.* Leningrad.

Watkins, N. D., and H. G. Goodell (1967) Confirmation of the reality of the Gilsa Geomagnetic Polarity Event. *Earth and Planetary Science Letters,* 2, 123.

Watkins, N. D., and H. G. Goodell (1967) Geomagnetic polarity and faunal extinction in the Southern Ocean. *Science,* 156(3778), 1083–1087.

Watkins, N. D., and H. G. Goodell (1967) Inconsistencies resulting from a limited study of magnetic polarity variation in submarine sedimentary cores. *Trans. Amer. Geophys. Union,* 48(1), 90.

Wiseman, J. D. H. (1954) The determination and significance of past temperature changes in the upper layer of the equatorial Atlantic Ocean. *Proc. Roy. Soc. (A),* 222, 296–323.

*Zhivago, A. V., and A. P. Lisitzin (1957) New data on bottom relief and marine sediments of the eastern Antarctic. *Izv. Akad. Nauk, Ser. Geograph.,* No. 1, 19–35.

*Unable to check.

Brian Funnell

SCHOOL OF ENVIRONMENTAL SCIENCES,
UNIVERSITY OF EAST ANGLIA, NORWICH, ENGLAND

OCEANIC MICROPALEONTOLOGY OF THE SOUTH PACIFIC

INTRODUCTION

This paper attempts to review and summarize existing knowledge on the micropaleontology of the South Pacific, from the equator southward to the position of the Antarctic Convergence. Consideration is restricted to oceanic occurrences, although pelagic microfossils are of course also found in marine sequences on islands and other land masses in the South Pacific (see, for example, Cole, 1960; Jenkins, 1967; McTavish, 1966).

HISTORY OF INVESTIGATION

Ehrenberg (1855, 1872) listed a few diatoms, radiolarians, and foraminifers from South Pacific sediments, but, as with so many other aspects of oceanographic investigation, the first substantial data on the micropaleontology of the South Pacific were obtained by H.M.S. *Challenger* during her circumnavigation of the World Ocean in 1873–1876. Some of the most voluminous reports of that voyage were those compiled by G. S. Brady (1880), H. B. Brady (1884), Haeckel (1887), and Murray and Renard (1891) on the potential microfossils, represented by the ostracodes, foraminifers, radiolarians, and sediments, respectively, of the deep-sea floor. For the most part the skeletal parts of microorganisms that they describe can be assumed to have accumulated only recently at the ocean bottom. In some instances, however, their illustrations indicate that microfossils of some antiquity were obtained. In the South Pacific, as Riedel (1957) has pointed out, the Radiolaria of *Challenger* station 272 (03°48'S, 152°56'W; 4,755 m) include reworked Tertiary forms.

Later the Swedish Deep Sea Expedition of 1947–1948 recovered more examples of Tertiary sediments and microfossils from the South Pacific, as exemplified by cores 62, 64, 69, 73, 88–91, and 93. Some of these contain only re-

worked Tertiary microfossils, whereas others consist, at least in part, of actual Tertiary accumulations (Riedel, 1952, 1957, etc.).

Following this expedition a variety of investigations were mounted into various aspects of the micropaleontology of the South Pacific, ranging from the occurrence of skeleton-bearing organisms in the plankton, through a study of their accumulation and distribution in bottom sediments, to their occurrence and interpretation in both Quaternary and Tertiary sediments. The principal contributions in these fields are referred to in the following pages.

DISTRIBUTION OF POTENTIAL MICROFOSSILS IN THE PLANKTON AND BENTHOS OF THE SOUTH PACIFIC

The principal planktonic organisms that secrete skeletons likely to accumulate on the deep-sea floor as microfossils are the Bacillarophyceae (diatoms), the Silicoflagellatae (silicoflagellates), the Coccolithophorales (coccolithophores), the Radiolaria (radiolarians), the Globigerinacea (planktonic foraminifers) and the Euthecosomata (euthecosomatous pteropods).

The principal benthic organisms contributing microfossil remains to the deep-sea floor are the Foraminiferida (benthic foraminifers) and the Ostracoda (ostracodes).

The distribution of these groups in the South Pacific as living organisms has been studied to varying extents.

Diatoms and silicoflagellates are commonly associated with one another as microfossils. Diatoms comprise a class of planktonic (neritic and freshwater) algae producing a skeleton of opaline silica; silicoflagellates are a group of planktonic protozoans producing a skeleton of opaline silica somewhat similar in size to that of diatoms.

Contributions to knowledge of diatoms in the South Pacific have been made by Castracane degli Antelminelli

(1886) and Hendey (1937). Haeckel (1887) included some consideration of silicoflagellates in his treatment of radiolarians, but generally speaking, their distribution in the plankton of the South Pacific seems to all intents and purposes to be unknown.

Coccolithophores constitute a coccolith-bearing order of the planktonic algae Chrysophyceae. They have been little studied in the living state in the South Pacific. Hasle (1960) recorded some 33 species from the equatorial and subantarctic South Pacific, and Norris (1961) made qualitative studies on some 31 species from surface waters between New Zealand and the Tonga Islands. Uschakova (in press) studied coccoliths in suspension along a north–south section at 140°W between the equator and approximately 20°S, but intact coccolithophore cells, as opposed to isolated coccoliths, were rare in the collections made. Marshall (1933) recorded five species from inshore Australian waters, and knowledge of the distribution of coccolithophores in the oceans at large has been reviewed by Gaarder (in press).

Radiolarians constitute an opaline-skeleton-bearing subclass (the Acantharia are here considered to be a separate subclass) of the planktonic protozoan class Reticulosia. In the order Tripylea the opaline silica is intimately mixed with organic material, and the skeletons rarely survive into the fossil state. In the order Polycystina the opaline silica skeletons are frequently preserved.

Haeckel (1887) recorded radiolarians, including living abyssal forms, from bottom-sediment samples obtained by H.M.S. *Challenger* at numerous stations in the South Pacific, but no plankton samples appear to have been examined from this region (Haeckel, 1887, clix–clxiii). Recently Petrushevskaya (1966a, in press a) has described polycystine radiolarians from the plankton between the equator and 13°S.

Planktonic foraminifers comprise a planktonic superfamily with a calcite skeleton, within the otherwise benthic order Foraminiferida of the protozoan class Granuloreticulosia.

The earliest studies of this group in the plankton of the South Pacific were made by H. B. Brady, who listed species found at several stations throughout the region (Brady, 1884, pp. 111–116). Bradshaw (1959) made quantitative studies of planktonic foraminifers in the equatorial South Pacific, and Parker (1960) followed this with a similar examination of plankton tows from the southeastern Pacific. Boltovskoy (1966) studied the planktonic foraminifers from plankton tows taken mainly between 90° and 160°W astride the Antarctic Convergence.

Euthecosomatous pteropods are planktonic gastropod mollusks bearing a shell of aragonite. Their occurrence in the plankton of the South Pacific was first studied by Pelseneer (1888) on the basis of the collections made by H.M.S. *Challenger* in 1873–1875. The classic works of Meisenheimer (1905) and Tesch (1948) also refer to collections from the South Pacific.

Benthic foraminifers comprise the residuum of the order Foraminiferida (i.e., excluding the superfamily Globigerinacea), generally possessing a calcareous or agglutinated test. Descriptions of collections from bottom sediments made by H. B. Brady (1884), Cushman (1932, 1933, 1942), Todd (1965), and others do not distinguish between dead tests and specimens alive at the time of collection.

Ostracodes are a subclass of crustacean arthropods. Descriptions of collections from bottom sediments made by G. S. Brady (1880) do not distinguish between dead valves and specimens alive at the time of collection. [N.B. skeletal remains of species of the planktonic ostracode genus *Halocypris*, recorded by Brady (1880, pp. 21–28) from plankton tows in the South Pacific, do not appear to survive in the bottom sediments.]

Benson (1964) has recently reviewed the distribution of Recent marine ostracodes in the Pacific, although most of the occurrences cited refer to shallow-water faunas.

DISTRIBUTION IN BOTTOM SEDIMENTS

The occurrence of microfossils in deep-sea sediments is affected both by conditions controlling their production in the overlying waters (or on the deep-sea floor) and by solution or destruction in the overlying water, at the sediment surface, or within the accumulating sediment.

Production in the overlying waters affects both the amounts of microfossils contributed to the bottom sediments and the specific composition of the assemblages. The productivity of the overlying waters clearly controls the amounts of microfossils contributed, and higher concentrations and rates of accumulation of microfossils are noted under both the Equatorial Current system and the Antarctic Convergence in the South Pacific, with lowest rates of accession under the South Pacific Central Water Mass. The specific composition and ultimately the absolute concentrations of microfossils, however, may be profoundly modified by post-mortem solution and destruction.

Solution of the opaline silica of which the skeletons of diatoms, silicoflagellates, and radiolarians are composed is a well-known phenomenon. Many of the more delicate diatom species seem never to reach the deep-sea floor, and many even of the more delicate polycystine radiolarian species are not preserved in the bottom sediment. Berger (1968b) has directly observed the solution of radiolarian skeletons suspended at varying depths in an experiment carried out in the central Pacific. He concludes that solution proceeds more rapidly at shallower depths, with a maximum at 250 m (the shallowest observed), and less rapidly at greater depths, especially below 3,000 m. He suggests that high rates of solution may be related to the lower-dissolved silicate content and higher temperatures of the oceanic water less than 1,000 m deep, and that, if this is so, solution rates should be least in the Antarctic and in the deep Pacific, where accumulations of siliceous skeletons are best known. Riedel (1959b) has pointed out that siliceous skeletons normally only survive in bulk to contribute to bottom sediments under waters where the productivity is high (e.g., the Equatorial Current system and the Antarctic Convergence), and that they may disappear from originally sparsely diatomaceous or radiolarian sediments a few tens of centimeters below the sediment surface (Riedel and Funnell, 1964).

Solution of the calcite of which the skeletons of coccolithophores, planktonic foraminifers, most benthic foraminifers, and ostracodes are composed has also been investigated by Berger (1967). He found that solution of planktonic foraminiferal skeletons suspended at various depths in the central Pacific increased markedly below 3,000 m, and even more below 5,000 m. This last increase he attributed to the pronounced undersaturation of abyssal water with $CaCO_3$. The phenomenon of solution of calcite skeletons in the deeper parts of the oceans is well known, of course, and was described by Murray (1889) and Murray and Renard (1891). The level at which the rate of solution exceeds supply is variable within limits in different parts of the oceans and is known as the "critical depth" or "calcium carbonate compensation depth." In the South Pacific it is generally between 4,000 and 5,000 m (Bramlette, 1961), but in the Ross Sea, Antarctica (Kennett, 1966), it rises to 500 m. Beneath this depth calcareous skeletal remains are rarely found, but near it the more robust coccoliths and foraminifers persist while the others are partially or completely destroyed by solution; benthic foraminifers usually outlast the planktonic forms. Some coccolithophores and planktonic foraminifers are rarely found in bottom sediments, although they may be abundant in the plankton (e.g., coccoliths such as the holococcoliths, and delicate planktonic foraminifers such as *Hastigerina pelagica*). Many authors have pointed out that solution strongly affects the quantitative and qualitative composition of planktonic foraminiferal assemblages even well above the calcium carbonate compensation, and recently, Berger (1968a) has made this effect the subject of quantitative analysis. It is generally assumed that most of the solution of calcitic skeletons occurs while they are at the sediment–water interface and not during the relatively short time when they are falling through the water column; rapid burial in sediments, caused for example by density currents, may lead to entombment of calcitic microfossils well below the "critical" or "compensation" depth.

Solution of aragonite of which the skeletons of euthecosomatous pteropods are composed occurs in general at much shallower depths than does the solution of calcite. In the South Pacific it probably occurs below about 500 m.

Different combinations of conditions of production and solution of siliceous and calcareous skeletons lead to the distinctive varieties of siliceous, calcareous, calcareous-siliceous and unfossiliferous deep-sea sediment. In the South Pacific, the high productivity and moderate depths beneath the Equatorial Current system generally produce calcareous-siliceous oozes, the high productivity but aggressive bottom waters beneath the Antarctic Convergence produce siliceous oozes, and the low productivity and moderate depths beneath the Central Water Mass produce calcareous oozes, except where the depths are in fact somewhat greater where largely unfossiliferous clays accumulate.

Diatoms and Silicoflagellates

The distribution of diatom frustules in bottom sediments of the South Pacific has been described by Jousé *et al.* (1967, in press a and b), and references to silicoflagellates can be found in Haeckel's account (1887) of the radiolarians.

Jousé *et al.* show that the maximum concentrations of diatoms in bottom sediments occur in a belt either side of 60°S, i.e., corresponding approximately to the Antarctic Convergence. In this belt, diatom valves exceed 100×10^6 per gram of sediment. Concentrations fall off rather rapidly either side of this belt and are reduced to 5 to 10×10^6 valves per gram of sediment within 10° of latitude. Concentrations decrease with distance northward, and north of 40°S, diatoms are generally absent from bottom sediments until within 10° of the equator. At the equator, concentrations of 5 to 25×10^6 valves per gram of sediment are common east of 180° longitude, with peaks of 25 to 50×10^6 valves per gram. Details of these distributions can be seen in Jousé *et al.* (1967, in press a and b).

Jousé *et al.* recognize four diatom complexes (associations) in South Pacific sediments. The equatorial complex, which extends to about 5°S, and westward to 180° of longitude, contains the following species among others: *Asteromphalus imbricatus, Coscinodiscus africanus, Triceratium cinnamomeum, Asterolampra marylandica.* The tropical diatom complex, which generally reaches at least 10°S, with a further extension toward New Zealand at about 180° longitude, contains *Coscinodiscus crenulatus, C. nodulifer, Ethmodiscus rex, Hemidiscus cuneiformis, Nitzshia marina, Planktoniella sol, Rhizosolenia bergonii, Thalassiosira oestrupii,* etc. The central part of the South Pacific is essentially devoid of diatoms in bottom sediments, and the subantarctic complex, between about 50° and 65°S, contains *Fragilariopsis antarctica, Coscinodiscus lentiginosus, Thalassiothrix antarctica, Schimperiella antarctica,* etc. The Antarctic complex, which extends from about 65°S to the coast of Antarctica, contains *Fragilariopsis curta, F. cylindricus, Eucampia balastrium, Thalassiosira gracilis, Charcotia actinochilus, C. oculoides, C. symbolophorus, Biddul-*

phia weissflogii, etc. Complexes of transitional type are found over about 5° of latitude between the subantarctic and Antarctic complexes.

In general, the concentrations of diatom valves in the bottom sediments appear to reflect productivity conditions in the overlying surface waters. The species associations in the bottom sediments also seem to directly reflect distinctive associations in the plankton above, although they may differ markedly in the proportions of species (many being absent altogether) because of the solution of the delicate skeletons after the death of the organisms.

Coccolithophores (in the form of coccoliths) have been described from bottom sediments of the South Pacific by Shumenko and Uschakova (1967) and Uschakova (in press).

Uschakova (in press) has examined coccoliths in the surface layer of the bottom sediment from the equator to approximately 10°S along a section at 140°W. Two complexes (associations) were recognized. The first, an equatorial one, extending from the equator to 7°S, is dominated by *Cyclococcolithus leptoporus* (35%), *Gephyrocapsa oceanica* (34%), *Coccolithus* aff. *C. obtusus* (12%), along with *Umbilicosphaera mirabilis, Coccolithus huxleyi,* and some 13 other forms. The second, a south equatorial one, south of 7°S, is dominated by *Cyclococcolithus leptoporus* (50%), "*Ellipsoplacolithus*" *productus* (10%), "*Cycloplacolithus*" *laevigatus* (10%), along with *Umbilicosphaera mirabilis, Ceratolithus cristatus,* and seven other forms. "*C*". *laevigatus* and *C. cristatus* are particularly characteristic.

Radiolarians from the bottom sediment samples recovered by H.M.S. *Challenger* were given a full taxonomic treatment by Haeckel (1887; see pp. clix–clxiii for station list). Recently, Petrushevskaya (1966a, in press a) and Nigrini (1968) have given further consideration to the accumulation of Polycystine radiolarians in bottom sediments in the equatorial zone of the South Pacific, and Riedel (1958), Hays (1965), and Petrushevskaya (1966b, 1967, in press b) have considered their distribution in bottom sediments of the Antarctic zone.

Petrushevskaya's (1966a, in press a) data are based on only four stations between the equator and 13°S, and between 154° and 175°W. At the station on the equator (actually at about 00°30'N) there are more than 5×10^5 radiolarians per gram of sediment, which corresponds to a high observed content of radiolarians in the overlying plankton at that station. Specimens of some species, including *Phormacantha hystrix,* were found to be absent or rare in the sediment although abundant in the overlying plankton; species such as *Artostrobus annulatus, Stylatractus neptunus,* and *Spongodiscus resurgens,* were abundant in the bottom sediment although rare or unobserved in the overlying plankton; yet other species, such as *Psilomelissa calvata* and *Peromelissa phalacra,* are equally abundant in both bottom sediment and plankton. Generally speaking, representatives of the Discoidea (Spongodiscidae), Larcoidea, and Botryoidea are more abundant in the bottom sediments, and representatives of the Sphaeroideae, Plectoidae, Stephoidae, and Cyrtoidae (Dicyrtida), with skeletons consisting of thin bars or thin-walled shells, are more abundant in the plankton.

Nigrini (1968) described the occurrence of 10 radiolarian species in bottom sediments at 12 stations in the equatorial South Pacific and concluded that their distribution corresponded to near-surface current patterns.

Hays (1965) recognized two radiolarian faunas in the bottom sediments of high southern latitudes, a warm-water fauna north of the Antarctic Convergence and an Antarctic fauna south of it. Notable species in the warm-water fauna included *Cenosphaera nagatai, Echinomma leptodermum, Axoprunum stauraxonium, Heliodiscus astericus, Androcyclas gamphonycha, Calocyclas amicae, Lamprocyclas maritalis, Stichopilium annulatum,* and *Eucyrtidum tumidulum?,* whereas the Antarctic fauna included such species as *Spongoplegma antarcticum* and *Peromelissa denticulata.* Petrushevskaya (in press b) shows radiolarian concentrations of 1,000 to 10,000 radiolarians per gram of sediment in the vicinity of the Antarctic Convergence south of New Zealand (with occasional peaks of 10,000 to 100,000 per gram). Concentrations fall to 100 to 1,000 per gram north and south, but there is another zone of high concentrations (10,000 to 100,000 per gram) midway between Australia and New Zealand.

Planktonic foraminifers were recorded extensively from South Pacific bottom sediments by Brady (1884) on the basis of samples taken by H.M.S. *Challenger,* and an up-to-date taxonomic treatment of Pacific species was given by Parker (1962). In 1965, Todd published descriptions of the planktonic foraminifers obtained from bottom sediments in the tropical Pacific by the *Albatross* in 1899–1900. Kustanowich (1963) described the distribution of planktonic foraminifers in bottom sediments of the southwestern Pacific, Blackman (1966; Blackman and Somayajulu, 1966) has defined faunal groups for planktonic foraminifers in bottom sediments in the southeastern Pacific, and Parker (in press) has reviewed their distribution in the oceans at large. See also Beliaeva (1966).

Kustanowich (1963), basing his conclusions on a number of bottom-sediment samples between 15° and 55°S, and between 165°E and 170°W, indicated five different planktonic foraminiferal faunas in the southwestern Pacific. These were:

1. Subequatorial fauna—between about 15° and 26°S, with abundant *Globigerinoides conglobatus, G. ruber,* and *G. sacculifer.*

2. Northern fauna—between about 26° and 32°S, with the subequatorial species accompanied by numerous *Globo-*

rotalia punctulata (= *G. crassaformis*), *G. eggeri* (= *Globo-quadrina dutertrei*), *Globigerinella aequilateralis* (= *G. siphonifera*), and *Globigerina subcretacea*.

3. North Central fauna—between about 32° and 37°S, essentially transitional.

4. Central fauna—between about 37° and 44°S, with abundant *Globigerina bulloides, Globorotalia truncatuli-noides,* and *G. inflata*.

5. South Central fauna—between about 44° and 54°S, with the central species accompanied by numerous *Globigerina pachyderma*.

Blackman (see Blackman and Somayajulu, 1966) has recognized four faunal groups in the southeastern Pacific on the basis of vector analysis of assemblages from surface-sediment samples. These are as follows:

1. Equatorial group—consisting of *Globoquadrina duter-trei, Globorotalia tumida, G. cultrata, Pulleniatina obliqui-loculata, Globigerinella siphonifera,* and *Globigerinoides quadrilobatus sacculifer*. This group occupies a belt gener-ally between the equator and 6°S, but swings south as far as 20°S, east of 100°W.

2. Tropical group—consisting of *G. quadrilobatus sac-culifer, G. ruber, Globoquadrina conglomerata, G. sipho-nifera, G. tumida, G. cultrata, Globigerinoides conglobatus,* and *G. dutertrei*. This group occupies a belt between ap-proximately 6° and 22°S but is not observed to continue east of 100°W.

3. Midlatitude group—consisting of *G. conglobatus, G. ruber, Globorotalia truncatulinoides, G. quadrilobatus sac-culifer, G. siphonifera, G. conglomerata,* and *Globorotalia inflata*. This group occupies a belt generally between ap-proximately 22° and 35°S, but is not observed to continue east of about 85°W.

4. High-latitude group—consisting of *Globigerina bul-loides, G. inflata,* and *"Orbulina universa."* This group occu-pies a belt generally south of 35°S, but the northern bound-ary swings northeast of 85°W, to abut against the equatorial group at about 20°S.

Parker (in press) has shown the existence of two coiling provinces for *Globorotalia truncatulinoides* in bottom sedi-ments in the South Pacific. North of a rather sinuous boundary, varying between approximately 20° and 40°S, populations tend to be right-coiling; south of that boundary they tend to be left-coiling. For *Globigerina pachyderma* in the South Pacific, Parker shows a right-coiling province north of about 50°S, and a left-coiling province south of about 50°S but expresses some doubt that a single species is involved.

Euthecosomatous pteropods were described from a few *Challenger* bottom samples from the South Pacific by Pelseneer (1888), but their remains are not widespread.

Benthic foraminifers make up the majority of forms in-cluded in H. B. Brady's work on the *Challenger* samples (1884), a taxonomic revision of which was issued by Barker (1960). Cushman (1932, 1933, 1942) and Todd (1965) de-scribed foraminifers obtained by the *Albatross* from the tropical Pacific in 1899-1900. More recently, Bandy and Rodolfo (1964) described benthic foraminifers from sam-ples taken in the Peru-Chile Trench area, and Beliaeva and Saidova (1965) have commented on the correlation between benthic and planktonic foraminifers in the bottom sedi-ments of the Pacific.

Ostracodes from the South Pacific floor are recorded by G. S. Brady (1880); their occurrence mainly at shallow-water localities in the Pacific has recently been reviewed by Benson (1964), who is currently studying their distribution in ocean-floor sediments.

QUATERNARY SEQUENCES

Although the sequences of Quaternary microfossils in the South Pacific have probably not received as much attention as have those in some other parts of the World Ocean, the results of a considerable variety of studies in the region are now either available or becoming available.

At this stage it is not possible to reconcile the rather diverse views that have been expressed following these stud-ies. In point of fact, it is probably in this field (i.e., com-parison of results obtained on different microfossil groups) that advances could be most fruitfully pursued. Unfortu-nately, with very few exceptions, work on different groups has hitherto been undertaken on different cores, often from different parts of the ocean, so that few direct comparisons are possible. (Those that are possible do not instill confi-dence in *in*direct comparisons.)

In the following account the Quaternary occurrences of the groups are treated in turn. Attention is drawn to the few in-stances in which direct comparisons may be made on present data. Figure 1 shows the positions of all the Quaternary cores that have so far been studied and published in any detail.

Diatoms and silicoflagellates Arrhenius (1952), Kolbe (1954), and Olausson (1960) made a variety of observations on diatoms in the eastern equatorial Pacific cores obtained by the Swedish Deep-Sea Expedition of 1947-1948. More recently, Muhina (in press) has studied some 24 equatorial Pacific cores for diatoms and silicoflagellates; 10 of these cores come from the South Pacific side of the equator. Donahue (1967) has described Quaternary diatom assem-blages from the Antarctic sector.

Arrhenius (1952), Kolbe (1954), and Olausson (1960), between them, examined seven cores from the equatorial South Pacific for diatoms, namely,

S D-S E   38   (02°52'S, 89°50'W; 3,225 m)
S D-S E   39   (02°44'S, 92°45'W; 3,617 m)
S D-S E   61   (00°06'S, 135°58'W; 4,475 m)
S D-S E   62   (03°00'S, 136°26'W; 4,520 m)
S D-S E   74   (00°29'S, 151°33'W; 4,545 m)
S D-S E   91   (02°56'S, 171°18'E; 4,120 m)
S D-S E   93   (01°20'S, 167°23'E; 3,965 m)

Of these, S D-S E 91 proved to be devoid of diatoms. No general Quaternary stratigraphy of diatoms was attempted, but interesting observations were made (Arrhenius, 1952) on the variation in size of *Coscinodiscus nodulifer* in Quaternary sequences, and Kolbe (1954) gave a taxonomic treatment of the species encountered. *Coscinodiscus nodulifer* was shown to have produced large-diameter frustules of around 120 μm at intervals corresponding to periods of high productivity, intensified water circulation, and presumed glacial episodes deduced from the sedimentological evidence. The more usual size of *C. nodulifer* is about 40 μm, and the increase in proportions of larger individuals was attributed to auxospore formation during periods of higher productivity.

Muhina (in press) recognizes up to seven "Quaternary" horizons based on diatoms and silicoflagellates:

- Horizon I, said to correspond to the Post-Glacial or Holocene epoch, contains a high proportion of equatorial-tropical diatom species, including, *Asteromphalus imbricatus, Asterolampra marylandica, Actinocyclus ellipticus* var. *lanceolata, A. ellipticus* var. *elongata, Coscinodiscus africanus, C. nodulifer, C. crenulatus, Ethmodiscus rex, Hemidiscus cuneiformis, Nitzschia marina, Planktoniella sol, Rhizosolenia bergonii,* and *Triceratium cinnamomeum.*
- Horizon II, said to correspond to glacial conditions (sometimes with an indication of a median interstadial), contains up to about 50 percent of such subtropical diatom species as *Pseudoeunotia doliolus, Roperia tesselata, Thalassiosira lineata, T. oestrupii, Thalassionema nitzschioides, T. parva, Nitzschia sicula, N. bicapitata, N. braarodii,* and *N. kolaschezkii.*
- Horizon III is similar to Horizon I, and is said to correspond to interglacial conditions.
- Horizon IV is similar to Horizon II, and is said to correspond to glacial conditions; it is distinguished by the presence of the silicoflagellate *Mesocena elliptica,* and by the extinction of the diatom *Rhizosolenia praebergonii* just above the base.
- Horizon V is similar to Horizons I and III and is said to correspond to interglacial conditions; it is distinguished by the maximum abundance of *Rhizosolenia praebergonii* and by the extinction of *Thalassiosira convexa* about halfway through the horizon.
- Horizon VI is similar to Horizons II and IV, and is said to correspond to glacial conditions; it is distinguished by

the entry of *R. praebergonii* part way through the horizon, by abundant *T. convexa,* and by the presence of *Nitzschia praemarina,* which becomes extinct at the end of the horizon.
- Horizon VII is similar to Horizons I, III, and V and is said to correspond to interglacial conditions.
- Horizon VIII corresponds to the first indication of cooling.

These horizons have been recognized in four of the South Pacific cores examined: USSR 5098 (05°03'S, 139°56'W; 4,400 m) shows all eight horizons in 3.37 m of calcareous-siliceous sediment; USSR 3802 (03°17'S, 172°52'W; 5,329 m) shows Horizons III to VIII in 2.66 m of siliceous sediment; USSR 5100 (07°08'S, 140°13'W; 4,076 m) shows Horizons I to VI in 2.40 m of calcareous sediment; and apparently Swedish Deep-Sea Expedition core 62 (03°00'S, 136°26'W; 4,511 m) shows Horizons I to V in 14.60 m of calcareous-siliceous sediment. In five of the other South Pacific cores no Quaternary stratigraphy could be distinguished on the basis of diatoms and silicoflagellates. These were: USSR 3921 (00°55'S, 142°30'E; 3,075 m), consisting of 2.88 m of calcareous sediment; CAP 8 BP (13°39'S, 174°58'E; 2,570 m), consisting of 7.62 m of calcareous sediment; USSR 5145 (08°01'S, 175°57'W; 5,520 m), consisting of 8.00 m of red clay; USSR 5110 (12°59'S, 154°06'W; 5,222 m), consisting of 2.08 m of red clay; and USSR 5103 (11°08'S, 140°10'W; 4,136 m), consisting of 2.60 m of calcareous sediment. In USSR 5113 (05°01'S, 154°15'W; 5,046 m), consisting of 2.90 m of siliceous sediments, the midpart of the core from 85 to 210 cm, seemingly corresponding to Horizons III to V inclusive, is devoid of diatoms and silicoflagellates. Horizons I–II and VI are recognizable at the top and bottom of the core, respectively. It would appear that this barren interval in the Quaternary is more extensive in time and thickness away from the equator.

Coccolithophores Uschakova (1966) has given some data on the occurrence of coccolithophores in the South Pacific core USSR 5100 (07°08'S, 140°13'W; 4,076 m), which was also studied by Muhina (in press) from the point of view of diatoms and silicoflagellates (see above). Uschakova distinguishes four horizons:

- Horizon I (approximately 0 to 20 cm) is shown as dominated by increasing proportions of *"Calcidiscus."*
- Horizon II (approximately 20 to 140 cm) is shown as containing abundant *"Calcidiscus"* with high proportions of *Cyclolithus* in the lower part and *Tremalithus* in the upper part. *Discolithus, Zygolithus,* and *Ceratolithus* are indicated consistently; *Rhabdolithus,* sporadically throughout Horizons I and II.
- Horizon III (approximately 140 to 220 cm) is shown as

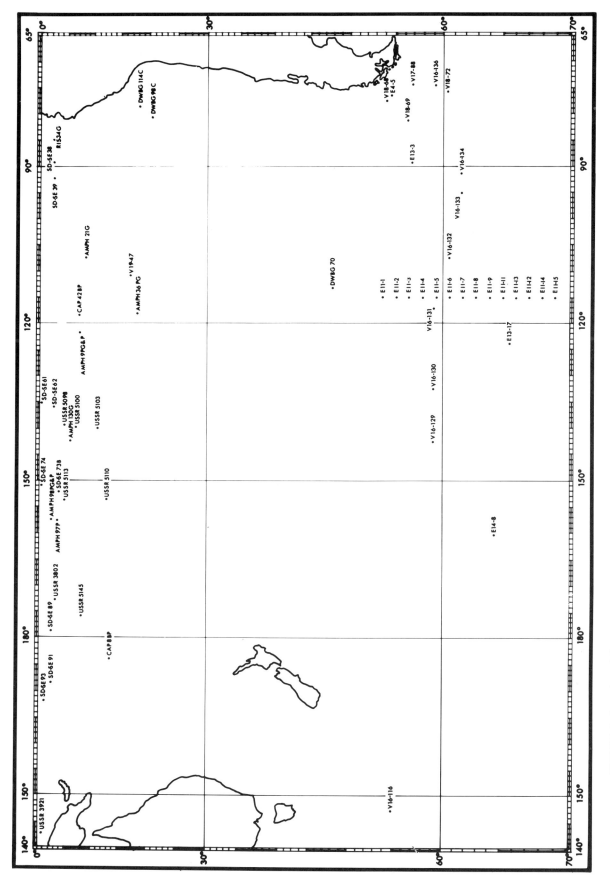

FIGURE 1 Location of Quaternary cores in the South Pacific (only cores that have been studied and published in some detail are included).

containing abundant *Coccolithus* cf. *C. pelagicus* and *Discoaster*, both of which continue sporadically in Horizon II above, in the lower part, and abundant *Ceratolithus* in the upper part. *Cyclolithus,* "*Calcidiscus,*" and *Tremalithus* are also present.

● Horizon IV (approximately 220 to 240 cm) is shown as having a similar flora to Horizon III.

It would seem that Uschakova's Horizon I corresponds approximately to Muhina's Horizon I, her Horizon II approximately to Muhina's Horizons II to IV, her Horizon III approximately to Muhina's Horizon V, and her Horizon IV to Muhina's Horizon VI. The presence of abundant *Discoaster* (including forms recorded as *D. brouweri* and "*D. tribrachiatus*") in the lower part of the core (below 180 cm) suggests that it, and by implication Muhina's Horizons V (pars) and VI, etc., may be Pliocene in age.

Radiolarians from five Quaternary equatorial South Pacific cores were examined by Riedel (1957), but no Quaternary stratigraphy was attempted. These cores were:

S D-S E	73B	(04°04'S, 152°53'W;	5,200 m)
S D-S E	74	(00°29'S, 151°33'W;	4,575 m)*
S D-S E	89	(02°48'S, 178°57'W;	5,465 m)
S D-S E	91	(02°50'S, 171°18'E;	4,096 m)*
S D-S E	93	(01°20'S, 167°23'E;	3,942 m)*

The cores designated by asterisks were also studied for diatoms by Kolbe (1954). All the cores except S D-S E 74 contained reworked Tertiary radiolarians in the Quaternary sequences.

More recently, further radiolarians from Quaternary cores in the equatorial South Pacific have been studied by Nigrini (in press; see also Nigrini, 1968), and from cores taken astride the Antarctic Convergence by Hays (1965, 1967). The latter cores are of particular interest in connection with their paleomagnetic stratigraphy, subsequently determined by Opdyke *et al.* (1966). Additional studies on the relation between Quaternary (and Tertiary) radiolarians and paleomagnetic reversals in cores taken astride the Antarctic Convergence in the South Pacific were made by Watkins and Goodell (1967) and Hays and Opdyke (1967).

Nigrini (in press) has investigated the radiolarians of 17 equatorial Pacific cores; 10 of these lie south of the equator. Nigrini recognizes a fourfold Quaternary sequence based on the appearance and disappearance of radiolarian species, which do not appear to be related to climatic variations. The zonation she proposes is as follows:

Buccinosphaera invaginata Range Zone (Zone 1; uppermost Quaternary) is defined by the range of *B. invaginata*. *Collosphaera tuberosa* and *Theocorythium trachelium trachelium* are also present, and *Amphirhopalum ypsilon* is rare.

Collosphaera tuberosa Concurrent Range Zone (Zone 2) is defined by the appearance of *C. tuberosa* at the base. *T. trachelium trachelium* and *A. ypsilon* are also present.

Amphirhopalum ypsilon Assemblage Zone (Zone 3) is defined by the disappearance of *Anthocyrtidium angulare* at the base. *A. ypsilon* and *T. trachelium trachelium* are consistently present, and *Lithopera bacca* is sometimes present.

Anthocyrtidium angulare Concurrent Range Zone (Zone 4; lowermost Quaternary) is defined at the top by the disappearance of *A. angulare* and at the base by the disappearance of *Pterocanium prismatium. A. ypsilon, T. trachelium trachelium,* and *L. bacca* are also present.

The underlying deposits containing *P. prismatium* are considered to be Tertiary (Pliocene).

The ten South Pacific cores in which these zones have been recognized are:

AMPH 97P	(03°42'S, 157°40'W; 5,228 m)	Zones 1 to 4, plus Pliocene, in 5.92 m
AMPH 98PG & P	(02°50'S, 157°13'W; 5,225 m)	Zones 1 to 4, plus Pliocene, in 1.81 and 5.69 m
AMPH 130G	(05°58'S, 142°43'W; 4,450 m)	Zones 1 to 2, plus Pliocene, in 1.14 m
S D-S E 61	(00°06'S, 135°58'W; 4,437 m)	Zones 1 to 2, in 10.29 m
S D-S E 62	(03°00'S, 136°26'W; 4,510 m)	Zones 1 to 4, plus Pliocene, in 14.79 m
AMPH 9 PG & P	(07°31'S, 121°56'W; 4,410 m)	Zones 1 to 4, plus Pliocene, in 1.40 and 5.94 m
CAP 42 BP	(07°19'S, 118°40'W; 4,200 m)	Zones 1 to 4, plus Pliocene, in 8.03 m
AMPH 21 G	(08°29'S, 107°26'W; 3,120 m)	Zones 1 to 2, in 0.79 m

It is interesting that Swedish Deep-Sea Expedition core 62 has also been examined by Muhina (in press; see above) for diatoms and silicoflagellates. Muhina's horizons appear to occupy approximately the following intervals in this core: Horizon I (Post-Glacial or Holocene) 9 to 90 cm; Horizon II (glacial) 90 to 380 cm; Horizon III (interglacial) 380 to 740 cm; Horizon IV (glacial) 740 to 1,090 cm; Horizon V (interglacial) 1,090 to 1,480 cm. Nigrini's zones occupy the following intervals: Zone 1 (uppermost Quaternary), 0 to 200 cm; Zone 2, 218 to 460 cm; Zone 3, 508 to 810 cm; Zone 4 (lowermost Quaternary), 868 to 1,170 cm; Pliocene, 1,170 to 1,470 cm. (However, see section on Coccolithophores above—it would seem that part, in this case the greater part, of Muhina's Horizon V is Pliocene.)

Hays (1965, 1967; Hays and Opdyke, 1967) established five faunal zones based on radiolarians from cores taken in the Antarctic sector of the Pacific astride the Antarctic Convergence. These were, from latest to earliest, Ω, ψ, χ, ϕ, and γ. The first three of these have subsequently been referred to the Quaternary.

Zone Ω contains only species found in recent Antarctic sediments.

Zone ψ is defined at its top by the last occurrence of the radiolarian *Stylatractus* sp. and its base (in sediments south of the Antarctic Convergence) by the last occurrence of the species *Saturnulus planetes* and *Pterocanium trilobum* (both of which are still living north of the Antarctic Convergence). Their disappearance south of the Antarctic Convergence probably indicates a change to colder conditions.

Zone Χ is defined at its top by the disappearance of *S. planetes* and *P. trilobum* as described for Zone ψ, and its base by the last occurrence of *Eucyrtidium calvertense*. At or near the base of Zone Χ there is a change to highly siliceous sediments, which may reflect upwelling engendered by the glaciation of Antarctica.

Opdyke *et al.* (1966) and Hays and Opdyke (1967), by paleomagnetic measurements on the same cores, show that the Brunhes normal polarity epoch contained the ψ/Ω boundary at or slightly below its middle, corresponding to an age of approximately 0.4 to 0.5 m.y. The Matuyama reversed-polarity epoch contained the Χ/ψ boundary just below its top, corresponding to an age of approximately 0.7 m.y. The φ/Χ boundary fell just below the base of the Olduvai normal polarity event, corresponding to an age of about 2.0 m.y. The Olduvai event is regarded as closely approximating the international definition of the base of the Quaternary (Pleistocene) (see Berggren *et al.*, 1967). Hays (unpublished; *fide* Hays and Berggren, in press) has found that the radiolarian *Pterocanium prismatium* disappears consistently, in some equatorial Pacific cores, just above the top of the Olduvai event, thereby confirming the assumption of previous authors, including Nigrini (in press), that its disappearance corresponds with (actually it very slightly postdates) the true base of the Quaternary in that region.

Planktonic Foraminifers in cores from the equatorial Pacific were included in the studies of Arrhenius (1952) and Brotzen and Dinesen (1959). More recently, Blackman (1966) has investigated planktonic foraminifers in Quaternary cores from the southeastern Pacific, some details of which have been published by Blackman and Somayajulu (1966). Blackman recognized five Holocene and late Pleistocene zones, based on the proportional representation of faunal groups (defined by vector analysis of surface sediments) and variations of coiling ratios in certain species of planktonic foraminifer. The faunal groups (equatorial, tropical, mid-latitude, and high-latitude) are given above in the section on planktonic foraminifers in bottom sediments.

The zones established are as follows:

● Zone 1 (Holocene and latest Pleistocene, extending back to 8,000–12,000, average 11,000 years B.P.) contains representatives of faunal groups in similar proportions to those found at the same site in surface sediment samples, indicating similar oceanographic conditions to those found there at the present day.

● Zone 2 (approximately 11,000 to 51,000–85,000 years B.P.) contains higher proportions of representatives of higher latitude faunal groups, indicating displacement of faunas toward the equator, presumably corresponding to a cooler phase.

● Zone 3 (51,000–85,000 to 110,000–140,000 years B.P.) contains representatives of faunal groups in similar proportions to those in Zone 1, indicating conditions similar to those at the present day.

● Zone 4 (110,000–140,000 to 120,000–180,000 years B.P.) is similar to Zone 2.

● Zone 5 (beyond 120,000–180,000 years B.P.) is similar to Zones 1 and 3.

Ages were determined by the ionium:thorium (^{230}Th: ^{232}Th) method (Blackman and Somayajulu 1966).

Details for the two cores DWBG 98C (20°49'S, 81°08'W; depth 2,300 m, length 0.68 m) and DWBG 114 (18°20'S, 79°21'W; depth 3,090 m, length 1.34 m) were given by Blackman and Somayajulu (1966). Both cores come from a position at which the equatorial and high-latitude faunal groups closely approach one another, and are therefore particularly well suited to showing Quaternary faunal fluctuations. The proportional contribution of the end-members of the different faunal groups has been plotted for different levels in each core. Zones 1, 3, and 5 contain higher proportions of the equatorial end-member and distinctly lower proportions of the high-latitude end-member than do Zones 2 and 4. *Globorotalia crassaformis* and *Globorotalia truncatulinoides* tend to right-coiling in Zones 1, 3, and 5 and to left-coiling in Zones 2 and 4, whereas Zone 2 is distinguished by a tendency to left-coiling in *Globoquadrina dutertrei*, which at other levels is almost 100 percent right-coiling. Both *G. truncatulinoides* and *G. dutertrei* tend to left-coiling in surface sediments in higher latitudes, and *G. crassaformis* seems to have a left-coiling province south of the core area, so that these tendencies support the concept of an equatorward displacement of faunas during Zones 2 and 4.

Wiles (1967), who examined pore concentrations in *Globigerina eggeri* (= *Globoquadrina dutertrei*) from Lamont Geological Observatory core V-19-47 (17°00'S, 111°12'W; 3,422 m), north of Easter Island, commented that the planktonic foraminifers in the core did not show marked faunal variation and suggested that conditions were warm throughout the Pleistocene. The usual variation in pore concentrations, which elsewhere seem to be related to interglacial (high counts) and glacial (low counts) conditions, was found, however. Wiles suggested that the pore concentration might be a response to some environmental change other than temperature, but it does not seem that this is necessarily so.

Euthecosomatous Pteropods have not so far been described from Quaternary sequences in the South Pacific. Herman (in press) reports that Quaternary cores from the Pacific examined so far are devoid of pteropods (see also Arrhenius, 1952, p. 224).

Benthic Foraminifers from Quaternary sequences in the South Pacific have not so far been given much attention, although some general descriptions of assemblages are given by Saidova (1967), and others are referred to by Arrhenius (1952) and Brotzen and Dinesen (1959).

Ostracodes from some of the Quaternary cores investigated by Blackman (1966) have been described by Swain (in press a). He examined ostracodes from the following cores:

RIS 34 G	(02°46'S, 85°28'W; 3,210 m)
AMPH 36PG	(18°17'S, 118°17'W; 3,550 m)
DWBG 70	(48°28'S, 113°17'W; 2,580 m)

Ten species were found, five of which are believed to represent described forms, namely, *Bradleya dictyon* (Brady), *Echinocythereis echinata* (Brady), *Krithe* cf. *K. tumida* Brady, *K.* cf. *K. glacialis* Brady, Crosskey and Robertson, and *"Cythere" sulcatoperforata* Brady.

The Plio-Pleistocene Boundary

Considerable confusion has attended the recognition of this boundary in deep-sea sediments in recent years. The observations of Berggren et al. (1967) and Phillips et al. (1968) on the relationships between paleomagnetic reversals, coccolithophore and planktonic foraminiferal successions in some North Atlantic cores, and those of Hays and Berggren (in press) on the relationship between paleomagnetic reversals and radiolarian successions in the equatorial Pacific, may have considerably clarified the situation.

One particular South Pacific core has been of special interest ever since attempts were made to define this boundary in deep-sea sediments. It is Swedish Deep-Sea Expedition core 62. In 1952, Arrhenius suggested, on the basis of sedimentological changes that he took to reflect the onset of Pleistocene climatic changes, that the boundary should be placed at 1,080 cm. Details of the different interpretations put on the stratigraphy of S D-S E 62 are compared in Figure 2.

At present it appears that the Plio-Pleistocene boundary should be placed at, or slightly below, 1,170 cm in S D-S E 62. This level corresponds to the base of Nigrini's radiolarian Zone 4 (*Anthocyrtidium angulare* Concurrent Range Zone) and to the disappearance of the radiolarian species *Pterocanium prismatium*, which Hays and Berggren (in press) state occurs consistently just above the top of the Olduvai (normal) polarity event in equatorial Pacific cores.

(The *Globoratalia tosaensis–Globorotalia truncatulinoides* transition, which is taken as the criterion of the base of the Pleistocene because of its occurrence at the base of the Calabrian of the Italian succession, occurs within the Olduvai event.) This correspondence between the disappearance of *P. prismatium* and the Plio-Pleistocene boundary confirms a view long held by micropaleontologists (Riedel, 1957, Riedel et al., 1963, Riedel and Funnell, 1964).

According to Uschakova (1966), the *Discoaster* content in S D-S E 62 falls abruptly between 1,335 and 1,230 cm (the forms identified as *D. brouweri* and *"D. tribrachiatus"* disappear altogether), although *D.* cf. *woodringi* and *Discoaster* spp. are shown as continuing to 1,060 cm in reduced quantities. Bramlette (in Riedel et al., 1963) has also recorded a marked decrease in the abundance of *Discoaster* between 1,300 and 1,250 cm in S D-S E 62; he indicates the presence of *Discoaster brouweri* up to about 1,210 cm and of *D.* aff. *woodringi* up to about 1,170 cm. He records the disappearance of *Coccolithus* cf. *C. pelagicus* at about 1,250 cm (cf. Uschakova's Horizon III in USSR 5100). This also corresponds fairly well with experience in North Atlantic cores (Berggren et al., 1967) where *Discoaster* disappears just above the Olduvai event, with a brief reappearance within the Jaramillo normal event at about 0.90–0.95 m.y.

PRE-QUATERNARY DEPOSITS

As noted above, the first pre-Quaternary microfossils were obtained from the South Pacific by H.M.S. *Challenger* (1873–1876), although they were not recognized as such until much later. More pre-Quaternary microfossils were obtained by the Swedish Deep-Sea Expedition (1947–1948), which also obtained the first examples of actual pre-Quaternary deposits from the South Pacific floor. Subsequently, an increasing number of samples of pre-Quaternary sediment have been obtained, initially mainly by expeditions mounted by the Scripps Institution of Oceanography, but latterly also by ships of the Lamont Geological Observatory, the USSR Institute of Oceanology, and others. Twelve primary drilling sites have been proposed for the South Pacific by the Pacific advisory panel to the U.S. Deep-Sea Drilling Project (JOIDES, 1967).

Diatoms and silicoflagellates. Some of the first pre-Quaternary microfossils to be recorded from the Pacific as a whole were specimens of *"Coscinodiscus diophthalmus"* (= *C. excavatus* var. *quadriocellatus*) described by Castracane (1886, Plate 16, Figure 4) [?from *Challenger* station 268 (07°35'N, 149°49'W; 5,304 m); see Kolbe (1954, p. 29)]. Kolbe (1954) recognized and described further late-Tertiary diatoms in several Swedish Deep-Sea Expedition cores from the equatorial Pacific, and Kanaya (in Olausson, 1961a, and in press) has also listed and described Tertiary diatoms from

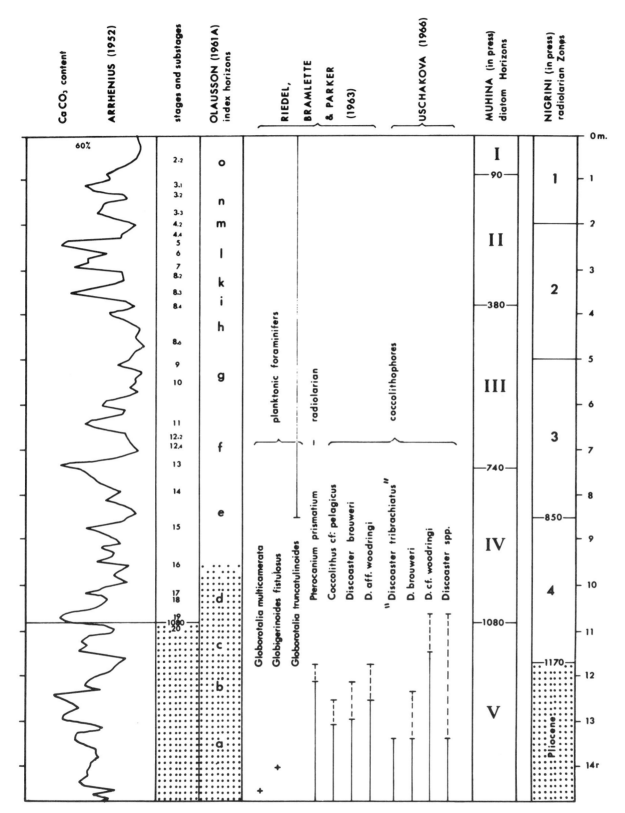

FIGURE 2 Comparison of interpretations of Swedish Deep-Sea Expedition core 62. (Stipple indicates portion of core regarded as Pliocene by various authors.)

143

both equatorial and South Pacific cores. In addition, it appears that the low-latitude diatoms and silicoflagellates of part of Horizon V and of Horizon VI to VIII of Muhina (in press) may be referable to the late Tertiary. In the Antarctic sector, Donahue (1967) has described diatoms from the latest Tertiary portions of cores also dated by radiolarians and paleomagnetic reversals.

Coccolithophores have been widely used in the age determination of South Pacific cores since Bramlette and Riedel (1954), and it is difficult to provide a complete catalog of their application. Pre-Quaternary age assignments, based at least in part on determinations of coccoliths by Bramlette, are contained in papers by Bramlette and Riedel (1959), Riedel and Bramlette (1959), Riedel (1957), Olausson (1961a,b), Riedel *et al.* (1963), Riedel and Funnell (1964), Riedel (in press), and others. Few of the taxa utilized in these determinations have been listed or described in these papers, although papers by Martini (1965, in press) refer to forms from the South Pacific. Uschakova (1966) has provided an account of the latest Tertiary coccolithophores of USSR core 5100.

Radiolarians. Pre-Quaternary radiolarians appear in at least one of the *Challenger* sediment samples illustrated by Murray and Renard (1891, Plate 15, Figure 4, station 268), and they were included among those described by Haeckel (1887). The occurrence of Tertiary radiolarians in west Pacific sediments was first recognized by Riedel (1952). Since then he has described numerous instances of *in situ* and reworked pre-Quaternary radiolarians from the Pacific deep-sea floor: *viz.* Riedel (1957, 1959a, in press), Riedel *et al.* (1963), Riedel and Funnell (1964), Friend and Riedel (1967). In his latest account (Riedel, in press), 132 occurrences of pre-Quaternary radiolarians are recorded from the South Pacific alone. Nigrini (in press) has described several instances of latest Tertiary radiolarians from the equatorial South Pacific.

In the Antarctic sector Hays (1965, 1967; and Hays and Opdyke, 1967, Opdyke *et al.*, 1966) has described latest Tertiary radiolarians from numerous cores for which the paleomagnetic stratigraphy has also been determined.

Planktonic foraminifers have been considered in the determination of the age of sediments and their reworked constituents in several papers concerning pre-Quaternary and contaminated Quaternary cores from the South Pacific, including Riedel (1959a), Olausson (1961 a,b), Riedel *et al.* (1963), Riedel and Funnell (1964), Parker (1965, 1967). Parker (1965, 1967) is the only one of these to actually describe and illustrate the forms recorded.

Euthecosomatous pteropods have not been recorded from pre-Quaternary sediments in the South Pacific; neither are pre-Quaternary forms known to be reworked into later sediment.

Benthic foraminifers occur in pre-Quaternary sediments in the South Pacific, and there are almost certainly reworked forms in later sediments, but these have yet to be described.

Ostracodes of pre-Quaternary age have so far been recorded from only one South Pacific core of Pliocene age (Swain, in press b). Like the benthic foraminifers they are probably present much more widely.

No account of pre-Quaternary microfossils from the South Pacific could be complete without at least a brief reference to the problems that these remains may help to resolve. They fall into three categories: those concerning the history and evolution of life, those related to the history of the water masses of the South Pacific, and those concerning the structural evolution of the ocean basin.

Riedel and Funnell (1964), on the basis of an observed relationship between the limits of siliceous sediments and the Equatorial Current system, speculated on a possible northward broadening of that system during the earlier parts of the Tertiary. On the other hand, they drew attention to the overall similarity of microfossil distributions during the Tertiary to those of the present day, implying an overall similarity in the general oceanic circulation pattern, and probably also in the gross depth pattern of the ocean floor. Whether this will also prove to be true of the pre-Tertiary, likely to be revealed by the Deep-Sea Drilling Project, remains to be seen. Geitzenauer *et al.* (1968) recently presented evidence of the presence of glacially attrited quartz grains in Eocene sediments from 58°S, and Parker (1967) noted a possible cooling in the equatorial region toward the end of the Pliocene, as shown by the appearance of cooler water species of planktonic foraminifera. Hays (1965, 1967) has shown a striking change in microfossil production south of the Antarctic Convergence at the Tertiary-Quaternary boundary, where comparatively sparsely fossiliferous radiolarian red clays give place to diatom oozes —presumably a response to substantial changes in the water-mass characteristics in those latitudes at that time.

Evidence of the structural evolution of the basin can be obtained by the use of microfossils in dating the volcanic floor or "basement" on which the sediments rest. According to the current theory of ocean-floor spreading, the age of this igneous "basement" should vary systematically from place to place. By present methods the minimum age of this floor can be set by the oldest microfossils, or more reliably the oldest sediments, recovered from above it. It can be a minimum age only, because the chances of permanent deposition commencing immediately after its formation and the chances of recovering the very oldest sediment of microfossils in the vicinity by current methods of gravity, piston and free-fall coring, are small. Nevertheless Burckle *et al.* (1967) and Riedel (1967) have been able to show that the distribution of recovered Tertiary sediments and microfossils across and on the flanks of the East Pacific Rise, are consistent with the hypothesis of spreading. When the Deep-Sea Drilling Project samples are available, better estimates of

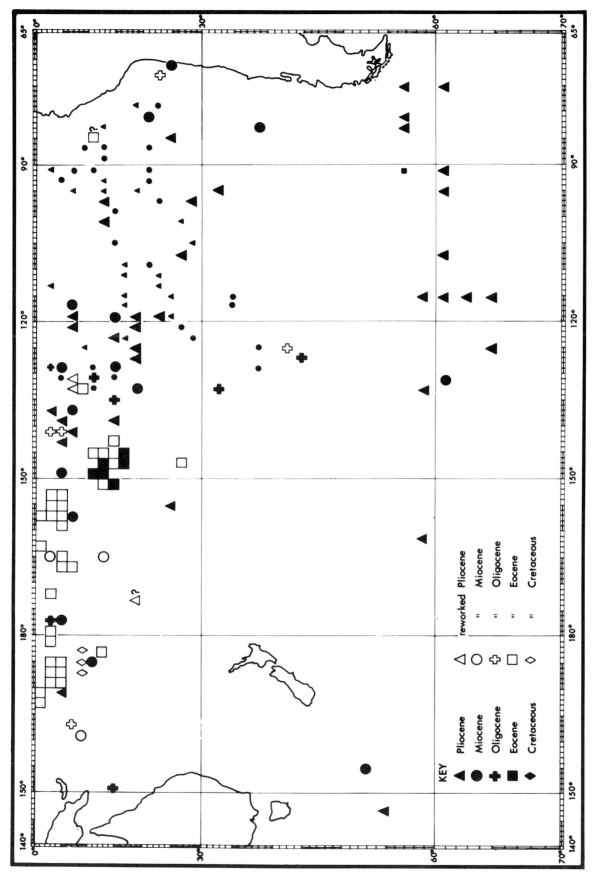

FIGURE 3 Occurrence of pre-Quaternary microfossils in the South Pacific. [In each case the oldest occurrence in each 2° square has been plotted. Details will be found either in the Appendix to this paper or in Tables 1 and 2 of Riedel (in press) for the larger symbols. The smaller symbols are after Figure 2 of Burckle et al. (1967).]

KEY

	Pliocene	reworked Pliocene
◢	Pliocene	◺ reworked Pliocene
●	Miocene	○ = Miocene
✚	Oligocene	✚ = Oligocene
■	Eocene	☐ = Eocene
◆	Cretaceous	◇ = Cretaceous

145

the age of the oldest sediment, and its relation to the age of the underlying "basement" at those points, should be obtained. Smaller scale vertical movements may be detected as a result of the sensitivity of microfossils to depth during life. Burckle and Saito (1966) have inferred changes in elevation of the South Pacific Tuamotu Ridge from Eocene sediments and microfossils recovered from it.

A comprehensive index of pre-Quaternary occurrences in the South Pacific is given in the Appendix to this paper.

PROBLEMS AND PROSPECTS

It would seem that the South Pacific has some unique possibilities for further micropaleontological investigation and interpretation.

The rather broad, seemingly uncomplicated distribution of the main water masses might be expected to lead to relatively broad, uncomplicated distributions of potential microfossils, although so far only the barest start has been made on determining whether this is so. Only scattered and rather localized information is available for the radiolarians and planktonic foraminifers; the distribution of diatoms, silicoflagellates, coccolithophores, and pteropods in relation to physical, chemical, and biological conditions is practically unknown. Unless further progress is made in this field, the necessary basis of understanding for interpretation of fossil occurrences will remain lacking.

Compared with conditions in the North Pacific, those in the South Pacific are generally favorable for the accumulation of skeletal remains on the deep-sea floor. Calcareous microfossils occur widely in the considerable tracts of the South Pacific floor that lie above the "calcium carbonate compensation depth." Siliceous microfossils are also widespread, although few persist beneath the low-productivity regions of the South Pacific Central Water Mass. The Quaternary record should therefore be good over broad areas and its interpretation not unduly frustrated by differential solution of floras and faunas. Future investigations of microfossils in Quaternary sequences ought to avail themselves more of the opportunity to study different groups in parallel, preferably from the same samples.

The pre-Quaternary microfossil record of the South Pacific will probably continue to enjoy some of the advantages of the Quaternary record, at least through the Tertiary, providing an opportunity to study well-preserved microfossils of different groups comparatively. The Deep-Sea Drilling Project should provide up to four long sequences approximately along the 140°W meridian, and a further seven across the East Pacific Rise. In the longer term, it will be interesting to see whether the history of the South Pacific differs in any significant respect, or extends over a longer period than that of the South Atlantic and Indian oceans, which are currently thought to owe their origin to the post-Paleozoic disruption and separation of the southern continents.

ACKNOWLEDGMENTS

M. Black, F. L. Parker, W. R. Riedel, and W. H. Berger kindly read and commented on the draft manuscript, and M. N. Bramlette assisted with some aspects of the coccolithophores. Compilation of data was assisted in part by a United Kingdom N.E.R.C. research grant, and the illustrations were drafted by Mr. J. Lewis at the Department of Geology, Cambridge.

APPENDIX. INDEX OF PRE-QUATERNARY MICROFOSSIL OCCURRENCES IN THE SOUTH PACIFIC, ARRANGED ACCORDING TO LATITUDE AND LONGITUDE

[N.B. This index does not include the pre-Quaternary occurrences reported by Burckle *et al.* (1967), or those listed for the first time by Riedel (in press).]

A. *0° to 10°S* (listed W to E)

01°20'S, 167°23'E. 3,965 m. Swedish Deep-Sea Expedition 93. *Reworked early, middle and upper Tertiary.* Kolbe, 1954, p. 14; Riedel, 1957, pp. 63, 75; Olausson, 1960, pp. 192–193.

04°31'S, 168°02'E. 3,208 m. LSDH 78P. *lower Pliocene.* Riedel *et al.*, 1963, p. 1240; Parker, 1965, pp. 151–152; Friend and Riedel, 1967, p. 219, etc.; Ross and Riedel, 1967, pp. 287–288; Parker, 1967, p. 129, etc.: Swain (in press b).

02°50'S, 171°18'E. 4,120 m. Swedish Deep-Sea Expedition 91. *lower Miocene (approx. G. insueta zone).* Riedel, 1952, pp. 5, 12; Kolbe, 1954, pp. 4, 14; Riedel, 1957, pp. 63, 74–76; Riedel, 1959a, pp. 286–288, etc.; Olausson, 1960, pp. 190–192;

Olausson, 1961a, p. 32; Olausson, 1961b, pp. 299–393; Parker (unpublished).

08°44'S, 173°24'E. 5,397 m. PROA 62PG & P. *Quaternary with Pliocene, Miocene and Cretaceous admixture, or Pliocene with Miocene and Cretaceous admixture.* Ross and Riedel, 1967, p. 288.

03°21'S, 174°12'E. 4,830 m. Swedish Deep-Sea Expedition 90. *upper Oligocene-lower Miocene (approx. C. stainforthi zone).* Olausson, 1960, p. 190; Olausson, 1961a, pp. 14, 30; Olausson, 1961b, pp. 299–303.

09°03'S, 174°52'E. 4,960 m. CAP 5 BG. *Mixture of Cretaceous to late Tertiary.* Riedel and Funnell, 1964, pp. 321, 361–362; Ross and Riedel, 1967, p. 288. 09°03'S, 174°52'E. 4,960 m. CAP 5

BP. *Mixture of Cretaceous to Pliocene.* Riedel and Funnell, 1964, pp. 322, 361–362.

02°04'S, 178°16'E. 5,519 m. MSN 16G. (?) *Pliocene with early and middle Tertiary admixture.* Riedel and Funnell, 1964, pp. 339, 361–363.

02°48'S, 178°57'W. 5,480 m. Swedish Deep-Sea Expedition 89. *(? middle Tertiary, with early Tertiary admixture).* Riedel, 1957, pp. 63, 74–75; Olausson, 1960, pp. 188–189.

02°37'S, 177°45'W. 5,770 m. Swedish Deep-Sea Expedition 88. *middle Tertiary.* Riedel, 1957, pp. 63, 73–75; Olausson, 1960, pp. 187–188.

03°17'S, 172°52'W. 5,329 m. USSR 3802 [(?) *Pliocene.*] Muhina (in press).

04°38'S, 158°10'W. 5,140 m. AMPH 96P. *upper lower Miocene.* Friend and Riedel, 1967, p. 219.

03°42'S, 157°40'W. 5,228 m. AMPH 97P. *upper Pliocene.* Friend and Riedel, 1967, p. 219; Nigrini (in press).

06°02'S, 157°28'W. 5,177 m. AMPH 91P. *middle Miocene (approx. G. fohsi barisanensis zone).* Friend and Riedel, 1967, p. 219.

02°50'S, 157°13'W. 5,225 m. AMPH 98PG & P. *Pliocene.* Nigrini (in press).

01°52'S, 156°42'W. 4,983 m. AMPH 99PG & P. *Quaternary, with Miocene and Pliocene admixture.* Ross and Riedel, 1967, p. 287.

03°51'S, 155°48'W. 5,161 m. AMPH 101GV. *middle Miocene.* Friend and Riedel, 1967, p. 219.

03°52'S, 155°53'W. 5,050 m. AMPH 105G. *middle Miocene (approx. G. fohsi barisanensis zone).* Friend and Riedel, 1967, p. 219.

03°52'S, 155°43'W. 5172 m. AMPH 102P. *upper Miocene.* Friend and Riedel, 1967, p. 219.

04°49'S, 155°19'W. 5,265 m. AMPH 109P. *upper lower Miocene.* Friend and Riedel, 1967, p. 219.

05°01'S, 154°15'W. 5,046 m. USSR 5113. [(?) *Pliocene.*] Muhina (in press).

03°48'S, 152°56'W. 4,755 m. *Challenger* 272. (?) *reworked middle Tertiary.* Murray and Renard, 1891, pp. 120–121; Riedel, 1957a, p. 80.

04°04'S, 152°53'W. 5,200 m. Swedish Deep-Sea Expedition 73. *upper Tertiary, with lower and middle Tertiary admixture.* Riedel, 1952, p. 6; Riedel, 1957, pp. 63, 66–68, 76; Olausson, 1960, pp. 174–175; Riedel *et al.*, 1963, p. 1240; Riedel and Funnell, 1964, p. 311; Ross and Riedel, 1967, p. 286.

04°26'S, 149°24'W. 4,600 m. MSN 135PG & P. *lower Miocene (G. insueta zone) and Quaternary with lower Miocene admixture.* Riedel and Funnell, 1964, pp. 341–342; Ross and Riedel, 1967, p. 286.

05°58'S, 149°33'W. 5,115 m. MSN 132PG & P. *(?) early, middle to late Miocene and Pliocene, with Miocene admixture.* Riedel and Funnell, 1964, pp. 341, 364; Friend and Riedel, 1967, p. 219.

05°58'S, 142°43'W. 4,450 m. AMPH 130G. *lower Pliocene.* Friend and Riedel, 1967, p. 219; Nigrini (in press).

07°08'S, 140°13'W. 4,076 m. USSR 5100. *Pliocene and Plio-Pleistocene boundary.* Uschakova, 1966, pp. 136–143; Muhina (in press).

05°03'S, 139°56'W. 4,400 m. USSR 5098. [(?) *Pliocene.*] Muhina (in press).

03°00'S, 136°26'W. 4,510 m. Swedish Deep-Sea Expedition 62. *Pliocene; Plio-Pleistocene boundary.* Arrhenius, 1952, pp. 179–185, 194–195; Kolbe, 1954, p. 14; Riedel, 1957, p. 65; Brotzen and Dinesen, 1959, pp. 52–53; Emiliani, 1961a, p. 144; Olausson, 1961a, pp. 5–33; Riedel *et al.*, 1963, pp. 1239–1240; Uschakova, 1966, pp. 136, 141; Muhina (in press); Nigrini (in press).

06°23'S, 136°11'W. 4,410 m. RIS 102G. *upper Miocene and lower Pliocene.* Friend and Riedel, 1967, p. 220.

09°59'S, 133°23'W. 4,340 m. DWHH 13. *Quaternary with Eocene and later Tertiary admixture.* Riedel and Funnell, 1964, p. 331.

07°31'S, 121°56'W. 4,410 m. AMPH 9P. *upper Pliocene.* Friend and Riedel, 1967, p. 219; Nigrini (in press).

07°19'S, 118°40'W. 4,200 m. CAP 42BP. *upper Pliocene.* Friend and Riedel, 1967, p. 219; Nigrini (in press); Kanaya (in press).

B. *10° to 20°S*

14°45'S, 151°14'E. 4,400 m. MSN 21G. *Quaternary or Pliocene, with reworked middle Tertiary.* Riedel and Funnell, 1964, pp. 339–340.

11°33'/11°51'S, 174°17'/174°27'E. 30 to 41 m. (Alexa Bank). *Miocene.* Bramlette (unpublished); Riedel and Funnell, 1964, p. 322.

11°05'S, 175°10'E. 2,560 m. CAP Dredge No. 2. (Alexa-Penguin Bank). *Miocene (N17) with Quaternary admixture.* Parker, 1967, p. 130, etc.

18°41'S, 172°17'W. 860 m. N.Z.O.I. Station B79. Capricorn Seamount. *(?) Plio-Pleistocene.* Kustanowich, 1962, pp. 427–434.

13°53'S, 150°35'W. 3,623 m. MSN 128G. *Pliocene with Eocene admixture.* Riedel and Funnell, 1964, p. 340–341.

14°29'S, 150°01'W. 1,628 m to 1,884 m. V18-RD29. (Tuamotu Ridge). *Middle and Upper Eocene.* Burckle and Saito, 1966, pp. 1207–1208.

13°25'S, 149°30'W. 4,300 m. Swedish Deep-Sea Expedition 69. *Middle Eocene overlain by Upper Oligocene-Lower Miocene.* Olausson, 1960, pp. 169–171; Olausson, 1961a, pp. 29–32.

16°23'S, 146°02'W. 1,380 m. DWBG 25. (Tuamotu Ridge). *Middle Eocene.* Riedel and Funnell, 1964, pp. 328, 362.

16°47'S, 146°15'W. 980 m. DWBD 4. from side of unnamed seamount on Southwest flank of the Tuamotu Ridge. *upper Eocene (Tertiary 'b').* Cole, 1959, p. 10–11; Menard and Hamilton, 1964, p. 201, Table 1.

16°42'S, 145°48'W. approx. 2,200 m. DWBG 23B. *Middle Eocene.* Riedel and Funnell, 1964, pp. 327, 362; Black, 1964, pp. 307, 315; Martini, 1965, p. 395 (in press).

17°17'S, 145°45'W. 1,470 m. Albatross H3866. *reworked Pliocene.* Todd, 1964, p. 1084.

12°00'S, 144°20'W. 5,008 m. Swedish Deep-Sea Expedition 64. *Mio-Pliocene, with Eocene admixture.* Bramlette (unpublished).

12°00'S, 144°21'W. 5,000 m. Dolphin 2. *Mixture of early, middle and late Tertiary.* Wiseman and Riedel, 1960, pp. 215–216; Riedel and Funnell, 1964, pp. 324, 361–362.

12°10'S, 144°25'W. 4,332 m. Dolphin 1. *Pliocene or Quaternary with Pliocene admixture.* Riedel and Funnell, 1964, p. 324.

15°54'S, 143°06'W. 2,868 m. Albatross H3881. *reworked Pliocene.* Todd, 1964, p. 1084.

16°13'S, 143°48'W. 1,805 m. Albatross H3878. *reworked Pliocene.* Todd, 1964, p. 1084.

15°15'S, 142°27'W. 3,675 m. RIS 84G. *Quaternary with admixture of Pliocene and Eocene.* Parker, 1967, p. 129, etc.

14°03'S, 139°35'W. 3,900 m. RIS 82G. *Pliocene (lower N19).* Parker, 1967, p. 129, etc.

19°45'S, 139°54'W. 2,732 m. Albatross H3919. *reworked Pliocene.* Todd, 1964, p. 1084.

14°28'S, 135°29'W. 4,400 m. DWHH 14. *Oligocene.* Riedel and Funnell, 1964, pp. 331, 363; Martini, 1965, p. 394 (in press).

14°02'S, 128°29'W. 3,985 m. RIS 77G. *late Miocene (?N18).* Parker, 1967, p. 129, etc.

14°01'S, 122°28'W. 3,790 m. RIS 75G. *Pliocene (N21).* Parker, 1967, p. 129, etc.

14°16'S, 119°11'W. 3,400 m. CAP 38BP. *Miocene, overlain by Pliocene (N17–19, N21).* Parker, 1967, p. 129, etc.

C. 20° to 30°S

24°41'S, 154°45'W. 4,500 m. MSN 126G. *Pliocene.* Riedel and Funnell, 1964, p. 340; Parker, 1967, p. 129, etc.

26°19'S, 147°07'W. 3,680 m. DWBG 36. *Quaternary with late Eocene or early Oligocene admixture.* Riedel and Funnell, 1964, pp. 331–332.

23°37'S, 118°14'W. 3,440 m. DWHG 79. *Pliocene.* Riedel and Funnell, 1964, pp. 333–334.

27°54'S, 106°53'W. 3,200 m. DWBP 119. *Pliocene.* Riedel and Funnell, 1964, p. 330.

28°02'S, 96°20'W. 3,400 m. DWBG 118C. *Pliocene.* Riedel *et al.*, 1963, p. 1240; Riedel and Funnell, 1964, p. 330.

25°31'S, 85°14'W. 920 m. DWHD 72. *Late Tertiary (?lower Pliocene).* Bramlette (unpublished).

approx. 20°S, 80°W. DWHD 72. Nazca Ridge seamount. *?Miocene.* Fisher, 1958, pp. 22–23; Wilson, 1963, p. 536; Allison *et al.*, 1967, p. 4.

23°29'S, 72°59'W. 3,710 m. DWBG 94. *Quaternary with Oligocene and late Miocene to Pliocene admixture.* Riedel and Funnell, 1964, p. 329.

D. 30° to 40°S

38°49'S, 83°21'W. 4,080 m. DWHG 54. *middle Miocene.* Riedel and Funnell, 1964, p. 333.

44°13'S, 127°20'W. 4,600 m. DWHG 34. *early Oligocene.* Riedel and Bramlette, 1959, p. 106; Riedel and Funnell, 1964, p. 332; Martini (in press).

44°21'S, 127°14'W. 4,675 m. DWHH 35. *late Tertiary.* Riedel and Funnell, 1964, pp. 332–333.

42°50'S, 125°32'W. 4,560 m. DWBG 57B. *Quaternary with middle Tertiary admixture.* Shipek, 1960, Figure 2; Riedel and Funnell, 1964, pp. 328–329.

43°07'S, 125°23'W. 4,640 m. DWBG 58. *Quaternary with middle Tertiary admixture.* Riedel and Funnell, 1964, p. 329.

F. 50° to 60°S

55°06'S, 147°29'E. 3,296 m. V-16-116. *Pliocene.* Hays, 1965, pp. 156, 164.

51°08'S, 152°56'E. 4,356 m. RC8-67. *Tertiary.* L.G.O. (unpublished).

53°29'S, 155°56'E. 4,303 m. RC8-69. *Miocene.* L.G.O. (unpublished).

59°40'S, 160°17'W. 3,875 m. E14-8. *Pliocene.* Hays and Opdyke, 1967, pp. 1001–1011.

59°22'S, 132°46'W. 3,910 m. V-16-130. *Pliocene.* Hays, 1965, pp. 157, 164; Donahue, 1967, Figure 3.

58°–60°S, 115°W. c. 4,500 m. E11-5, E11-6. *Pliocene,* Hays, 1967, pp. 126–127; Donahue, 1967, p. 137; Watkins and Goodell, 1967b, p. 1084.

57°46'S, 90°48'W. E13-4. *middle Eocene.* Burckle *et al.*, 1967, Figure 2, p. 537; Geitzenauer *et al.*, 1968, p. 173.

57°00'S, 82°29'W. 5,090 m. E13-3. *Pliocene.* Hays and Opdyke, 1967, pp. 1001–1011.

56°33'S, 81°45'W. 5,000 m. V-18-69. *Pliocene.* Hays, 1965, pp. 155, 164.

57°02'S, 74°29'W. 4,064 m. V-17-88. *Pliocene.* Hays, 1965, pp. 157, 164.

G. 60° to 70°S

65°41'S, 124°06'W. 4,720 m. E13-17. *Pliocene.* Watkins and Goodell, 1967a, p. 123; Hays and Opdyke, 1967, pp. 1001–1011.

60°–66°S, 115°W. c. 4,500 m. E11-8, E11-9, E11-11. *Pliocene.* Hays, 1967, pp. 126–127; Donahue, 1967, p. 137; Watkins and Goodell, 1967b, p. 1084.

60°45'S, 107°29'W. 4,898 m. V-16-132. *Pliocene.* Hays, 1965, pp. 153, 164; Opdyke *et al.*, 1966, p. 350; Donahue, 1967, p. 136.

61°57'S, 95°03'W. 4,062 m. V-16-133. *Pliocene.* Hays, 1965, p. 164; Opdyke *et al.*, 1966, p. 350.

61°54'S, 91°15'W. 5,145 m. V-16-134. *Pliocene.* Hays, 1965, pp. 150, 164; Opdyke *et al.*, 1966, pp. 350, 355.

60°29'S, 75°57'W. 4,695 m. V-18-72. *Pliocene.* Hays, 1965, p. 164; Opdyke *et al.*, 1966, p. 350.

REFERENCES

Allison, E. C., J. W. Durham, and L. W. Mintz (1967) New southeast Pacific echinoids. *Occ. Pap. Calif. Acad. Sci., 62*, 1–23.

Arrhenius, G. (1952) Sediment cores from the East Pacific. *Rep. Swed. Deep-Sea Exped., 5*, 1–227.

Bandy, O. L., and K. S. Rodolfo (1964) Distribution of foraminifera and sediments, Peru-Chile Trench area. *Deep-Sea Res., 11*, 817–837.

Barker, R. W. (1960) Taxonomic notes on the species figured by H. B. Brady in his report on the foraminifera dredged by *H.M.S. Challenger* during the years 1873–1876. *Spec. Publ. Soc. Econ. Paleont. Mineral., 9*, i-xxiv, 1–238.

Beliaeva, N. V. (1966) Climatic and vertical zonality in the distribution of planktonic foraminifera in the sediments of the Pacific Ocean (abstr.). *2nd. Int. Oceanogr. Cong. Abst. Pap.*, 35–36.

Beliaeva, N. V., and Kh. M. Saidova (1965) Correlation of benthonic and planktonic foraminifera in the surface layers of the sediments of the Pacific Ocean. *Okeanologiia, 5*, 1010–1014.

Benson, R. H. (1964) Recent marine podocopid and platycopid ostracodes of the Pacific. *Publ. Sta. Zool. Napoli, 33*, 387–420.

Berger, W. H. (1967) Foraminiferal ooze: Solution at depths. *Science, 156*, 383–385.

Berger, W. H. (1968) Planktonic foraminifera: Selective solution and palaeoclimatic interpretation. *Deep-Sea Res., 15*, 31–43.

Berger, W. H. (in press) Radiolarian skeletons: Solution at depths.

Berggren, W. A., J. D. Phillips, A. Bertels, and D. Wall (1967) Late Pliocene-Pleistocene stratigraphy in deep sea cores from the south-central North Atlantic. *Nature, Lond., 216*, 253–254.

Black, M. (1964) Cretaceous and Tertiary coccoliths from Atlantic seamounts. *Palaeontology, 7*, 306–316.

Blackman, A. (1966) *Pleistocene stratigraphy of cores from the southeastern Pacific Ocean.* University of California, San Diego (unpubl. PhD thesis).

Blackman, A., and B. L. K. Somayajulu (1966) Pacific Pleistocene: Faunal analyses and geochronology. *Science, 154*, 886–889.

Boltovskoy, E. (1966) Zonacion en las latitudes altas del Pacifico sur segun los foraminiferos planctonicos vivos. *Revta. Mus. Argent.*

Cienc. Nat. Bernardino Rivadavia Inst. Nac. Invest. Cienc. Nat. (hydrobiol.), 2, 1-56.

Bradshaw, J. S. (1959) Ecology of living planktonic foraminifera in the north and equatorial Pacific Ocean. *Contr. Cushman Fdn., 10,* 25-64.

Brady, G. S. (1880) Report on the Ostracoda dredged by H.M.S. *Challenger,* during the years 1873-1876. *Rep. Voy. Challenger, Zool., 1,* 1-184, pls. 1-44.

Brady, H. B. (1884) Report on the foraminifera dredged by H.M.S. *Challenger* during the years 1873-1876. *Rep. Voy. Challenger, Zool., 9,* i-xxi, 1-814, pls. 1-115.

Bramlette, M. N. (1961) Pelagic sediments, pp. 345-366 in: *Oceanography.* M. Sears, editor. American Association for the Advancement of Science, Pub. No. 67. Washington, D.C.

Bramlette, M. N., and W. R. Riedel (1954) Stratigraphic value of discoasters and some other microfossils related to recent coccolithophores. *J. Paleont., 28,* 385-403.

Bramlette, M. N., and W. R. Riedel (1959) Stratigraphy of deep-sea sediments of the Pacific Ocean (abstr.). Preprints Int. Oceanogr. Congr., 1959, 86-87. American Association for the Advancement of Science, Washington, D.C.

Brotzen, F., and A. Dinesen (1959) On the stratigraphy of some bottom sections from the central Pacific. *Rep. Swed. Deep-Sea Exped., 10,* 43-55.

Burckle, L. H., J. Ewing, T. Saito, and R. Leyden (1967) Tertiary sediment from the East Pacific rise. *Science, 157,* 537-540.

Burckle, L. H., J. Ewing, and T. Saito (1966) An Eocene dredge haul from the Tuamotu ridge. *Deep-Sea Res., 13,* 1207-1208.

Castracane degli Antelminelli, A. F. (1886) Report on the Diatomaceae collected by H.M.S. *Challenger,* during the years 1873-1876. *Rep. Voy. Challenger, Bot., 2,* 1-178, pls. 1-30.

Cole, W. S. (1959) *Asterocyclina* from a Pacific seamount. *Contr. Cushman Fdn., 10,* 10-14.

Cole, W. S. (1960) Upper Eocene and Oligocene larger foraminifera from Viti Levu, Fiji. *Prof. Pap. U.S. Geol. Surv., 374-A,* 1-7.

Cushman, J. A. (1932) The foraminifera of the tropical Pacific collections of the *Albatross,* 1899-1900. Pt. 1. Astrorhizidae to Trochamminidae. *Bull. U.S. Natl. Mus., 161*(1), 1-88.

Cushman, J. A. (1933) The foraminifera of the tropical Pacific collections of the *Albatross,* 1899-1900, Pt. 2. Lagenidae to Alveolinellidae. *Bull. U.S. Natl. Mus., 161*(2), 1-79.

Cushman, J. A. (1942) The foraminifera of the tropical Pacific collections of the *Albatross,* 1899-1900, Pt. 3. Heterohelicidae and Buliminidae. *Bull. U.S. Natl. Mus., 161*(3), 1-67.

Donahue, J. G. (1967) Diatoms as indicators of Pleistocene climatic fluctuations in the Pacific sector of the southern ocean. *Prog. Oceanogr., 4,* 133-140.

Egger, J. G. (1893) Foraminifera aus Meeresgrundproben, gelothet von 1874 bis 1876 von S.M. Sch. Gaselle. *Abh. K. bayer, Akad. Wiss. Munchen,* II, Cl. 18. Bd. Z. abt. 2, 195-458, pls. 1-21.

Ehrenberg, C. G. (1855) Uber die wietere Entwicklung der Kenntniss des Grünsandes als grüner Polythalamien-Steinkerne, uber braunrothe und corallrothe Steinkerne der Polythalamien-Kreide in Nord-Amerika, und über den Meeresgrunde aus 12,900 Fuss Tiefe. *Monatsber. Kgl. Preuss. Akad. Wiss. Berlin,* Jahrg. 1855, 172-178.

Ehrenberg, C. G. (1872) Microgeologischen Studien über das Kleinste Leben des Meeres-Tiefgrunde aller Zonen und dessen geologischen Einfluss. *Abh. Preuss. Akad. Wiss.,* Jahrg. 1872, 131-399.

Emiliani, C. (1961) The temperature decrease of surface sea water in high latitudes and of abyssal-hadal water in open ocean basins during the past 75 million years. *Deep-Sea Res., 8,* 144-147.

Fisher, R. L. (1958) Downwind investigation of the Nasca Ridge. In: *Preliminary Report on Expedition Downwind,* R. L. Fisher (ed.).

IGY Gen. Rep. Ser., 2, p. 23. National Academy of Sciences, Washington, D.C.

Friend, J. K., and W. R. Riedel (1967) Cenozoic orosphaerid radiolarians from tropical Pacific sediments. *Micropaleontol., 13,* 217-232.

Funnell, B. M. (in press) The occurrence of pre-Quaternary microfossils in the oceans. In: *The micropalaeontology of oceans.* B. M. Funnell and W. R. Riedel, editors. Cambridge Univ. Press, London.

Gaarder, K. R. (in press) Comments on the distribution of coccolithophorids in the oceans. In: *The micropalaeontology of oceans.* B. M. Funnell and W. R. Riedel, editors. Cambridge Univ. Press: London.

Geitzenauer, K. R., S. V. Margolis, and D. S. Edwards (1968) Evidence consistent with Eocene glaciation in a South Pacific deep sea sedimentary core. *Earth Planetary Sci. Letters, 4,* 173-177.

Haeckel, E. (1887) Report on the radiolaria collected by H.M.S. *Challenger* during the years 1873-1876. *Rep. Voy. Challenger, Zool., 18,* i-clxxxviii, 1-1803.

Hasle, G. R. (1959) A quantitative study of phytoplankton from the equatorial Pacific. *Deep-Sea Res., 6,* 38-59.

Hasle, G. R. (1960) Plankton coccolithophorids from the subantarctic and equatorial Pacific. *Nytt. Mag. Bot., 8,* 77-88.

Hays, J. D. (1965) Radiolaria and late Tertiary and Quaternary history of Antarctic Sea. *Amer. Geophys. Union Antarc. Res. Ser., 5,* 125-184.

Hays, J. D. (1967) Quaternary sediments of the Antarctic Ocean. *Progr. Oceanogr., 4,* 117-131.

Hays, J. D., and W. A. Berggren (in press) Quaternary boundaries. In: *The micropalaeontology of oceans.* B. M. Funnell and W. R. Riedel, editors. Cambridge Univ. Press, London.

Hays, J. D., and N. D. Opdyke (1967) Antarctic radiolaria, magnetic reversals and climatic change. *Science, 158,* 1001-1011.

Hendey, N. I. (1937) The plankton diatoms of the Southern Seas. *Discovery Rep., 16,* 151-364, pls. 6-13.

Herman, Y. (in press) Vertical and horizontal distribution of pteropods in Quaternary sequences. In: *The micropalaeontology of oceans.* B. M. Funnell and W. R. Riedel, editors. Cambridge Univ. Press, London.

Jenkins, D. G. (1967) Planktonic foraminiferal zones and new taxa from the Lower Miocene to the Pleistocene of New Zealand. *N. Z. J. Geol. Geophys., 10,* 1064-1074.

Joides (1967) Deep-sea Drilling Project. *Bull. Amer. Assoc. Petrol. Geol., 51,* 1787-1802.

Jousé, A. P., O. G. Koslova, and V. V. Muhina (1967) Species composition and zonal distribution of diatoms in the surface sedimentary layer of the Pacific Ocean. *Dokl. Akad. Nauk SSSR, 172,* 1183-1186.

Jousé, A. P., O. G. Koslova, and V. V. Muhina (in press a) Diatom algae in the surface sedimentary layer of the Pacific Ocean. In: *Monograph on the Pacific Ocean.* Nauka, Moscow.

Jousé, A. P., O. G. Koslova, and V. V. Muhina (in press b) Distribution of diatoms in the surface sediment layer of the Pacific Ocean. In: *The micropalaeontology of oceans.* B. M. Funnell and W. R. Riedel, editors. Cambridge Univ. Press, London.

Kamptner, E. (1962) Tertiare und nach-tertiare Coccolithineen-Skeletreste aus Tiefsee-ablagerungen des ostlichen Pazifischen Ozeans. *Palaontol. Z., Stuttgart, 36,* 13.

Kamptner, E. (1963) Coccolithineen-Skeletreste ans Tiefsee-ablagerungen des Pacifischen Ozeans. *Ann. Nat. Mus. Wien., 66,* 139-204.

Kanaya, T. (in press) Some aspects of pre-Quaternary diatoms in the oceans. In: *The micropalaeontology of oceans.* B. M. Funnell and W. R. Riedel, editors. Cambridge Univ. Press, London.

Kennett, J. P. (1966) Foraminiferal evidence of a shallow calcium carbonate solution boundary, Ross Sea, Antarctica. *Science, 151*, 191–193.

Kolbe, R. W. (1954) Diatoms from equatorial Pacific cores. *Rep. Swed. Deep-Sea Exped., 6*, 1–49.

Kustanowich, S. (1962) A foraminiferal fauna from Capricorn Seamount, south-west equatorial Pacific. *N. Z. J. Geol. Geophys., 5*, 423–434.

Kustanowich, S. (1963) Distribution of planktonic foraminifera in surface sediments of the south-west Pacific Ocean. *N. Z. J. Geol. Geophys., 6*, 534-565.

Marshall, S. M. (1933) The production of microplankton in the Great Barrier Reef region. *Sci. Rep. Gr. Barrier Reef Exped., 2*, 111–157.

Martini, E. (1965) Mid-Tertiary calcareous nannoplankton from Pacific deep-sea cores, pp. 393–410 in: *Submarine geology and geophysics*. W. F. Whittard and R. B. Bradshaw, editors. Butterworths, London.

Martini, E. (in press) The occurrence of pre-Quaternary calcareous nannoplankton in the oceans. In: *The micropalaeontology of oceans*. B. M. Funnell and W. R. Riedel, editors. Cambridge Univ. Press, London.

McTavish, R. A. (1966) Planktonic foraminifera from the Malaita Group, British Solomon Islands. *Micropaleontol., 12*, 1–36.

Meisenheimer, J. (1905) Pteropoda. *Wiss. Ergebn. Dtsch. Tiefsee-Exped., 9*, 1–314.

Menard, H. W., and E. L. Hamilton (1964) Paleogeography of the tropical Pacific. *Proc. 10th Pac. Sci. Cong., Hawaii, 1962.*

Muhina, V. V. (in press) Problems of diatom and silicoflagellate Quaternary stratigraphy in the equatorial Pacific Ocean. In: *The micropalaeontology of oceans*. B. M. Funnell and W. R. Riedel, editors. Cambridge Univ. Press, London.

Murray, J. (1889) On marine deposits in the Indian, southern and Antarctic Oceans. *Scot. Geogr. Mag., 5*, 405–436.

Murray, J., and A. F. Renard (1891) Deep sea deposits. *Rep. Voy. Challenger*, i–xxvii, 1–525, pls. 1–29.

Nigrini, C. A. (1968) Radiolaria from eastern tropical Pacific sediments. *Micropaleontol., 14*, 51–63.

Nigrini, C. (in press) Radiolarian zones for the Quaternary of the equatorial Pacific Ocean. In: *The micropalaeontology of oceans*. B. M. Funnell and W. R. Riedell, editors. Cambridge Univ. Press, London.

Norris, R. E. (1961) Observations on phytoplankton organisms collected on the N.Z.O.I. Pacific cruise, September 1958. *N. Z. J. Sci., 4*, 162–188.

Olausson, E. (1960) Description of sediment cores from the central and western Pacific with the adjacent Indonesian region. *Rep. Swed. Deep-Sea Exped., 6*, 163–214.

Olausson, E. (1961a) Remarks on some Cenozoic cores from the central Pacific, with a discussion of the role of coccolithophorids and foraminifera in carbonate deposition. *Göteborgs. Vetensk. Samh. Handl. Följ., 7, (B), 8*, 1–35.

Olausson, E. (1961b) Remarks on Tertiary sequences of two cores from the Pacific. *Bull. Geol. Inst. Univ. Uppsala, 40*, 299–303.

Opdyke, N. D., B. Glass, J. D. Hays, and J. Foster (1966) Paleomagnetic study of Antarctic deep-sea cores. *Science, 154*, 349–357.

Parker, F. L. (1960) Living planktonic foraminifera from the equatorial and southeast Pacific. *Sci. Rep. Tohoku Univ., 2nd Ser. (geol.), spec. vol. 4*, 71–82.

Parker, F. L. (1962) Planktonic foraminiferal species in Pacific sediments. *Micropaleontol. 8*, 219–254.

Parker, F. L. (1965) A new planktonic species (Foraminiferida) from the Pliocene of Pacific deep-sea cores. *Contr. Cushman Fnd., 16*, 151–152.

Parker, F. L. (1967) Late Tertiary biostratigraphy (planktonic foraminifera) of tropical Indo-Pacific deep-sea cores. *Bull. Amer. Paleontol., 52*, 111–208.

Parker, F. L. (in press) Distributions of planktonic foraminifera in recent deep-sea sediments. In: *The micropalaeontology of oceans*. B. M. Funnell and W. R. Riedel, editors. Cambridge Univ. Press, London.

Pelseneer, P. (1888) Report on the Pteropoda. II. Thecosomata. *Rep. Voy. "Challenger", Zool., 23*, 1–132.

Petrushevskaya, M. G. (1966a) Radiolarii v planktone i donnykh osadkakh, pp. 219–245 in: *Geokhimiya kremnesema, Moscow.*

Petrushevskaya, M. G. (1966b) Radiolaria in Antarctic bottom sediments, pp. 1–8 in: *Symposium on Antarctic Oceanography*, Santiago.

Petrushevskaya, M. G. (1967) Radiolarii otryadov Spumellaria i Nasselaria Antarticheskoy oblasti (po materialam Sovetskoi Antartichesksoi Expeditsii). Issledovanya faunymorei, t. IV (Xii). *Resutaty biologicheskikh issledovaniy Sovetskoi Antartich. Eksped., 1955-58, 3*, 5–186.

Petrushevskaya, M. G. (in press a) Radiolaria (Spumellaria and Nassellaria) in the plankton and in bottom sediments of the central Pacific. In: *The micropalaeontology of oceans*. B. M. Funnell and W. R. Riedel, editors. Cambridge Univ. Press, London.

Petrushevskaya, M. G. (in press b) Radiolaria in recent sediments from the Indian and Antarctic oceans. In: *The micropalaeontology of oceans*. B. M. Funnell and W. R. Riedel, editors. Cambridge Univ. Press, London.

Phillips, J. D., W. A. Berggren, A. Bertels, and D. Wall (1968) Paleomagnetic stratigraphy and micropaleontology of three deep sea cores from the central North Atlantic Ocean. *Earth Planetary Sci. Letters, 4*, 118–130.

Riedel, W. R. (1952) Tertiary radiolaria in western Pacific sediments. *Göteborgs Vetensk Samh. Handl. Följ. 7, (B), 6*, 1–18.

Riedel, W. R. (1957) Radiolaria: A preliminary stratigraphy. *Rep. Swed. Deep-Sea Exped., 6*, 59–96.

Riedel, W. R. (1958) Radiolaria in Antarctic sediments. *BANZ Antarc. Res. Exped. Rep., (B), 6*, 217–255.

Riedel, W. R. (1959a) Oligocene and Lower Miocene Radiolaria in tropical Pacific sediments. *Micropaleontol., 5*, 285–302.

Riedel, W. R. (1959b) Siliceous organic remains in pelagic sediments. In: *Silica in sediments*. H. A. Ireland, editor. Spec. Publ. S. E. P. M., 7, 80–91.

Riedel, W. R. (1967) Radiolarian evidence consistent with spreading of the Pacific floor. *Science, 157*, 540–542.

Riedel, W. R. (in press) Occurrence of pre-Quaternary Radiolaria in deep-sea sediments. In: *The micropalaeontology of oceans*. B. M. Funnell and W. R. Riedel (eds.). Cambridge Univ. Press, London.

Riedel, W. R., and M. N. Bramlette (1959) Tertiary sediments in the Pacific Ocean basin (abstr.). *Preprints Int. Oceanogr. Congr., 1959*, American Association for the Advancement of Science, 105–106, Washington, D.C.

Riedel, W. R., M. N. Bramlette, and F. L. Parker (1963) 'Pliocene-Pleistocene' boundary in deep-sea sediments. *Science, 140*, 1238-1240.

Riedel, W. R., and B. M. Funnell (1964) Tertiary sediment cores and microfossils from the Pacific Ocean floor. *Quart. J. Geol. Soc., Lond., 120*, 305–368.

Ross, D. A., and W. R. Riedel (1967) Comparison of upper parts of some piston cores with simultaneously collected open-barrel cores. *Deep-Sea Res., 14*, 285–294.

Saidova, H. M. (1967) Sediment stratigraphy and paleogeography of the Pacific Ocean by benthonic foraminifera during the Quaternary. *Progr. Oceanogr., 4*, 143–151.

Semina, G. (1963) Phytoplankton of the central part of the Pacific Ocean, along a section at 174°W. *Trud. Inst. Okeanol., 6*, 131-143.

Shipek, C. J. (1960) Photographic study of some deep-sea environments in the eastern Pacific. *Bull. Geol. Soc. Amer., 71*, 1067–1074.

Shumenko, S., and M. G. Uschakova (1967) Coccoliths of bottom sediments of the Pacific Ocean. *Dokl. Akad. Nauk SSSR, 176*, 932–934.

Swain, F. M. (in press a) Pleistocene ostracoda from deep-sea sediments in the southeastern Pacific Ocean. In: *The micropalaeontology of oceans*. B. M. Funnell and W. R. Riedel, editors. Cambridge Univ. Press, London.

Swain, F. M. (in press b) Pliocene ostracodes from deep-sea sediments in the southwest Pacific and Indian oceans. In: *The micropalaeontology of oceans*. B. M. Funnell and W. R. Riedel, editors. Cambridge Univ. Press, London.

Tesch, J. J. (1948) The Thecosomatous pteropods II. The Indo-Pacific. *Dana Rep., 30*, 1–44.

Todd, R. (1964) Planktonic foraminifera from deep-sea cores off Eniwetok Atoll. *Prof. Pap. U.S. Geol. Surv.,* 260 CC, 1067–1100.

Todd, R. (1965) The foraminifera of the tropical Pacific collections of the *Albatross,* 1899–1900, Pt. 4. Rotaliform families and planktonic families. *Bull. U.S. Natl. Mus., 161*(4), 1–139, pls. 1–28.

Uschakova, M. G. (1966) Biostratigraphic significance of coccolithophorids on the basis of bottom sediments from the Pacific Ocean. *Okeanologiia, 6*, 136–143.

Uschakova, M. G. (in press) Coccoliths in suspension and from the surface sedimentary layer of the Pacific Ocean. In: *The micropalaeontology of oceans*. B. M. Funnell and W. R. Riedel, editors. Cambridge Univ. Press, London.

Watkins, N. D., and H. G. Goodell (1967a) Confirmation of the reality of the Gilsa Geomagnetic Polarity Event. In: *Earth planetary science letters, 2*, 123.

Watkins, N. D., and H. G. Goodell (1967b) Geomagnetic polarity change and faunal extinction in the Southern Ocean. *Science, 156*, 1083–1087.

Wiles, W. W. (1967) Pleistocene changes in the pore concentrations of a planktonic foraminiferal species from the Pacific Ocean. *Progr. Oceanogr., 4*, 153–160.

Wilson, J. T. (1963) Evidence from islands on the spreading of ocean floors. *Nature, 197*, 536–538.

Wiseman, J. D. H., and W. R. Riedel (1960) Tertiary sediments from the floor of the Indian Ocean. *Deep-Sea Res., 7*, 215–217.

John Ewing

Robert Houtz

Maurice Ewing

SOUTH PACIFIC STRUCTURE AND MORPHOLOGY

About 200,000 km of profiler data from the South Pacific have been used to compile a sediment isopach map and to study the marginal tectonics and geological history of the area.

Profiler crossings at about 30° to the coastline of northern New Zealand and near Valparaiso, Chile, may be interpreted as showing that the deep-sea sediments are intensely folded near the continental slope. Nearby crossings made at right angles to the coastline show only confused, overlapping reflection hyperbolas. Folding such as that along the east coast of the north island of New Zealand does not occur along the east coast of the south island. A recently active, left-lateral fault was profiled on the northern edge of the Chatham Rise, near the south island. Despite the absence of intense seismic activity, the fault might be attributed to large differences in response to sea-floor spreading between northern and southern New Zealand. Recent folding of the sediments off the east coast of the north island apparently diverted the Hikurangi submarine canyon.

The sediment cover (up to 1 km thick) over most of the Solomons plateau diminishes rapidly wherever water depth exceeds the calcareous compensation level. This implies general stability during the period of deposition. However, anomalous sediment thicknesses (in relation to water depth) suggest relative vertical movements of up to 4 km along the southwestern edge of the plateau, where it is separated from the Melanesian block by a major fault. Southwest of the fault, in the Solomon Sea and the Santa Cruz basin, the sea floor is devoid of sediment. It is proposed that these sediment-free regions were formerly well below the calcareous compensation level and have only recently been elevated.

The fault bounding the northern edge of Melanesia extends at least 10° of longitude east of the Tonga Trench and bounds the southern edge of the "opaque layer" thought by Ewing and co-workers to be Cretaceous in age. The almost complete lack of sediment to the south of the fault (it is about 700 m thick at the northern edge of the fault) is difficult to reconcile with simple sea-floor spreading, unless the sea floor south of the fault has remained below the calcareous compensation level during most of Cainozoic time. The most recent movement on this fault (here, as elsewhere) seems to be vertical.

The amount of sediment in the Peru-Chile Trench appears to be related simply to present-day rainfall, but such complicating effects as lateral migration of turbidites, the existence of barriers in the trench, and variability of runoff may actually control the sediment thickness. The southern half of the trench has filled, and the overflow has given rise to a small offshore abyssal plain. About one fourth of the volume of sediments south of 50°S in the trench is acoustically transparent and may be deep-sea pelagic materials.

Sediments near the northern limit of turbidites derived from Antarctica are distorted in the region where transform faults are believed to extend into the Bellingshausen basin. To the west, in the Balleny basin, the sediments over a wide area appear to be disturbed; disturbance has probably occurred since Antarctica was frozen over.

The sediment isopach map of the South Pacific shows that sediments from the New Zealand plateau have streamed to the north, probably under the influence of bottom currents. A thick band of sediment between 50° and 55°S extends across the entire South Pacific and is 5° north of present-day zones of maximum productivity. The anomalous band may represent a former northerly location of the polar front. Sediment distribution in Antarctic waters has also been strongly modified by bottom currents and by recent activity, possibly sea-floor spreading, near the crest of the Pacific Antarctic Ridge.

IV
BIOTA OF
THE SOUTH PACIFIC

George A. Knox

DEPARTMENT OF ZOOLOGY,
UNIVERSITY OF CANTERBURY,
CHRISTCHURCH, NEW ZEALAND

BIOLOGICAL OCEANOGRAPHY OF THE SOUTH PACIFIC

INTRODUCTION

I presume that one of the reasons for my being invited to participate in this symposium is to present a southern viewpoint of the biological problems of the South Pacific. My own country, New Zealand, is the only land area of any considerable size in the vast expanse of ocean forming the South Pacific. As an island nation whose sphere of political influence extends from the Tokelau Islands (8°S) to the Ross Sea (the most southerly part of the Southern Ocean), we have long been orientated toward the sea, and New Zealand marine scientists have made significant contributions to our understanding of the southwest Pacific region. I have often noted the differences in viewpoint between the southern worker viewing the problems of the South Pacific from the south and the northern worker viewing the same problems from the north; these different viewpoints and some suggestions for resolving them are dealt with later in this paper.

My task, as I see it, is to introduce the section on Biota of the South Pacific by briefly surveying the overall status of our knowledge of the region, by pointing out the noticeable gaps, and by dealing with some of the interesting and exciting problems that await investigation. As I have not had prior knowledge of the contents of the other contributors to this section, there may be some duplication, but I am sure that if there is, the treatment will differ. Since none of the contributions deal with benthic floras and faunas, I propose to deal with them in some detail. I must apologize for any bias toward problems of the southwest Pacific and for the emphasis on the southern section of the South Pacific.

LIMITS OF THE "SOUTH PACIFIC"

The limits of the region under discussion have been variously defined according to whether they are based on geog-

raphy, physical oceanography, or biological distributions. On a geographic basis, Viglieri (1966) defines it as that part of the Pacific south of the equator with the western limit to the south on a line from Tasmania down the meridian 146°53'E to the Antarctic Continent, and with the eastern limit to the south on a line from the meridian of Cape Horn (67°14'W) to the Antarctic Continent, but excluding the marginal seas. Fairbridge (1966) would include the marginal seas, the Coral and Tasman seas, as well as the smaller Bismarck and Solomon seas (see Table 1). These arbitrary geographic limits deny the reality of the South Ocean (Deacon, 1957, 1963). Physical oceanographers generally exclude this southern ocean region from the South Pacific proper, but as Fairbridge (1966) has pointed out, the position at which the southern boundary is placed varies according to the viewpoint adopted. According to Kort *et al.* (1965), it has been established with certainty that the West Wind Drift (Figure 4) is the boundary to the north of which the waters of the Atlantic, Indian, and Pacific oceans form individual systems, having their characteristic dynamic, physical, chemical, and biological properties that differ from the corresponding systems and processes south of the current. This "Southern Ocean," characterized by "independent systems of sea and air currents, and independent salinity and temperature distribution at the surface" (Shokal'skiy, 1959), is limited to the north by the Subtropical Convergence (Figure 5). This feature, although it may fluctuate markedly in position from season to season and year to year and may not always be well defined, can perhaps best mark the southern boundary of the South Pacific. One difficulty, however, concerns the boundary on the eastern sector where the cold Peru Current, composed largely of water derived from the region of the West Wind Drift, penetrates to equatorial latitudes.

While the location of the southern boundary may be a satisfactory one when surface waters only are concerned, there are complications when subsurface waters are considered. Antarctic Intermediate Water (Figure 6) originating

TABLE 1 Extent of the Pacific Ocean

	Pacific	Pacific (Excluding Marginal Seas)	North Pacific	South Pacific[a]
Area	$179.7 \times 10^6 km^2$	$165.2 \times 10^6 km^2$	$70.8 \times 10^6 km^2$	$94.4 \times 10^6 km^2$
Mean Depth	4,028 m	4,282 m	4,753 m	3,928 m
Volume	$727.7 \times 10^6 km^3$	$707.5 \times 10^6 km^3$	$336.7 \times 10^6 km^3$	$370.8 \times 10^6 km^3$

[a]Includes the southern ocean sector.

south of the Antarctic Convergence is recognizable well into subtropical latitudes (Brodie, 1965; Garner, 1962; Rotschi and Lemasson, 1967; Wyrtki, 1962). Further complications arise when biological distributions are concerned. While the Subtropical Convergence is undoubtedly a significant faunal boundary (Bary, 1959; David, 1958; Mackintosh, 1960; Tebble, 1960), many plankton organisms are found to both the north and south of this feature. In the northern tropical regions the flora and fauna are part of a central Pacific fauna with a strong Indo-Pacific element. Again, when benthic faunas are considered, the position of the Subtropical Convergence may have little significance apart from shallow inshore species.

It has proved impossible to limit the discussion that follows to the region from the equator to the Subtropical Convergence since the South Pacific thus defined cannot be understood without considering the equatorial waters to the north and the subantarctic waters to the south. Both profoundly affect processes within the arbitrary limits of the South Pacific as defined on primarily geographic criteria.

PRESENT STATE OF OUR KNOWLEDGE OF THE SOUTH PACIFIC

The South Pacific is a vast expanse of ocean even if the Pacific sector of the Southern Ocean is excluded (see Table 1), and it is certainly the least known from the biological viewpoint. If one starts to survey the distribution of any taxonomic group, species, or other biological parameter in the South Pacific, one is struck by the considerable gaps that exist in the information available. Figure 1 is taken from a recent review by Koblentz-Mishke (1965) of primary production in the Pacific. At the time her data were assembled (1964), there was not a single sample in the central South Pacific from 20°S. Since that date, many additional measurements have been made, mostly in the Southern Ocean sector by U.S. vessels, especially during cruises of the U.S. research vessel *Eltanin* (Sandved, 1966); the picture for the central portion of the South Pacific, however, remains

substantially the same. Figure 8, showing the Pacific distribution of the euphausiid *Thysanoessa gregaria* (Johnson and Brinton, 1963), illustrates a common distributional pattern contrasting the completeness of the information for the North Pacific with the paucity of information for the South Pacific. Many other examples could be cited.

When the results of the investigations of the *Eltanin* are published, considerable additions will have been made to our knowledge of the biological oceanography of the South Pacific south of the region of the Subtropical Convergence. Eighteen cruises have been carried out involving 13 traverses of the Pacific between South America and New Zealand and two traverses between New Zealand and the United States (Figure 2). On these voyages a great variety of biological observations have been made, including bacterial studies, primary productivity studies, chlorophyll and ^{14}C analyses, phytoplankton and zooplankton collections, midwater and bottom trawling, bottom photography, and a variety of physiological investigations (Sandved, 1966; Anon. 1967). A vast amount of information has been collected, and I await the publication of the results with considerable interest. The problem of receiving, sorting, accessioning, and distributing the large biological collections that have been assembled has been solved by channeling the collections through the Smithsonian Oceanographic Sorting Center of the Smithsonian Institution. After processing, the specimens are sent into the original collecting agency, distributed to specialists for study, or deposited in the U.S. National Museum. While this is an admirable procedure, the point I am concerned with is that these collections may suffer the fate of those of the great expeditions of the past. With few exceptions, these have resulted in monographic treatment of the separate taxonomic groups of animals, but no one has taken it upon himself to write a general account of the communities of organisms that were collected. As an example, in spite of the considerable amount of research carried out in Antarctic waters and the publication of large series of expedition reports, information as to the communities of organisms to be found in a particular depth with a specific type of sediment is not available. Information as to

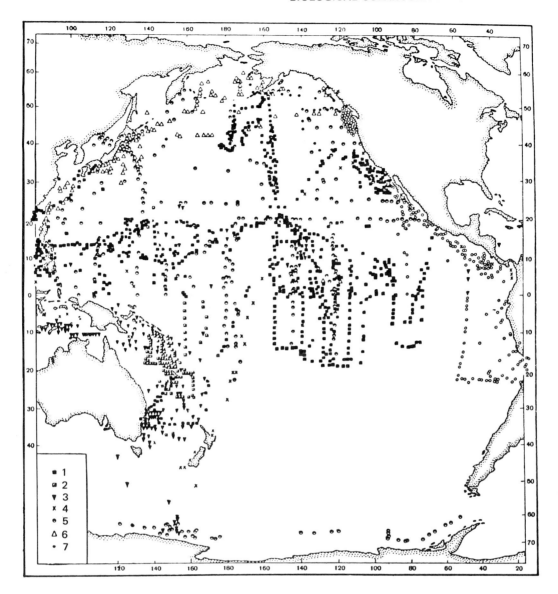

FIGURE 1 Map of localities at which primary production has been measured by workers from various countries: (1) U.S.A. (Hawaiian group), (2) Nova Scotia, (3) Australia, (4) Denmark, (5) U.S.S.R., (6) Japan, (7) U.S.A. (Scripps Institution and University of Washington.) Data up to 1964. (Reproduced with permission from Koblentz-Mishke, 1965, p. 104, Figure 1.)

the relative densities of different species is lost when each taxonomic group is reported on separately. I trust that this will not be the fate of more recent expeditions.

While I have been engaged in compiling this survey, I have often been struck by the confidence with which some workers have drawn isolines separating regions of like concentration of some variable, such as numbers of organisms per cubic meter or milligrams of carbon per square meter, to give an appearance of completeness, which, upon close inspection, is seen to be based on extremely meager data.

Another fault is the lumping of information gained at different times of the year or obtained by different sampling methods. I realize that in many cases we need to draw upon all the information available and that generalizations based on meager data often have been substantiated by more detailed sampling, but I do make a plea for authors to make clear the limitations of their data.

The only areas from which more than scattered observations are available comprise the margins of the South Pacific, the tropics, the more northerly coasts of South

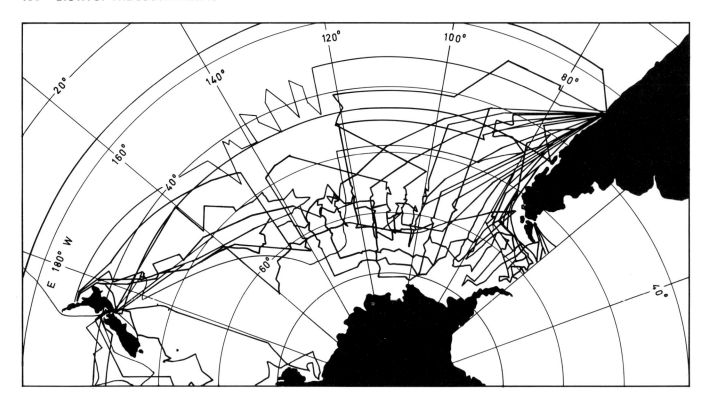

FIGURE 2 The tracks of the cruises of the U.S. research vessel *Eltanin* in South Pacific waters.

America, the Coral and Tasman seas and the Southern Ocean sector. In the center there is an immense area that is still virtually unknown.

A further indication of the comparative lack of work in this region is shown in Table 2, which analyzes the papers published on the biological oceanography of the Pacific in three major oceanographic journals. The ratio of papers dealing exclusively with the North Pacific to those dealing exclusively or in part with the South Pacific is 11 to 1.

This preponderance of work carried out in the North Pacific can be traced to two main factors. First, the extremely well developed exploitation of the marine resources of the North Pacific, and second, the fact that this region is ringed by countries with advanced technological development, resulting in the allocation of considerable resources to marine research. Figure 3 shows the location of the marine and fisheries laboratories of the South Pacific region. As far as I can ascertain, there are eight marine laboratories with inshore shallow-water facilities only, only four with ocean-going facilities, and five fisheries research laboratories. This contrasts with 67 marine laboratories and 30 fisheries research laboratories in the North Pacific (data from Shilling and Tyson, 1964). In the whole central tropical Pacific there is not a single well-equipped marine laboratory. I could suggest that the development of such a laboratory is an urgent requirement for the future exploration and study of this important region. This would have to be developed as an

international venture, and I suggest that it might appropriately be associated with the newly established University of the South Pacific in Fiji.

CURRENTS, BOUNDARIES, AND WATER MASSES

In order to discuss the biological parameters of the region, it is necessary first to outline briefly the principal current systems, the significant hydrological boundaries, and the main water masses. As a nonspecialist in this area I have found the present situation somewhat confusing, especially as regards terminology. The following is an attempt to interpret the situation from the viewpoint of a biologist interested in the distribution patterns of marine plants and animals and the factors determining such patterns. It is based on recent reviews by Burkov (1966), Deacon (1963), Knauss (1963), Kort *et al.* (1965), Radzikhovskaya (1966), Reid (1966), Rotschi and Lemasson (1967), Stepanov (1965), Wooster and Reid (1963), Wooster and Cromwell (1958), and Wyrtki (1966).

Currents

Figure 4 is a diagrammatic idealized representation of the principal surface currents of the South Pacific; data for the southwest Pacific is from Knox [(1963a, Figure 3), based

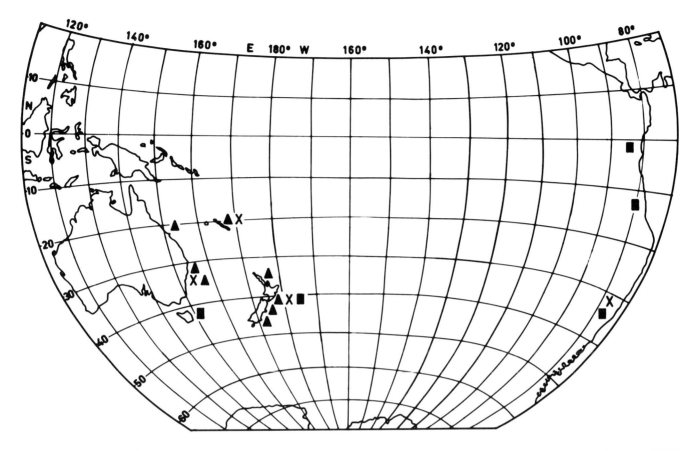

▲ LABORATORIES WITH
INSHORE FACILITIES ONLY

✕ LABORATORIES WITH
OCEAN-GOING FACILITIES

■ FISHERIES RESEARCH
LABORATORIES

FIGURE 3   Marine laboratories of the South Pacific.

TABLE 2   Published Papers on Biological Oceanography of the Pacific in Selected Journals

Name of Journal	Number of Papers Dealing with Pacific Ocean	Number of Papers Including South Pacific Observations
Journal of Marine Research 1950–1968	21	0
Journal of Limnology and Oceanography 1956–1968	38	3
Pacific Science 1949–1968	143	15
Totals	202	18

on data from Brodie (1960), and Wyrtki (1960)], and data for the eastern tropical region is from Wyrtki (1966).

According to Knauss (1963), the classic picture of the Pacific equatorial circulation has been that of three currents: the westward-flowing North and South Equatorial Currents and between them the eastward-flowing Equatorial Countercurrent. He stresses, however, that the real situation is much more complicated and that care must be taken in extrapolating the conditions from one region to another. The South Equatorial Current is developed as a result of the southeast trades; its northern boundary is the Equatorial Countercurrent at approximately 4°N, and it extends as a well-defined current to about 10°S. At its eastern end, according to Wyrtki (1966), about 2/5 of the flow is contributed by the Equatorial Countercurrent and the Peru Current in the ratio of 7:3 and the other 3/5 from the Equatorial Undercurrent, or Cromwell Current (Knauss, 1960, 1963), a fast, thin, eastward-flowing current along the equator at a depth of 50–100 m. To the west, the South Equatorial Current contributes to the Trade Wind Drift, from which the southward-flowing East Australian (off the New South Wales Coast) and East Cape Currents (off the New Zealand northeast coast) are derived. In the south the warm southward-moving waters meet the easterly flowing cold waters of the West Wind Drift along the Subtropical Convergence Zone. This

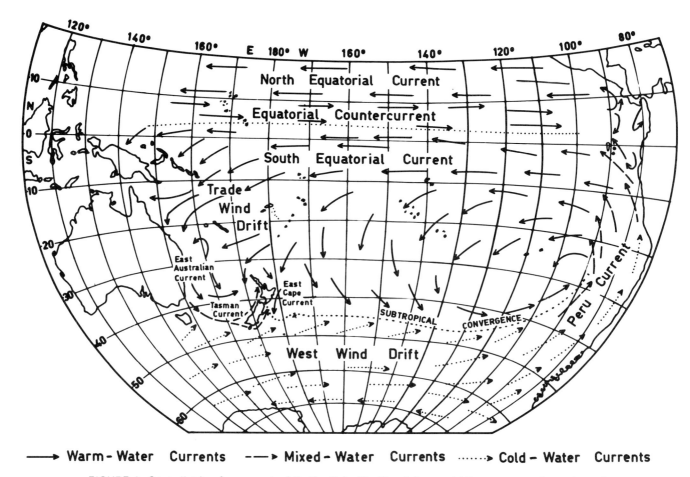

FIGURE 4 Generalized surface currents of the South Pacific. (South Equatorial Countercurrent is not shown.)

cold current between latitudes 45° and 59°S is deflected northward by the southerly projecting South American land mass to form the Peru Current. The latter, following Wyrtki (1966), is perhaps best termed the Peru Current system, as it consists of several more-or-less independent branches, the chief of which are the Peru Coastal Current, with the Peru Oceanic Current farther offshore (Gunther, 1936) and the Peru Countercurrent between, flowing south chiefly as a subsurface current (Wooster and Gilmartin, 1961; Wyrtki, 1963). Upwelling (Gunther, 1936; Schott, 1931; Schweigger, 1958; and Wooster, 1961) is a conspicuous feature in this current system. The combined effect of the northerly transport and upwelling is to produce coastal water conditions of a temperate nature well into the tropics (as far as 5°S), the average surface temperatures off the coast of Peru being something like 10°C lower than the theoretical values for the latitude. This has profound biological consequences.

The net effect of this circulation is to produce the Southern Subtropical Anticyclonic Gyre (Burkov, 1966). Burkov has proposed the name South Pacific Current for the eastward-flowing southerly component of this gyre to the north of the Subtropical Convergence. The generally ac-

cepted terminology for these currents is shown in Figure 4. Russian workers have proposed an alternative terminology (see Table 3).

Convergences and Divergences

Zones of convergence of surface currents are important oceanographic features from the biological viewpoint. The

TABLE 3 Principal Currents of the South Pacific

Terminology Used by British and U.S. Workers	Russian Terminology (Burkov, 1966)
Equatorial Countercurrent	Intertrade Countercurrent
South Equatorial Current	Southern Trade Currents
Trade Wind Drift	
East Australian Current	East Australian Current
	South Pacific Current
West Wind Drift	Antarctic Gyral Current
Peru Current	Peru Current

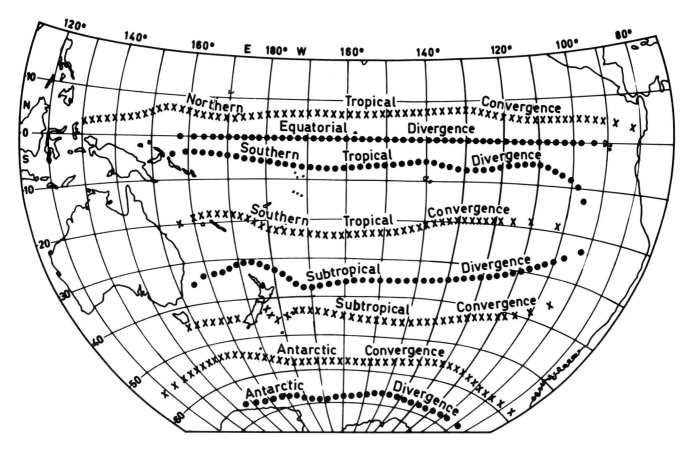

FIGURE 5 Convergences (xxx) and divergences (...) of the South Pacific. (The positions shown are only approximate, and for many regions they are largely speculative.)

characteristic features of these convergence zones are the coming together of currents carrying waters of different properties and the sinking of surface waters to depth (Burkov, 1966). Zones of divergence, on the other hand, occur where two bodies of water move away from each other with the ascent of deeper waters to the surface. This terminology has been confused by the use of the term "front" for either of these features. The generally recognized zones of convergence and divergence in the South Pacific are shown in Figure 5. (The positions shown are only approximate, and for many regions they are largely speculative.) Table 4 gives the terminology used in this paper and that proposed by Burkov (1966). The most important features in the south are the Antarctic Convergence (Deacon, 1957, 1963; Brodie, 1965; Garner, 1958; Ivanov, 1961; Wyrtki, 1960), the Subtropical Convergence (Deacon, 1963; Garner, 1959, 1962; Garner and Ridgway, 1965), and the Equatorial Divergence (Burkov, 1966).

The Antarctic Convergence forms one of the major fundamental boundary zones of the world's oceans. The principal property by which its location can be mapped is the steep temperature gradient at the sea surface (Mackintosh,

1946). Across the convergence, the temperature change in summer is from 4° to 8°C and in winter, from 1° to 3°C. Below the surface the convergence marks the area in which Antarctic Upper Water sinks beneath and mixes with the warmer south-moving Subantarctic Surface Water (Figure 6).

The Subtropical Convergence is a well-defined feature in the south Tasman Sea and off the east coast of New Zea-

TABLE 4 Convergences and Divergences in the South Pacific

Present Terminology	Burkov's (1966) Terminology
Equatorial Divergence	Equatorial Divergence
Southern Tropical Divergence	Southern Subtropical Divergence
Southern Tropical Convergence	Southern Subtropical Convergence
Subtropical Divergence	Subantarctic Divergence
Subtropical Convergence	Subantarctic Convergence
Antarctic Convergence	Southern Polar Front
Antarctic Divergence	Antarctic Divergence

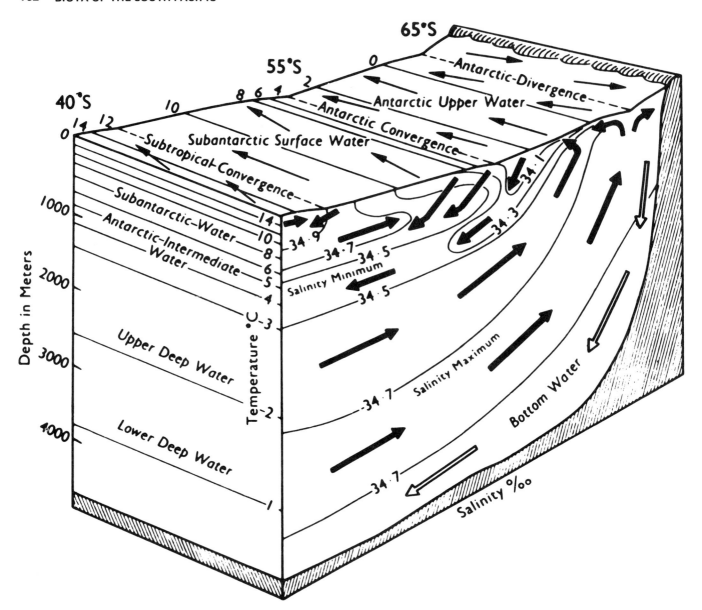

FIGURE 6 Schematic diagram of the meridional and zonal flow in the Southern Ocean (after Brodie, 1965; reproduced with permission). The diagram represents summer conditions with the average positions of convergences and divergences. In the Pacific the Upper Deep Water is not well developed as in the Atlantic sector, and the southward component of the Lower Deep Water is weak or may be reversed.

land. Across the convergence from south to north the temperature change in the summer is from 13° to 16°C and in the winter, from 8° to 11°C (Garner and Ridgway, 1965), while there is a sharp rise in salinity from about 34.3–34.9‰. There are marked seasonal fluctuations in the position and degree of development of this convergence. Its structure east of the Chatham Islands is imperfectly known, and there is very little information available as to what happens when the West Wind Drift turns north as the Peru Current.

The Equatorial Divergence at the equator is formed as a result of the divergence of drift currents to the north and south in the Northern and Southern hemispheres, respectively, leading to the upwelling of deeper water.

Burkov (1966) recognizes an additional convergence that he terms the Southern Subtropical Convergence (here termed the Southern Tropical Convergence) in the region of 20°S, with divergence zones to the north and south.

Water Masses

The net result of the oceanographic features discussed above is the formation at the surface of water masses of characteristic temperature and salinity structure, which can be distinguished on the basis of characteristic temperature-salinity (T-S) curves. The line of curves diagnostic of the Antarctic (polar) type of water structure extends along the *S*-axis, while that diagnostic of tropical types extends along the *T*-axis. Since the distributions of many planktonic animals (Bary, 1959; Johnson and Brinton, 1963; McGowan, 1963) are associated with water types falling within a defined range of T-S curves (see Figure 9), our knowledge of the distribution of the basic water masses of the region is important.

As I have previously pointed out, the terminology for the water masses in the southern oceans is in a confused state (Knox, 1960). Not only are different terms used by different workers (see Table 5), but the terminology does not correspond with that used for equivalent water masses in the Northern Hemisphere. Since many individual species or related species show similar distribution patterns in the North and South Pacific, it is of the utmost importance that as uniform a system of nomenclature as possible be agreed upon for the two regions. One of the problems has been the use of the term "subtropical" for water immediately north of the Subantarctic (Cold-Temperate) Water. This, in reality, corresponds with water masses termed "warm-temperate" elsewhere.

In the most recent reviews by Stepanov (1965) and Radzikhovskaya (1966), five broad zones are recognized from the Antarctic Continent to the equatorial regions. These, as shown in Figure 7 with the terminology adopted in this paper, are (1) *Antarctic* (Polar), south of the Antarctic Convergence; (2) *Subantarctic* (Cold-Temperate), between the Antarctic and Subtropical Convergences; (3) *Subtropical* (Warm-Temperate), from the Subtropical Convergence south to about the position of the Southern Tropical Convergence at 20° to 25°S; (4) *Tropical*, from the Southern Tropical Convergence to within 4° to 5°S; and (5) *Equatorial*, from 4° to 5°S to about 15° north.

As Garner (1959) points out, the terms "subtropical" and "subantarctic," in the hydrological sense, require qualification when coastal waters are concerned. He states:

Modifications of nearshore water properties may occur. Shallow or semi-enclosed waters may undergo large changes in temperature under local climatic effects, turbulent flows in tidal waters may induce intense vertical mixing of colder subsurface and nearsurface layers and the runoff of fresh water from the adjacent land will modify the temperature and salinity pattern in some areas. The formation of "slope" water over the continental shelf through mixing of oceanic and coastal shelf waters may be another source of complex surface gradients and areas of cold water may be formed by upwelling.

In a previous paper (1960), I have proposed the terms "cold temperate mixed waters" and "transitional warm temperate waters" for such coastal water types.

The different water masses may be characterized as follows (see also Table 6):

1. *Equatorial* Surface Water is found where the sea-surface temperature is high and its seasonal variation is small and where salinity is low due to excess of rainfall over evaporation (Wyrtki, 1966). Surface temperatures are generally greater than 25°C, and salinity is usually less than 34.0‰.
2. *Tropical* Surface Water is found where evaporation greatly exceeds precipitation; it is characterized by high salinities (35.5–36.5‰), but the temperature may vary over a wide range from 28° to 15°C.
3. *Subtropical* (Warm-Temperate) Surface Water, lying begion of the Subtropical Convergence and north to about the position of the Southern Tropical Convergence, is characterized by a winter temperature range of 12.8° to 20.0°C, a summer range of 15.0°C to about 25.0°C, and high salinities of 34.5–35.5‰.
4. *Subantarctic* (Cold-Temperate) Surface Water, lying between the Antarctic and Subtropical Convergences, is characterized by an even meridional temperature gradient,

TABLE 5 Terminology of Water Masses of the South Pacific

Present Terminology	Sverdrup *et al.* (1942)	Deacon (1957, 1963)	Wyrtki (1966)	Stepanov (1965)	Radzikhovskya (1965)	Ekman (1953)
Antarctic (Polar)	Circumpolar	Antarctic		Polar	Antarctic	Antarctic
Subantarctic (Cold-Temperate)	Subantarctic	Subantarctic	Temperate	Subantarctic	Subantarctic	Antiboreal
Subtropical (Warm-Temperate)	Western and Eastern South Pacific Central Water	Subtropical	Subtropical	Temperate-Tropical	Subtropical	Warm-Temperate
Tropical	Western and Eastern South Pacific Central Water	–	Subtropical	Tropical	Tropical	–
Equatorial	Equatorial	–	Equatorial	Equatorial	–	–

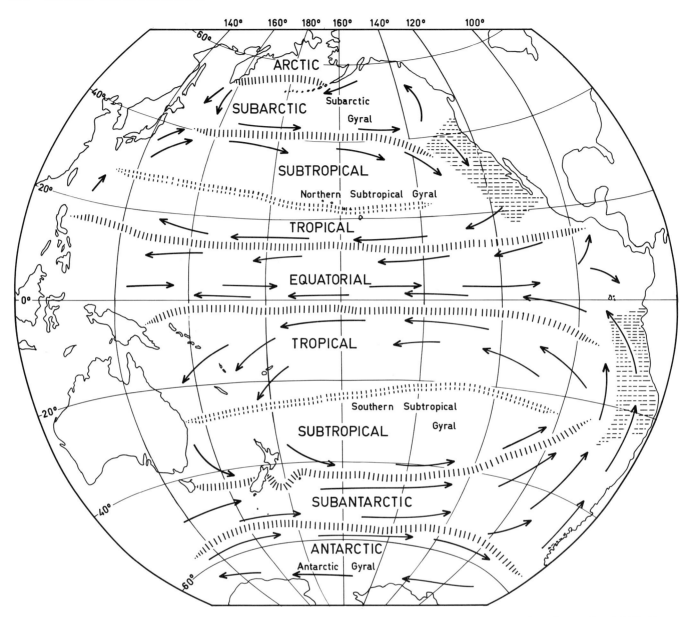

FIGURE 7 Water masses and biogeographical zones. Based on data from Burkov (1966); Johnson and Brinton (1963); Brodskiy (1965); Knox (1960, 1963a and b); Radzikhovskaya (1965); Stepanov (1965); Tully (1964). The position of the boundary zone between the subtropical and tropical zones is ill-defined, and the real situation is probably a gradient from the boundary with the equatorial zone to that with the cold-temperate zones.

with surface temperatures increasing from south to north from 3.0° to 11.5°C in the winter and from 5.5° to 14.5°C in the summer, and with a salinity of about 34.5‰, or somewhat higher.

5. *Antarctic* Surface Water is characterized by low temperatures and low salinities.

Radzikhovskaya (1966) also recognizes an Eastern Subtropical (Warm-Temperate) water mass off the South Amer-

ican coast between 20° and 40°S. This is the South Pacific equivalent of the "Transition Region" distinguished in the Californian Current by Sverdrup *et al.* (1942, Figure 209A, p. 740) on the basis of physical properties. It is a region where Subarctic (Cold-Temperate), Subtropical (Warm-Temperate), and Equatorial Waters converge.

With this background, it is possible to discuss some of the interesting problems concerning the distribution of marine organisms.

TABLE 6   Temperature and Salinity Characteristics of the Water Masses of the South Pacific

| Water Mass | Temperature Range | | Salinity Range |
	Winter	Summer	
Equatorial	$>25°$ C	$>25°$ C	$>34.0‰$
Tropical	$15.0°-23.0°$ C	$20.0°-28.0°$ C	$35.5-36.5‰$
Subtropical (Warm-Temperate)	$12.0°-20.0°$ C	$15.0°-25.0°$ C	$34.5-35.5‰$
Subantarctic (Cold-Temperate)	$3.0°-11.5°$ C	$5.5°-14.5°$ C	$34.0-34.5‰$
Antarctic	$-1.8°-0.5°$ C	$-1.0°-3.5°$ C	$33.0-34.0‰$

BIOGEOGRAPHY

Many of the immediate biological problems of the South Pacific are concerned with the distribution patterns of single species, related species groups, and communities of organisms. In the past, most marine biogeographical studies have, in the main, been concerned with the simple plotting of species distributions and have rarely been concerned with communities of organisms, the approach being systematic and historical, rather than ecological (Hedgpeth, 1957). However, in recent years new approaches, especially directed toward the North Pacific (Brinton, 1962; Brodskiy, 1961, 1962 a and b, 1965; McGowan, 1963), have resulted in some detailed analyses of the factors involved in determining distribution patterns.

Plankton Distributions

A tremendous amount of work must be done before we have anything like a general picture of the distribution of South Pacific plankton. The requirements are well expressed by Brodskiy (1965):

... two research trends must be distinguished: the characterization of water masses, fronts, currents and so on in terms of the fauna and the characterization of the distribution of species and their interspecific units followed by the establishment of the extent of agreement between the areas (of distribution) and the "hydrologic" provinces (in the sense used by Brinton, 1962, for the North Pacific).

For the North Pacific, certain species have been recognized as having geographical distributions that are in close agreement with the major water mass provinces, such as pelagic Foraminifera (Bradshaw, 1959), Chaetognatha (Alvarino, 1962; Bieri, 1959), Pteropod molluscs (McGowan, 1963), and euphausiid crustaceans (Brinton, 1962). Each of the above groups includes species limited to *Subarctic, Central,* and *Equatorial* (Cold-Temperate, Subtropical-Tropical, Equatorial). Studies of the distribution of polychaetous annelids (Dales, 1957; Tebble, 1962), some pelagic tunicates (Berner, 1960), and some copepods (Johnson, 1938) in

northeastern and northwestern Pacific waters are in general agreement with characteristic recurring patterns. Johnson and Brinton (1963) also recognize a fourth zone in temperate waters at the northern limit of the central region, a zone of transition between central and subarctic faunas occupying the water of the South Pacific Drift between 38° and 45°N. Toward its eastern limit, this "transition zone," as they term it, diverges with a northern branch entering the Gulf of Alaska and a southern branch occupying the Californian Current, terminating off Baja California.

Euphausiids are a very good group for illustrating distribution patterns and their relationship to water masses, boundaries, and currents because many species are widely distributed, occurring in equivalent zones in both the North and South Pacific. Table 7 summarizes some of these distributions in relation to the major water masses. It will be noted that a number of species form two isolated populations (so far indistinguishable on morphological grounds) in corresponding zones in the North and South Pacific; other species have their centers of concentration limited to a single water mass or part thereof. Similar patterns can be found in other groups of pelagic animals, especially polychaetes (Tebble, 1960, 1962), chaetognaths (Alvarino, 1964, 1965), and pteropods (McGowan, 1963). The actual situation, however, is complex in view of the uncertainty of the taxonomic status of many of the forms. In the words of Johnson and Brinton (1963),

The studies of distribution and systematics are interdependent in the development of biogeographical concepts. Oceanic populations found to be small in range sometimes prove to be ecotypically or genotypically distinct from other populations of the same morphological type. The apparently broad character of other ranges has prompted study also leading to the recognition of complexes of regional species or subspecies.

They cite the example of *Salpa fusiformis,* which was long recognized as cosmopolitan and variable in form, but which was shown by Foxton (1961) to consist of four species, two tropical-subtropical, one living south of the Subtropical Convergence, and one limited to the Pacific section of the

TABLE 7 Water Masses and Pacific Euphausiids

Water Mass	Species Occurring in Both North and South Pacific	Species with Center of Concentration Limited to One Water Mass
Subarctic (Cold-Temperate)	–	*Thysanoessa longipes, Euphausia pacifica, Tessarabrachion oculatus*
Transition (Warm-Temperate)	*Thysanoessa gregaria*	*Nematoscelis difficilis*
Central (Tropical)	*Euphausia brevis, Nematoscelis atlantica*	*Euphausia hemigibba, E. mutica, Thysanopoda aequalis* (W. Central); *T. subaequalis* (E. Central)
Equatorial	*Euphausia distinguenda, Nematoscelis gracilis*	*Euphausia exima* and *E. lamelligera* (E. region); *E. sibogae* (W. region)
Tropical	*Euphausia brevis, Nematoscelis atlantica*	*Euphausia gibba*
Subtropical (Warm-Temperate)	*Thysanoessa gregaria*	*Nematoscelis megalops, Thysanopoda aequalis, T. subaequalis.*
Subantarctic (Cold-Temperate)	–	*Euphausia vallentini, E. longirostris, E. luceas.*

Data from Brinton (1962); Johnson and Brinton (1963); John (1936); Sheard (1953).

Antarctic. A similar situation concerns the calanoid copepod *Calanus finmarchicus*, considered to be a cosmopolitan cold-water species but now subdivided into seven separate species, five in the Northern Hemisphere and two in the Southern Hemisphere—*C. astralis* and *C. chilensis*—both occurring in the Subantarctic zone of the South Pacific (Brodskiy, 1965). Many of the distribution records of planktonic species from the South Pacific are based on early determinations before modern approaches to the study of variability and population analysis were applied. Hence, the exact status of many forms remains in doubt, and some species recognized as having distinct allopatric populations may, on closer analysis, prove to be different species or subspecies. We need much more detailed systematic sampling and biometric analysis of populations in order to have a clear picture of speciation and the limits of distribution of the populations of the species or subspecific units.

Of particular interest are sibling species, such as the euphausiids *Nematoscelis difficilis* and *N. megalops*, the former occurring in the North Pacific "transition" zone, the latter, in warm-temperate waters of the South Pacific, and such species as the euphausiid *Thysanoessa gregaria*, which are widely separated geographically with populations occupying comparable zones in the North and South Pacific (Figure 8). As can be seen in Figure 9, the T-S habitats differ somewhat with the North Pacific populations, which occur at lower salinities and higher temperatures. It could be suspected that these allopatric populations differ physiologically, and studies of their physiology and reproductive cycles should prove rewarding.

In the southern oceans *Th. gregaria* has been considered to be a tropical–subtropical–temperate euphausiid that oc-

casionally occurs in subantarctic waters (Sheard, 1953; Boden, 1954). Bary (1959), however, has recorded breeding populations in waters believed to be of subantarctic origin to the south and east of New Zealand. He found that these populations reacted to increasing temperatures by a reduction in numbers in a manner similar to other subantarctic species and suggests that they may be a stock of *Th. gregaria* that is adapted to cold waters. As Bary points out, great interest would accrue from comparison of upper and lower temperature tolerances of this and subtropical stocks, both by experiments and through field data by means of temperature-salinity-plankton (T-S-P) diagrams.

There is no doubt that the Pacific offers unparalleled opportunities for the study of speciation processes in planktonic organisms. Johnson and Brinton (1963) have reported the effects of continental barriers and oceanwide warming and cooling in past periods on the exchange of genetic material across the tropics and the resultant speciation in euphausiids, and Knox (1963b) has given information on the possible effects of the migration in past geological periods of the cold-water–warm-water boundary (the Subtropical Convergence Zone) on speciation patterns in shallow-water species in the New Zealand region. Further detailed work on the problem of speciation is needed.

Biogeographical Terminology

Brodskiy (1965) questions whether biogeographical terminology can be the same as is employed in hydrology. In particular, he criticizes the use of the term "sub-Arctic" for what is in biogeographic terms the northern temperate Pacific region and puts forward the view that biogeographic

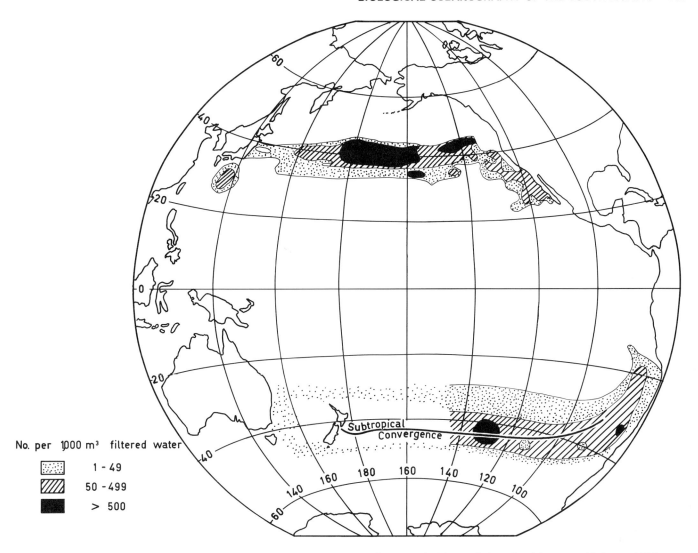

No. per 1,000 m³ filtered water

          1 - 49

          50 - 499

          > 500

FIGURE 8   The Pacific distribution of the euphausiid *Thysanoessa gregaria*. (Reprinted with permission after Johnson and Brinton, 1963, Figure 4, p. 341.) This has its greatest concentration as a warm-temperate (subtropical species), termed a "transition zone" species lying between the central (subtropical) and subarctic (subantarctic) water masses. The position of the Subtropical Convergence in the South Pacific is shown.

zoning should be based on work on the taxonomy, ecology, and distribution of organisms. While this criticism may be valid, I feel that there is much to be gained if the confusion of terminology is solved and the two sets of terms are in as close agreement as possible. In particular, a large number of names has been proposed for the Southern Cold-Temperate Zone, including "Subantarctic," "Antiboreal," "Notalian" (Bekker, 1964; Brodskiy, 1958), and "Austral" (Brodskiy, 1962b). I would, however, agree with Brodskiy on the use of the term "transition zone," as applied by Brinton (1962), for the region between his sub-Arctic and central zones. Johnson and Brinton (1963) have also suggested that a similar transition zone occurs in the South Pacific in the region of the Subtropical Convergence. As Brodskiy puts it, such a

biogeographic division is of little use to biogeographers and taxonomists because a transitional fauna can clearly be found in many regions of overlap.

Figure 7 and Table 8 give the major biogeographic zones of the Pacific Ocean, as based on the major water masses and distribution patterns. The general correspondence of these zones in the North and South Pacific can be seen. Whether a division of the warm waters into tropical and subtropical zones is justified will depend upon more detailed distributional analyses. It may be that there is a gradual change, with warm-temperate species being replaced by tropical species as the equatorial regions are approached. Reid (1962) suggests that the gyres themselves may offer a more effective classification for zooplankton. He suggests

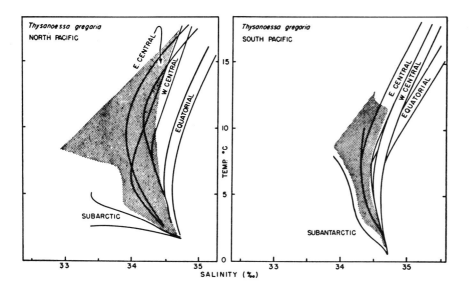

FIGURE 9   Temperature-salinity characteristics of water below 150 m for the North Pacific habitat (a) and the South Pacific habitat (n) of *Thysanoessa gregaria* compared with the T-S envelopes of the principal water masses. (Reprinted with permission after Sverdrup *et al.*, 1942.) The South Pacific T-S curves for the *T. gregaria* habitat are all from the eastern half of the ocean. (Reprinted with permission from Johnson and Brinton, 1963, Figure 4, p. 391.)

TABLE 8   Water Masses and Biogeographical Terminology

Gyral System	Water Mass	Biogeographical Zone
Subarctic Cyclonic Gyre	Subarctic	Northern Cold-Temperate
Northern Subtropical Anticyclonic Gyre	Northern Subtropical Northern Tropical	Northern Warm-Temperate Northern Tropical
Equatorial Zonal Flows	Equatorial	Equatorial
	Southern Tropical	Southern Tropical
Southern Subtropical Anticyclonic Gyre	Southern Subtropical	Southern Warm-Temperate
	Subantarctic	Southern Cold-Temperate
Antarctic Cyclonic Gyre	Antarctic	Antarctic

that particular features (surface convergences in the anticyclones, coastal boundaries along much of the cyclones) tend to confine individuals within the gyres.

Of particular interest are situations such as that found in the Subtropical Convergence Zone, where warm waters and cold waters converge. Bary (1959) has reported on some data from coastal waters off the east coast of New Zealand and has demonstrated the mixed nature of the fauna in the vicinity of the Subtropical Convergence. Recently, one of my students (Grieve, 1966) had confirmed and extended these results in a study of the annual cycle of plankton on the edge of the continental shelf off Kaikoura on the east coast of the south island of New Zealand. She found that the area was subjected to subtropical and subantarctic influences in varying amounts throughout the year. Certain species that were known from previous work to have distribu-

tions limited to either subantarctic or subtropical waters consistently occurred in her samples (Figure 10). Subtropical species included the euphausiids *Th. gregaria, Nematoscelis* sp. and *Nyctiphanes australis* (Sheard, 1953); the polychaete *Rhynchonerella angelini* (Tebble, 1960); and the tunicates *Pyrosoma* sp., *Ihlea magalhanica*, and *Thalia democratica* (Thompson, 1948). Subantarctic species included the euphausiid *Euphausia vallentini* (Sheard, 1953), the copepods *Candacia cheirura* and *Euchaeta bilobata* (Vervoort, 1957) and *Cyllopus magellanicus* and *Eucalanus longipes* = *E. acus* (Bary, 1959), and the tunicate *Salpa thompsoni* (Foxton, 1961). It is clear that further studies are required in order to understand the role of this important hydrological feature in the distribution of planktonic animals.

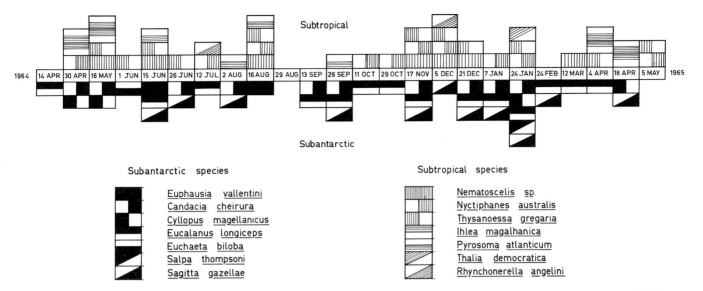

Subantarctic species

Euphausia vallentini
Candacia cheirura
Cyllopus magellanicus
Eucalanus longiceps
Euchaeta biloba
Salpa thompsoni
Sagitta gazellae

Subtropical species

Nematoscelis sp.
Nyctiphanes australis
Thysanoessa gregaria
Ihlea magalhanica
Pyrosoma atlanticum
Thalia democratica
Rhynchonerella angelini

FIGURE 10 Occurrences of selected Subtropical (Warm-Temperate) and Subantarctic (Cold-Temperate) species at a locality at the shelf edge (200 m) off Kaikoura, east coast, New Zealand. (Reprinted with permission from Grieve, 1967.)

BENTHIC ANIMALS

Over the greater part of the vast expanse of the South Pacific there is very scanty information concerning the distribution of benthic animals. A number of expeditions, starting with the *Challenger*, and culminating in recent times with the Danish Deep-Sea *Galathea* Expedition, the voyages of the USSR vessels the *Vityaz* and the *Ob*, and the cruises of the U.S. Antarctic research vessel *Eltanin*, have crossed and sampled parts of the region. However, the coverage is very scanty, and apart from very broad generalizations, biogeographic comparisons are certain to be misleading. This can best be illustrated by considering recent work on the shelf and slope benthic fauna of the New Zealand region.

Systematic study of the benthic fauna of the New Zealand region only commenced in 1954 with the Chatham Islands 1954 Expedition (Knox, 1957), which carried out a series of full oceanographic studies, including bottom sampling along the Chatham Rise and around the Chatham Islands. Prior to this expedition, only 11 bottom stations had been worked in New Zealand waters in depths between 100 and 200 fathoms, and only 5 in depths over 200 fathoms. Five of the former were carried out by the Danish Deep-Sea *Galathea* Expedition in the atypical deep basin of Milford Sound. The Chatham Islands Expedition carried out a further six stations in depths from 120 and 200 fathoms and ten in depths of over 200 fathoms. Table 9 gives an indication of the quantity of material obtained on this expedition.

One of the important results of the expedition was that it enabled the recognition of a well-marked slope (archiben-

thal) fauna, as distinct from the shelf fauna, for the first time (Dell, 1959). Another important result, from the paleoecological point of view, was the finding in large quantities of a number of genera of mollusca such as *Galeodea*,

TABLE 9 Species Recorded by the "Chatham Islands 1954 Expedition"[a]

Taxonomic Group	Number of Species	Number of New Species	Number of New Records for N.Z.
Porifera (Keratosa)	6	2	0
Porifera (Demospongiae)	22	5	3
Bryozoa	78	23	8
Hyroida	27	0	2
Scleractinia	3	1	0
Polychaeta Errantia	99	19	28
Polychaeta Sedentaria (estimated)	84	28	10
Bopyridae	1	1	0
Cumacea	3	1	1
Brachyura	16	2	1
Decapoda Natuntia (deep water)	6	4	0
Mollusca (slope species)	213	67	0
Mollusca (shelf species)	295	1	0
Sipunculoidea	3	0	0
Echinodermata	55	5	12
Tunicata	31	11	10
Fishes	60 (approx.)	1	7
Totals	1,002	171	82

[a]The list covers the groups so far reported on; a number of groups still await detailed study.

Comitas, Micantapex, Waipaoa, Teremelon, or *Scaphander* (Dell, 1959), which are well represented in the New Zealand Tertiary, but which had been considered to be extinct. In addition, one of the commonest species of sea urchin taken in the depth over 100 fathoms belonged to a species previously known only from the New Zealand Pleistocene of which it is very characteristic. Among the starfishes there were representatives of four families, the Zoroasteridae, Hippasteriidae, Peribolasteridae, and Solasteridae, not previously recorded from New Zealand waters. Fell (1957, 1960), in reporting on the echinoderms, considered the discovery of a starfish belonging to the genus *Hippasterias* to be of exceptional interest. The genus in the Australasian region was previously known only from a single fossil specimen from the Late Cretaceans of North Canterbury.

Since 1955, the New Zealand Oceanographic Institute has been engaged in a program of sampling and analyzing the bottom animals of the New Zealand region (Dawson, 1965, 1968). Some idea of the coverage is given in Figure 11: 400 shelf stations (10 to 100 fathoms) and 300 slope or archibenthal stations (100 to 1,000 fathoms and deeper) have been occupied. Material has been provided in quantity for systematic monographs of many groups of marine invertebrates, and distributional and ecological analyses of many species have been made. One of the better known groups is the echinodermata, which comprises 295 known species. Mapping and analysis of the physical variables of the environment, such as depth, sediments, water movements, and temperature, have helped to explain distribution patterns in this and other groups of invertebrates. A general picture is emerging of a widespread slope or archibenthal fauna with little regional differentiation throughout the region studied (Figure 12 shows examples from the echinodermata of the distribution of typical members of this fauna) and of a shelf fauna with two main elements, a northern warm-water element and a southern cold-water element (Figure 13 shows the distribution of typical representatives of these elements from the echinodermata).

As far as I am aware, this is the only example of systematic benthic work on a large scale in the whole of the South Pacific region, and it shows what can be accomplished with somewhat limited resources. There is a great need to extend this type of work to other parts of the South Pacific and into deeper waters. In particular, it is important to extend such sampling to areas such as the Lord Howe Rise, the Norfolk Ridge, and the Colville Ridge in order to determine how widespread the New Zealand slope fauna is and what its relationships are with similar faunas in other parts of the South Pacific. Dell (1959) has listed 21 examples of pairs of related species on the slope faunas of New Zealand and eastern Australia. Detailed comparative studies are needed to reveal the degree of relationship between these faunas. As Dell has pointed out, the nomenclature of the Australian slope species has not been revised since Hedley, together with Petterd and May, published a number of papers (1905–1908) on the molluscs from isolated deep-water stations (four in all). These papers comprise what is known of the slope molluscan fauna of the east Australian region. Dell suggests that there is a distinct possibility that in some cases the species pairs may be found to be end points in clines or perhaps two of a series of subspecies that extend across the Tasman Sea, since known depths on the complicated bottom topography of the northern Tasman Sea are not likely to be barriers to dispersal.

Comparable studies to those described above are needed for other regions of the South Pacific. In particular, it would be of considerable biogeographic interest to have a knowledge of the fauna of the major ridges, such as the Pacific-Antarctic Ridge, the South Chile Ridge, and the East Pacific Rise. This could provide evidence concerning the migration and evolution of the faunas of intermediate depths. Other areas of considerable interest are the numerous seamounts and guyots that cover the area. Menard (1959) has listed over 1,400 from the Pacific basin, mostly concentrated in the North Pacific, where the major exploratory work has been carried out, and has suggested that this represents about 10 percent of the true number. A recent report (Anon. 1968) on the voyage of the U.S. oceanographic survey ship *Oceanographer*, along 35°S from New Zealand to Chile, reveals that twenty-five new mountains were discovered, about half of them ranging in height from about 6,000 to 11,000 feet, as well as numbers of seamounts. Many parts of the sea floor were found to be up to 3,000 feet deeper than indicated on the stations. Further work of this kind will reveal areas that are likely to prove of interest for biological sampling. As an example of interesting finds yet to be made from these guyots and seamounts, Zullo and Newton (1964), reporting on the cirripede fauna from a guyot at a depth of 228 m 1,290 km off the Chilean coast, recorded four new species, including two "deep-water" verrucarids not previously reported from the eastern Pacific.

Role of Systematics

If we are going to carry out the biological exploration of the South Pacific outlined above, there will have to be a considerable increase in the number of taxonomists in order to study the material that will be collected (Dell, 1965). In this connection, it is pleasing to note that the grants for the training of taxonomists concerned with marine groups in the United States increased from $3,000 in 1937 to $10 million in 1963 (Brodskiy, 1965). However, with ecological research expanding rapidly on a worldwide basis, the still limited number of taxonomists will be in a position to help the ecologists only with the more difficult taxonomic problems, and the burden of the nomenclatural aspects of ecology will fall increasingly on the ecologist (Carriker, 1967). In the type of problem outlined above, it is imperative that

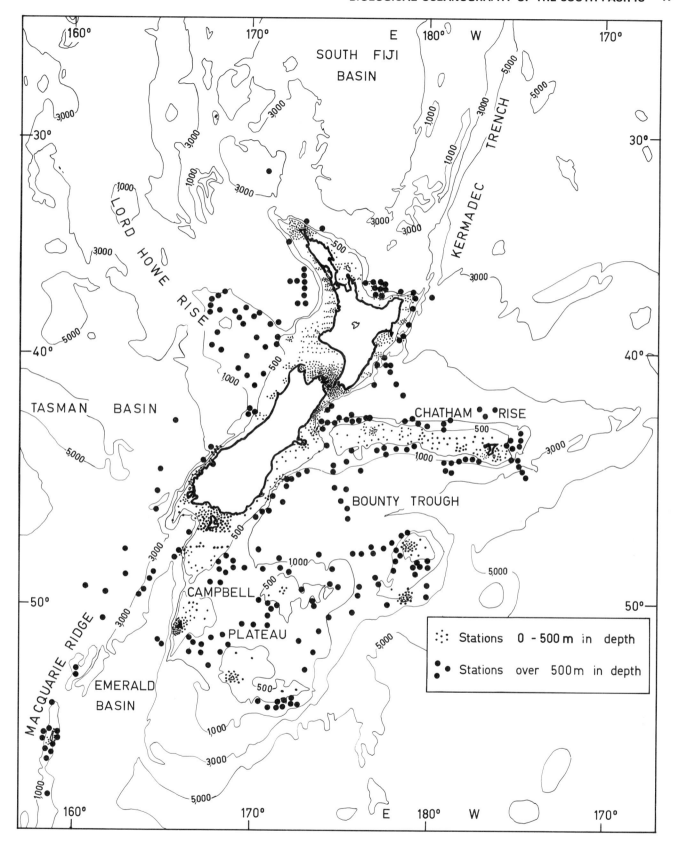

FIGURE 11   New Zealand Oceanographic Institute bottom stations, 1955–1967. (From data supplied by E. W. Dawson, N.Z.O.I.)

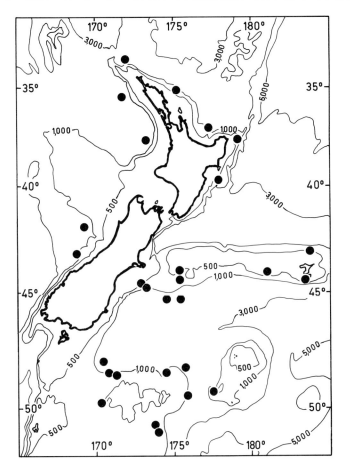

FIGURE 12 The distribution of the echinoid *Phormosoma bursarium*, a widespread slope (archibenthal) species of the New Zealand region.

PRODUCTIVITY

Since productivity will be dealt with in detail in a later contribution to this section, I propose to mention only a few general points. Koblentz-Mishke (1965) has summarized the information available at that time on the volume of primary production in the Pacific (as measured by variants of the ^{14}C method) and on its seasonal and geographic variations. As she points out,

The data contain systematic errors due to the conditions under which the samples were exposed and production at the surface may be underestimated in some instances and overestimated approximately one and a half times in others. In addition to the systematic errors specific to each of the methods used, the data on primary production have a random spread due in the first instance to irregularity in the spatial distribution of the phytoplankton. Random errors in the methods employed are also not insignificant factors.

A further source of error in making comparisons is the fact that the samples used for determinations of productivity have been collected at different times of the year. With the usual method of expressing productivity in terms of mg of C per m^3 per day, the season in which the samples were taken, especially in temperate regions, where the annual variation in productivity may be from quantities too small to be detected by the methods employed to more than 100 mg of C per m^3 per day. More meaningful figures would be based on annual production rates, but such information is available from only a handful of widely separated localities.

The general picture that emerges from these studies is that the regions of highest productivity (more than 100 mg of C per m^2 per day) are to be found in the neritic zone of the temperate waters of the North Pacific. However, inadequacy of sampling may account for the fact that values of this order have not yet been recorded from the South Pacific. Relatively high values (10–100 mg of C per m^2 per day) are characteristic of the northern part of the Peru Current and localized areas off Southwest Australia (Jitts, 1965) and off New Zealand (Grieve, 1967). A production of 5–10 mg of C per m^2 per day is found in the region of the Equatorial Divergence and in the subantarctic cold-temperate waters. In general, Subtropical and Tropical waters are characterized by low values, generally less than 5 mg of C per m^2 per day, while for large areas of tropical waters, values are unusually low, averaging less than 1 mg of C per m^2 per day.

The average rate of production of shallow temperate waters is, however, appreciably higher than that of the deep ocean (Ryther, 1963). This can be seen in the data presented by Burkenholder and Burkenholder (1967) on the chlorophyll *a* content of samples taken between South America and New Zealand on cruises of the research vessel *Eltanin* between May and December 1964 (Figure 14). In early December values of chlorophyll *a* (2.56 to 2.71 mg per m^2) in the area of 38°S latitude and 179° longitude

organisms be properly identified according to species, subspecies, or even races (Brodskiy, 1965; Carriker, 1967). It is therefore imperative that adequate monographic works be provided. A major bottleneck in taxonomic and ecological research is the scattered literature, much of it based on the reports of the great expeditions of the past. The type of work that is required for the Pacific as a whole is something like the "Fauna of the Ross Sea" series being sponsored by the New Zealand Oceanographic Institute. In order to achieve this, considerably larger numbers of taxonomists have to be trained, and much greater funds will need to be provided for their training and for the provision of employment for such trained people. Money can often readily be found to mount an expedition or to study animals in the field or laboratory, but it is often much more difficult to find support for people to study the collection obtained. I would like to see a budgeting requirement for such follow-up work built into the funds obtained for any collecting expedition (Knox, 1968).

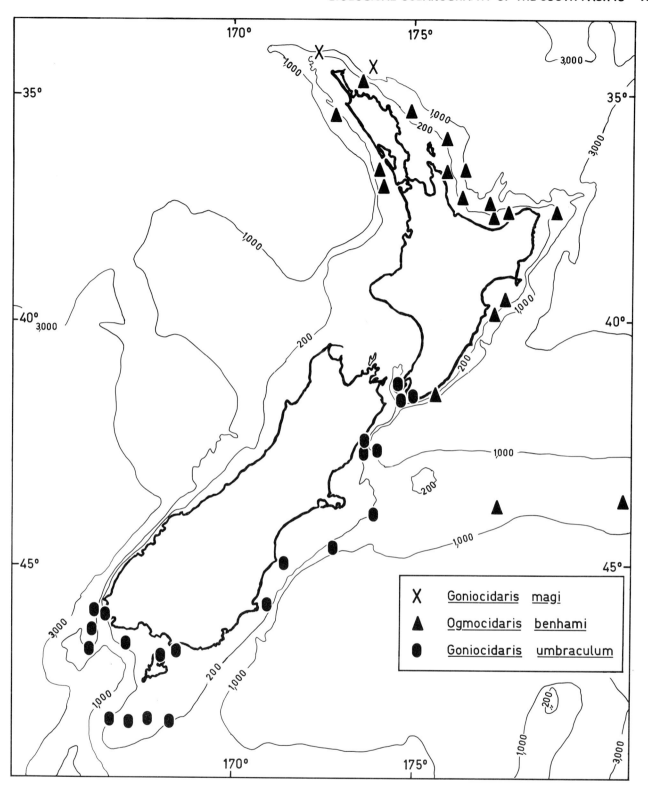

FIGURE 13 The distribution patterns of three New Zealand shelf echinoids.

  a) *Goniocidaris magi,*   a subtropical species

  b) *Ogmocidaris benhami,*   a northern warm-water species

  c) *Goniocidaris umbraculum,* a southern cold-water species

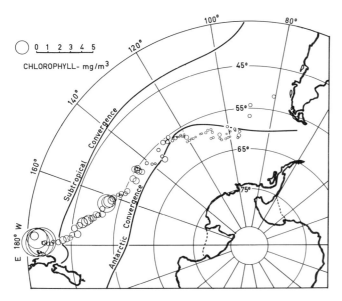

FIGURE 14   Chlorophyll *a* values in surface waters on Cruise 15 of the U.S. research vessel *Eltanin*, October 1–December 8, 1964. Diameters of circles indicate mg of chlorophyll per m^3. (Reprinted with permission from Burkenholder and Burkenholder, 1967, Figure 2, p. 608.)

were up to 10 times the average for stations farther to the east. Similar increases in waters off New Zealand are to be seen in the figures published by El-Sayed (1967) for the distribution of chlorophyll *a* and ^{14}C uptake in the Pacific Sector of Antarctica. Few detailed studies of productivity have been carried out in New Zealand waters, and only one investigator, Grieve (1967), has followed the annual cycle. At the edge of the shelf off Kaikoura on the east coast of the south island, she recorded an annual variation ranging from 0.41 mg of chlorophyll *a* per m^3 in June to a maximum of 4.45 mg per m^3 in September. For this locality Grieve has calculated the "gross" primary production from chlorophyll, radiation, and transparency data, enabling comparisons to be made with figures given by Ryther and Yentsch (1958). These comparisons are shown in Table 10.

TABLE 10   Comparisons of "Gross" Primary Production

Locality	Depth (m)	"Gross" Productivity (g of C per m^2 per year)
Long Island Sound (New York)	25	380
Continental Shelf	25–50	160
New York	50–1,000	135
	1,000–2,000	100
North Central Sargasso Sea (Bermuda)	>5,000	78
Kaikoura, New Zealand	200	157

It can be seen that the value for the New Zealand locality compares favorably with those calculated for similar waters off New York.

Another useful measure of productivity is zooplankton volume or biomass. Reid (1962) has calculated the distribution of zooplankton volume for the Pacific as a whole (Figure 15). The highest concentrations occur in Antarctic and Subantarctic (Cold-Temperate) waters in the vicinity of the Antarctic Convergence and where these cold waters turn north off the east coasts of New Zealand and along the west coast of South America in the Peru Current system. Moderately high values are found in the region of the Subtropical Convergence and in the Equatorial Zone, while values are low in tropical and subtropical waters (Bogorov and Vinogradov, 1960; Graham, 1941; King and Demond, 1953; King and Hida, 1957; Legand, 1961; Ponamareva and Lubini, 1958). As Tranter (1962) has pointed out, many of the early observations on zooplankton abundance, especially off Australian coasts, are contradictory, largely through failure to appreciate the relationship between depth of water and zooplankton abundance. For the open ocean areas of the Tasman Sea, Tranter's values (less than 50 mg per m^3) agree with other estimates for subtropical waters; values for the continental shelf, on the other hand, averaged 100 mg per m^3. On the edge of the shelf at Kaikoura on the east coast of New Zealand, Grieve (1967) recorded a variation in zooplankton biomass from 10.3 mg per m^3 in August to 403 mg per m^3 in December, with an average of 68 mg per m^3. More detailed sampling will probably alter details of this general picture of the South Pacific, as Reid's data for large areas were very scanty.

As Reid has demonstrated, zooplankton volume is distributed in very much the same way as inorganic phosphate phosphorus (PO$_4$ · P). Both are high in the cyclones and in the eastern boundary currents and low in the anticyclones. They are also relatively high in the region of the Equatorial Divergence and the zonal currents of the Equatorial system. However, nowhere in the South Pacific do we have the detailed information that is available for the Californian Current over a period of years (Reid, 1962, Figure 6, p. 303). It is in years of low temperatures within the Californian Current, with the presumably greater influx of Cold-Temperate Subarctic Water, that the plankton volumes are highest. High temperatures are usually accompanied by lower plankton volumes. Reid suggests that a similar shift from cold- to warm-water forms may occur near 20°S along the coast of South America. It would be of considerable interest to have details of such a shift. As mentioned previously, a similar shift, although of smaller magnitude, in the position of the warm-water–cold-water boundary marked by the Subtropical Convergence occurs off the New Zealand east coast (Garner, 1959). In some summers, Subtropical Water penetrates as far south as Dunedin (46°S); in other years it does not reach south of 43°S. These influxes of

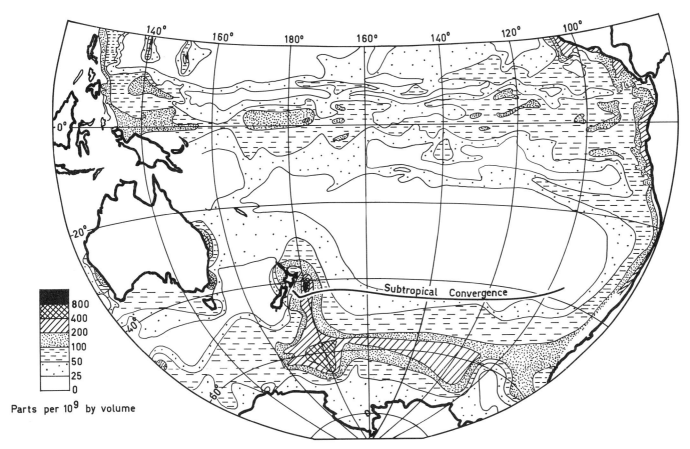

FIGURE 15 Distribution of zooplankton volume (parts per 10^9 by volume) in approximately the upper 150 m of the South Pacific. The position of the Subtropical Convergence is shown. (After Reid 1962, Figure 4 (b), p. 301; reproduced with permission.)

Subtropical Water are marked by dense swarms of the oceanic copepod *Calanus tonsus* and large numbers of whale sharks *Galcolamma* sp. Tranter (1962) has reported that salps reached swarm proportions off the Australian east coast most often in the region of mixing of warm (subtropical) and cool (subantarctic) waters. It is in the spring that these salp swarms dominate the plankton both on and off the shelf, coinciding with phytoplankton blooms (Dakin and Colfax, 1940; Humphrey, 1960) and increases in pelagic fish numbers. These increases coincide with the northward movement of subantarctic waters and their intrusion into the continental slope (Newell, 1961, Wyrtki, 1962). Bary (1959) and Grieve (1966) have reported similar salp swarms off the New Zealand east coast. As Tranter points out, it is important to establish the relationship between these phenomena and to determine whether there is a general increase in the crustacean biomass in the open ocean during this period.

Reid poses the question: "How much does the volume of zooplankton at any one place reflect that which can be sustained there by local growth and how much does it reflect the influx of plankton from other areas?" He points out that zooplankton volumes are consistently low in the large areas within the anticyclonic gyres and that it is inferred that the plankton is produced locally and represents an equilibrium between the nutrients available and the growth requirements of the species present. For many regions, however, the question cannot be readily answered. If large influxes occur in specific areas, primary productivity measurements will not give a true indication of the potential of the area to support fish populations.

MARINE RESOURCES

The South Pacific probably contains the largest unexploited marine resources of the world's oceans. These will be discussed under two headings: "Fish and Fisheries" and "Crustacean and Molluscan Resources."

Fish and Fisheries

Benthic or demersal fisheries in the South Pacific region are strictly limited in spite of the fact that it is the largest of the world's oceans. The total catch in 1966 was about 250,000 metric tons. The demersal fisheries of the world are concentrated in the shelf areas of temperate waters. Such shelf areas are strictly limited in the South Pacific, with restricted narrow continental shelves along the margins of the major land areas, eastern Austrialia and western South America. The most extensive shelf area, which is found around the coasts of New Zealand, yielded some 31,000 metric tons of fish in 1965. There are indications that this yield could probably be doubled. In addition to this shelf area, there are considerable areas of moderately deep water (up to 1,000 m) on the Challenger Plateau, the Chatham Rise, and the Campbell Plateau (Figure 11). These waters are unexploited and unexplored, and this would require larger vessels and more sophisticated techniques than are generally available in New Zealand. With the meager data available, it is impossible to estimate the fishing potential of these waters. The potential may be great, especially on the Chatham Rise and the Campbell Plateau, which lie in regions of moderately high productivity of surface waters and are known to support comparatively rich benthic faunas (Knox, 1957; Dawson, 1965, 1968). It is clear that exploration of these areas, especially the Campbell Plateau, would be worthwhile.

It is in the field of pelagic fisheries, however, that there are considerable prospects for expansion. Such fisheries in the South Pacific are based largely on small clupeoid plankton-feeding fishes and large predatory tunas. On the basis of available information, it would appear that the stocks of both these groups are underexploited in the South Pacific as a whole.

The most extensive clupeoid fishery is that centered on the Peruvian anchovy (*Eugraulus ringens*), with a production in 1966 of 8.5 million metric tons. In addition to this harvest by man, it had been estimated in 1956 that at least 4.3 million tons were removed by the cormorants, pelicans and other guano-producing birds (Wooster and Reid, 1963). Thus it can be estimated that at least 12 million tons are removed in a year from the inshore waters of Peru by birds and man; this is equivalent to about 24 percent of the total world fish landings (Food and Agricultural Organization, 1966). As Wooster and Reid (1963) point out, these fish are caught on a coastal strip less than 200 miles long and 30 miles wide, covering about 0.2 percent of the surface of the world's oceans, which attests to the great fertility of the Peru Current system, a feature of eastern boundary currents in general. Throughout the rest of the South Pacific there is relatively little exploitation of stocks of clupeoid fishes, although considerable populations are known to exist and may be much more extensive than hitherto reported. Off the New South Wales coast a standing stock of pilchards (*Sardinopsis neopilchardus*) of up to 50,000 tons has been estimated (Blackburn, 1960; Blackburn and Downie, 1955). In New Zealand waters, pilchards (*Sardinopsis neopilchardus*) and sprats (*Maugaclupea antipodum*), and to a lesser

degree, anchovies (*Austranchovia australes*), occur in quantity, although there are no detailed accurate quantitative estimates of their abundance.

In the southwest Pacific, apart from tunas, stocks of predatory fishes such as barracouta (*Leinoura atun*) and mackerel (*Trachurus* spp.) are very lightly exploited. Standing stock estimates of 100,000 tons for the latter species off Tasmanian coasts (Hynd and Robins, 1967) are likely to be underestimates. It is probable that comparable stocks are to be found on New Zealand shores because large shoals have been reported from time to time.

Throughout the South Pacific there are extensive tuna fisheries, the most important forming part of an extensive fishery embracing the whole of the central Pacific. As Manar (1966) has pointed out, it was not very many years ago that some responsible scientists regarded the central Pacific Ocean as a marine desert, whereas it has been shown to support a fishery resource of immensely large size. Because a later contribution to this volume (Kasahara, page 252) will deal in detail with the tuna resources of the South Pacific, I propose to consider only a few points of general interest.

In all probability, southern bluefin tuna (*Thunnus thynnus*) represents the least exploited of the tuna stocks. This species occurs between 30° and 45°S in waters off Chile, New Zealand (McKenzie, 1961), and the eastern and southern coasts of Australia (Blackburn, 1965). There is a moderate fishery for the species in Australian waters, the catch in 1966 being about 6,600 metric tons, but it is little-exploited elsewhere. It is believed that comparable stocks to the Australian ones are to be found in the Subtropical Convergence regions of the New Zealand southern coasts. Tagging experiments have revealed movements of bluefin across the North Atlantic and North Pacific (Blackburn, 1965). Similar movements probably occur in the South Pacific.

Albacore (*Thunnus alalunga*) form the basis of a long-line fishery that extends from the east coast of Australia to about 85°W at the eastern end of the fishery and from the equator to about 45°S at the western end. In the North Pacific, albacore are known to engage in transoceanic migrations, but no information is available concerning the migration of the South Pacific stocks. Otsu and Yoshida (1967) state that from a knowledge of the North Pacific albacore it may be postulated that the juvenile component of the South Pacific population migrates between the coastal waters of Australia and Chile and that it is possible that the volume of movement in an easterly direction is greater in the South Pacific than in the North Pacific. A study of such movements in the two species discussed above and in the other species found in the South Pacific would prove rewarding.

Many of the problems associated with tuna in the South Pacific are part of broader studies on tuna biology for the Pacific as a whole. For example, there is the general question of the existence of genetically distinct subpopulations

(Marr and Tester, 1966). A possible approach to the study of this problem is through investigations of inherited blood-group systems and other genetically controlled blood characteristics. Carefully designed tagging experiments would also be necessary to determine rates of exchange and migratory rates.

One of the areas requiring intensive study is the relationship between tuna and oceanic "fronts." Blackburn (1965) has recently described such phenomena as follows:

Fronts are widely considered to be very important in the ecology of tunas and other pelagic animals but they are seldom well defined and the effects on the biota are rather poorly understood. . . .Fronts are probably best considered as boundaries (lines of convergence) between surface waters of different densities, recognisable by strong horizontal gradients of temperate and/or salinity and accompanied by some sinking of one or both of the types of water involved. Drifting or weakly swimming biota (zooplankton) are believed to be carried towards a front from one or both sides and aggregated there and strongly swimming biota (nekton) are thought to move into the front to feed on the aggregations.

Fronts are, however, intermittent phenomena, which makes observational programs difficult. Such fronts have been described by Beebe (1926, Ch. 2), Cromwell (1953), Cromwell and Reid (1956), and Knauss (1957). However, the relationship between tuna and such fronts is not yet adequately documented. Murphy and Shomura (1968) have suggested that the abundance of tuna schools in certain zones of the open ocean may be explained by their association with fronts and also that phenomena similar to fronts may occur close to islands. As Ashmole and Ashmole (1967) put it, the overriding importance of fronts in producing local concentrations of forage animals probably accounts for the fact that, although zooplankton abundance in general is higher between the equator and 5°N (see Figure 15) than between 5° and 10°N and 0° and 10°S (King and Hida, 1957), surface schools of tuna are much more abundant in the latter zones. Murphy and Ikehara (1955) reported half as many schools of tuna in the latitudinal zone between 0° and 5°N as in the Equatorial Countercurrent (5°-10°N) and the South Equatorial Current (0° to 10°S). Ashmole and Ashmole also point out the importance of these fronts in providing favorable feeding grounds for marine birds. Schools of tuna and other pelagic fish are generally accompanied by flocks of seabirds. Many of these birds (especially tern species) are dependent on the presence of marine predators, such as tunas, to make potential prey available at the surface.

In two recent papers on studies of the feeding ecology of tropical birds on Christmas Island, Ashmole and Ashmole (1967, 1968) have drawn attention to the intriguing possibilities in using food samples from seabirds in the study of seasonal variations in the surface fauna of tropical oceanic areas. They point out that

. . . because of the difficulty of sampling mobile and patchily distributed animals and the cost of oceanographic investigations, few data are available on the extent of seasonal change. By regularly collecting regurgitations from sea birds, and identifying and measuring food items seasonal data could be obtained on the availability, size classes and perhaps reproductive cycles of the fish and squid characteristic of the surface layer of tropical seas. Flying fish (Exocoetidae), juvenile tunas (Scombridae) and squid of the family Ommastrephidae are especially easily obtainable.

It is of interest that few representatives of these groups were recorded in attempts to sample forage animals near the surface of the central Pacific by means of oblique trawls with Isaacs-Kidd and other trawl nets (King and Iversen, 1962). It is clear from the work of Ashmole and Ashmole that these groups are more efficiently sampled by seabirds. Of special importance is the effectiveness of these tropical seabirds in catching juvenile tunas. King and Iversen (1962, p. 301) said:

It was our hope that by means of mid-water trawls we would capture juvenile tunas of lengths above 12 mm which were able to elude the plankton nets. In 274 hauls made with the four mid-water trawls described in this report we captured only six juvenile tunas, which ranged from 18 to 60 mm in length.

In contrast, in 800 food samples from Christmas Island, Ashmole and Ashmole found 247 young tunas, of which 166 were in 243 samples from the sooty tern (*Sterna fuscata*). Of the latter group, 77 were examined in detail and proved to be Yellowfin Tuna. Since different species of birds range different distances from their island bases, from close inshore to hundreds of miles, large areas of ocean could be sampled. For inshore waters, sampling from seabirds could be supplemented by sampling with a neustron net and carrying out oceanographic observations from a small boat. Investigations of this kind could be carried out economically on any of a large number of tropical islands. The Kermadec Islands northeast of New Zealand would be an excellent site for such investigations. Such studies together with observations on tuna and the contemporaneous standing crop of food organisms in relation to fronts should add much to our understanding of tuna biology.

Crustacean and Molluscan Resources

The principal crustacean fisheries are for prawns, with a total catch of 35,200 metric tons in 1966, 76 percent of which was taken in Chilean waters; for crabs, largely from southern Chile, with a catch of 1,300 metric tons; and for spiny lobsters, with a total catch of about 12,000 metric tons. The lobster fishery is a very important one in terms of monetary value, and it presents a number of interesting problems. These center on the southern spiny lobsters (locally called crayfish) of the *Jasus "lalandei"* complex. Until recently, the general view was that there was a single species with a circum-southern distribution in cold waters. In 1963 Holthius recognized six geographically isolated species, three of which occur in the South Pacific: *J. novaehollandiae* from southeast Australia and Tasmania, *J. edwardsii* from New

Zealand, and *J. frontalis* from Juan Fernandez (Figure 16). These three species form the basis of a substantial fishery (Table 11). The fishing of a previously unexploited population of *J. edwardsii* near the Chatham Islands commenced in 1964 and has shown a spectacular increase, the landed catch increasing from about 8,000 cwt in 1965 to about 39,000 cwt in 1966 (Figure 17). There are indications that this figure will treble for 1967, an amount exceeding the total catch from the New Zealand mainland in 1964 (Kensler, 1968).

One of the unsolved problems of the biology of these southern spiny lobsters concerns duration of the larval life and the movements of larvae during this period. Batham (1967) has recently studied the length of larval life in laboratory conditions of the Stage I phyllosoma of *J. edwardsii* and found it to be 3 weeks. By comparison with information on the larval life of other palinurids (Johnson, 1960; Saisho, 1962), she suggests that the total duration is a period of 9 to 12 months or more. It is generally believed that the

TABLE 11   1966 Catch of Southern Spiny Lobsters

Southeastern Australia	New Zealand	South America
11,650,000 lb	14,445,872 lb	(?) 660,000 lb
5,300 metric tons	6,555 metric tons	(?) 300 metric tons

Totals   27,750,000 lb (approx.)
12,155 metric tons (approx.)

phyllosoma larvae are planktonic, but it has been suggested (Johnson, 1960; Batham, 1967) that they may be at least partially benthic in their habits. If they are pelagic, they could be carried for considerable distances during the larval period, a possibility that presents two questions. First, how do they return to the shallow-water benthic habitats where

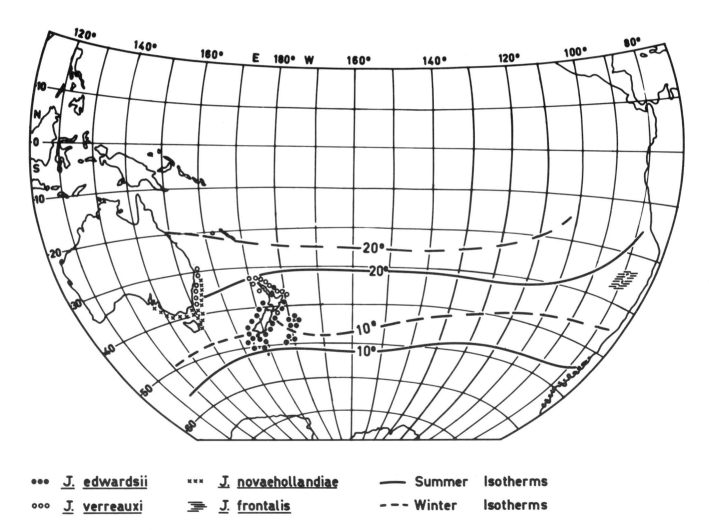

| ●●● | *J. edwardsii* | ××× | *J. novaehollandiae* | —— | Summer   Isotherms |
| ○○○ | *J. verreauxi* | ≡ | *J. frontalis* | - - - | Winter   Isotherms |

FIGURE 16   Distribution of the spiny lobster genus *Jasus* in the South Pacific. The 20° and 10°C summer and winter isotherms are shown.

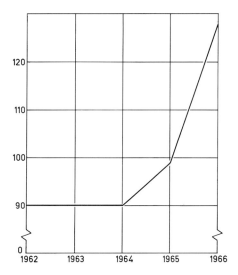

FIGURE 17 New Zealand crayfish production, 1962–1966.

they settle as transparent puerulus larvae (Kensler, 1967a), and second, how are the separate species populations maintained when a 12-month period is more than ample for larvae to drift considerable distances? Another mystery is the absence of *J. frontalis* from the mainland coast of South America. Temperature conditions would appear to be well within the range of tolerance of the South Pacific species in general. From Figure 16 it can be seen that apart from *Jasus verreauxi*, which is a warm-temperate species found in northern New Zealand and the coasts of New South Wales (Kensler, 1967b), they are, in general, limited to the north by the 20°C summer isotherm and to the south by the 10°C winter isotherm. The South American coast between about 55° and 15°S falls within these limits. Ecologically, the shallow waters off these coasts are similar to comparable waters in Australia and New Zealand, hence food supply is not likely to be a limiting factor. Information gained from a thorough study of the problems discussed above could, perhaps, lead to information that could lead to the stocking of areas that do not presently support populations of spiny lobsters. Kensler (1967a) has shown that it is possible to rear post-puerulus specimens of *Jasus edwardsii* successfully for periods up to 12 months. Further development of his technique on a large scale raises the possibility of the release of specimens at an age of 12 months or older when the normal early period of heavy mortality of young is past.

The principal molluscan resources are oysters, with a total production of 1,830 metric tons in 1966, and mussels, with a production in the same year of 29,500 metric tons (see Table 12). There are considerable areas that could be used for oyster cultivation. For example, oyster farming on a limited scale has only recently commenced in New Zealand. Apart from the northern coasts of Chile and localized areas in Australia and New Zealand, mussels, chiefly the

various subspecies of *Mytilus edulus*, are underexploited, and a considerable increase with possibilities of cultivation could be expected.

There are many other problems that I could have touched on, but most of these are general problems of the tropical and equatorial regions of the Pacific as a whole. For example, I have not mentioned the question of coral-reef biology; the majority of the detailed studies have been carried out in the North Pacific, and comparative studies in the South Pacific are required.

THE URGENT PROBLEMS FOR THE FUTURE

Based on this survey, the following are suggested as the most urgent and challenging problems on the biological oceanography of the South Pacific:

1. A detailed investigation of the Subtropical Convergence and the areas immediately north and south to determine the influence it has on biological distributions and productivity. Particular attention would have to be paid to seasonal observations.

2. Arising out of this, an intensive study of the biological characteristics of the eastern boundary currents, with particular reference to the Peru Current system, and directed especially toward an understanding of the factors involved in the high productivity of the waters and the biological and chemical conditions associated with the occurrence of El Niño.

3. An understanding of the relationship of the distribution of planktonic plants and animals and plankton communities to the major water masses and their boundaries (convergences and divergences). The factors underlying differences and similarities to North Pacific patterns should be explored, and particular attention should be paid to comparative studies of species populations and related species occurring in both oceans.

4. Systematic benthic investigations along the lines outlined above.

TABLE 12 1966 South Pacific Production of Oysters and Mussels (amounts in metric tons)

Area	Oysters	Mussels
Peru	700	4,400
Chile	900	22,700
New Zealand	6,800	2,200
Australia	9,900	200
Totals	18,300	29,500

5. Studies of tuna biology aimed at an understanding of population structure, migration patterns, and the factors involved in bringing about feeding concentrations.

ACKNOWLEDGMENT

The assistance of D. Cameron, Department of Zoology, University of Canterbury, in the preparation of the figures is gratefully acknowledged.

REFERENCES

Alvariño, A. (1962) Two new Pacific chaetognaths; their distribution and relationship to allied species. *Bull. Scripps Inst. Oceanogr., 8*, 1–50.

Alvariño, A. (1964) Bathymetric distribution of chaetognaths. *Pacif. Sci., 18*(1), 64–82.

Alvariño, A. (1965) Chaetognaths. *Oceanogr. Mar. Biol., 3*, 115–194.

Anonymous (1967) *Eltanin* Cruises 22–25. *Antarctic J. U.S., 2*(2), 41–44.

Anonymous (1968) New mountains in the South Pacific. *New Sci., 37*(584), 375.

Ashmole, N. P., and M. J. Ashmole (1967) Comparative feeding ecology of sea birds of a tropical ocean island. *Bull. Peabody Mus., 24*, 1–131.

Ashmole, N. P., and M. J. Ashmole (1968) Food samples from sea birds as indicators of changes in ocean surface fauna. *Pac. Sci., 12*(1), 1–10.

Batham, E. J. (1967) The first three larval stages and feeding behaviour of phyllosoma of the New Zealand palinurid crayfish *Jasus edwardsii* (Hutton, 1875). *Trans. Roy. Soc. N.Z., Zool., 9*(6), 53–64.

Bary, B. M. (1959) Species of zooplankton as a means of identifying different surface waters and demonstrating their movements and mixing. *Pac. Sci., 13*(1), 14–54.

Beebe, W. (1926) *The Arcturus Adventure*, G. P. Putman and Sons, New York and London, 493 pp.

Bekker, V. E. (1964) The moderate cold-water Myctophidae Pisces complex. *Okeanologiia, 4*(3), 469–476.

Berner, D. L. (1960) Unusual features of the distribution of pelagic tunicates in 1957 and 1958. *Rep. Calif. Coop. Oceanic Fish. Invest., 7*, 133–136.

Bieri, R. (1959) The distribution of planktonic Chaetognaths in the Pacific and their relationship to the water masses. *Limnol. Oceanogr., 4*, 1–28.

Blackburn, M. (1960) Synopsis of biological information on the Australian and New Zealand sardine, *Sardinops neopilchardus* (Steindacker), pp. 245–264 in: *Proceedings of World Scientific Meeting on the Biology of Sardines and Related Species*, Rome, September 14–21, 1959, editors H. Rosa, Jr., and G. Murphy, vol. 2. Species synopses. Subject synopses; issued also as FAO *Fish Bull. Synops.* No. 12.

Blackburn, M. (1965) Oceanography and ecology of tunas. *Oceanogr. Mar. Biol., 3*, 299–322.

Blackburn, M., and R. Downie (1955) The occurrence of oily pilchards in New South Wales waters. *Tech. Pap. Div. Fish. Oceanogr. C.S.I.R.O., 3*, 1–11.

Boden, B. P. (1954) The euphausiid crustaceans of southern South Africa waters. *Trans. Roy. Soc. S. Afr. 34*(1), 181–243.

Bogorov, V. G., and M. E. Vinogradov (1960) Distribution of zoo-plankton biomass in the central part of the Pacific Ocean. *Trans. Hydrobiol. Soc. U.S.S.R., 10*, 208–223.

Bradshaw, J. S. (1959) Ecology of living planktonic foraminifera in the North and Equatorial Pacific Ocean. *Contr. Cushman Found., 10*, 25–64.

Brinton, E. (1962) The distribution of Pacific euphausiids. *Bull. Scripps Inst. Oceanogr., 8*, 51–270.

Brodie, J. W. (1960) Coastal surface currents around New Zealand. *N.Z. J. Geol. Geophys., 3*, 235–252.

Brodie, J. W. (1965) Oceanography. In: *Antarctica*, T. Hatherton, editor. Methuen and Co., Ltd., London.

Brodskiy, K. A. (1961) Comparison of *Calanus* species (Copepoda) from the southern and northern hemispheres. *Inf. Ser. Dep. Sci. Ind. Res. N.Z., 33*, 1–22.

Brodskiy, K. A. (1962a) The species and distribution of Calanoidea in northwestern Pacific surface waters. *Issled. dal'nevost. Mor. SSSR, 8*.

Brodskiy, K. A. (1962b) On biogeographical division of pelagical zones of the Southern Hemisphere according to the distribution of *Calanus* species (Copepoda). *Polar Rec., 11*(72), 325–342.

Brodskiy, K. A. (1964) Plankton studies by the Soviet Antarctic Expedition (1955–58). *Sov. Antarct. Exped. Inf. Bull., 1*, 105–110. (Engl. trans. from *Sovet Antarkiticheskaia Eksped; Inform. Biull., 3*, 25–30).

Brodskiy, K. A. (1965) The taxonomy of marine plankton organisms and oceanography. *Oceanology, 5*(4), 1–11. (Engl. trans. from *Okeanologiia, 5*(4), 577–591).

Burkenholder, P. R., and L. M. Burkenholder (1967) Primary productivity in the surface waters of the South Pacific. *Limnol. Oceanogr., 12*(4), 606–617.

Burkov, V. A. (1966) Structure and nomenclature of Pacific Ocean currents. *Oceanology, 6*(1), 1–10. (Engl. trans. from *Okeanologiia, 6*(1), 3–14).

Carriker, M. R. (1967) Ecology of benthic invertebrates: A perspective, pp. 442–487 in: *Estuaries*, editor George H. Lauff. Amer. Ass. Adv. Sci., Publ. 83, Washington, D.C., 757 pp.

Cromwell, T. (1953) Circulation in a meridional plane in the Equatorial Pacific. *J. Mar. Res., 12*, 196–213.

Cromwell, T., and J. L. Reid, Jr. (1956) A study of oceanic fronts. *Tellus, 8*, 94–101.

Dakin, K. W., and A. N. Colfax (1940) The plankton of the Australian coastal waters off New South Wales. *Monogr. Dep. Zool. Univ. Sydney, 1*, 1–215.

Dales, D. P. (1957) Pelagic polychaetes of the Pacific Ocean. *Bull. Scripps Inst. Oceanogr., 7*, 99–168.

David, P. M. (1958) The distribution of Chaetognaths of the Southern Ocean. *Discovery Rep., 29*, 199–228.

Dawson, E. W. (1965) Oceanography and marine zoology of the New Zealand Subantarctic. *Proc. N.Z. Ecol. Soc., 12*, 44–57.

Dawson, E. W. (1968) The benthos of the New Zealand shelf and slope: an ecological and zoogeographical analysis. Paper read to Zoology Section, 40th ANZAAS Congress, Christchurch, N.Z.

Deacon, G. E. R. (1957) The hydrology of the Southern Ocean. *Discovery Rep., 15*, 1–123.

Deacon, G. E. R. (1963) The Southern Ocean, pp. 281–296 in: *The Sea:* Ideas and Observations on Progress in the Study of the Seas, vol. 2, editor M. N. Hill. Wiley-Interscience, New York, 554 pp.

Dell, R. K. (1959) The archibenthal Mollusca of New Zealand. *Bull. Dom. Mus. Wellington, 18*, 1–25.

Dell, R. K. (1965) Marine biology, pp. 129–152 in: *Antarctica*, editor T. Hatherton, A. H. and A. W. Reed, Wellington, 511 pp.

Ekman, S. P. (1953) *Zoogeography of the Sea*. Sidgwick and Jackson, London. 417 pp.

El-Sayed, S. Z. (1967) Biological productivity investigations of the Pacific sector of Antarctica. *Antarc. J. U.S., 2*(5), 200–201.

Fairbridge, R. W. (1966) Pacific Ocean: Introduction, pp. 653–658 in: *Encyclopaedia of oceanography*, editor R. W. Fairbridge, Reinhold, New York. 1021 pp.

FAO (1966) Yearbook of Fisheries Statistics. Rome.

Foxton, P. (1961) *Salpa fusiformis* Cuvier and related species. *Discovery Rep., 32*, 1–32.

Fell, H. B. (1957) Report on the Echinoderms, p. 33 in: Knox, G. A. "General account of the Chatham Islands 1954 Expedition." *Bull. N.Z. Dep. Sci. Ind. Res.*, 122 pp.

Fell, H. B. (1960) Archibenthal and littoral echinoderms. *Bull. N.Z. Dep. Sci. Ind. Res., 139*(2), 55–75.

Garner, D. M. (1958) The Antarctic Convergence south of New Zealand. *N.Z. J. Geol. Geophys., 1*, 577–594.

Garner, D. M. (1959) The Subtropical Convergence in New Zealand waters. *N.Z. J. Geol. Geophys., 2*(2), 315–337.

Garner, D. M. (1962) Analysis of hydrological observations in the New Zealand region. *Bull. N.Z. Dep. Sci. Ind. Res., 144*, 1–45.

Garner, D. M., and N. M. Ridgway (1965) Hydrology of New Zealand offshore waters. *Bull. N.Z. Dep. Sci. Ind. Res., 162*, 1–62.

Graham, H. W. (1941) Plankton production in relation to water character in the open Pacific. *J. Mar. Res., 4*, 189–197.

Grieve, J. A. (1967) The annual cycle of plankton off Kaikoura. Unpubl. Ph.D. thesis. University of Canterbury, Christchurch, New Zealand.

Gunther, E. R. (1936) A report on oceanographic investigations in the Peru Current. *Discovery Rep., 14*, 109–278.

Hedgpeth, J. W., Editor. (1957) Marine biogeography. In *Treatise on marine ecology and paleoecology*. Vol. 1. Ecology. *Mem. Geol. Soc. Amer., 67*(1), 461–534.

Hedley, C. (1905) Mollusca from one hundred and eleven fathoms, East of Cape Byron, New South Wales. *Rec. Aust. Mus., 6*, 41–54.

Hedley, C. (1907) The results of deep-sea investigations of the Tasman Sea: II. The Expedition of the *Way Way* 2; Mollusca from eight hundred fathoms. Thirty-five miles east of Sydney. *Rec. Aust. Mus., 6*, 356–364.

Hedley, C., and W. L. May (1908) Mollusca from one hundred fathoms, seven miles east of Cape Pillar, Tasmania. *Rec. Aust. Mus., 7*, 108–125.

Hedley, C., and W. F. Pettard (1906) Mollusca from three hundred fathoms, off Sydney. *Rec. Aust. Mus., 6*, 211–225.

Holmes, R. W., M. B. Schaefer, and B. M. Shimada (1957) Primary production, chlorophyll and zooplankton volumes in the tropical eastern Pacific. *Bull. Inter-Amer. Trop. Tuna Comm., 2*(4), 129–156.

Holthius, L. B. (1963) Preliminary description of some new species of Palinuridae (Crustacea, Decapoda, Macrura, Reptantia). *Proc. Ned. Akad. Wet.*, Ser. C., *66*, 54–60.

Humphrey, G. (1960) The concentrations of plankton pigments in Australian waters. *Tech. Pap. Div. Fish. Oceanogr., C.S.I.R.O., 9*.

Hynd, J. S., and J. P. Robins (1967) Tasmanian tuna survey report of first operational period. *Tech. Pap. Div. Fish. Oceanogr. C.S.I.R.O., 22*.

Ivanov, Y. U. A. (1961) Frontal zones in Antarctic waters. *Okeanol. Issled., 3*, 30–51.

John, D. D. (1936) The southern species of the genus *Euphausia*. *Discovery Reps., 14*, 165–180.

Johnson, M. W. (1938) Concerning the copepod *Eucalanus elongatus* Dana and its varieties in the northeast Pacific. *Bull. Scripps Inst. Oceanogr., 4*, 165–180.

Johnson, M. W. (1960) Production and distribution of larvae of the spiny lobster, *Palinurus interruptus* (Randall), with notes on *Palinurus gracilis* Streets. *Proc. Calif. Acad. Sci., 29*(1), 1–19.

Johnson, M. W., and E. Brinton (1963) Biological species, water-masses and currents, pp. 381–414 in: *The Sea: Ideas and Observations on Progress in the Study of the Seas*, editor N. M. Hill, Vol. 2, Wiley-Interscience, New York. 554 pp.

Kensler, C. B. (1967a) Notes on the laboratory rearing of juvenile spiny lobsters, *Jasus edwardsii* (Hutton). *N.Z. J. Mar. Freshwater Res., 1*(1), 71–75.

Kensler, C. B. (1967b) The distribution of spiny lobsters in New Zealand waters (Crustacea, Decapoda, Palinuridae). *N.Z. J. Mar. Freshwater Res., 1*(4), 412–420.

Kensler, C. B. (1968) Commercial landings of the spiny lobster *Jasus edwardsii* at the Chatham Islands. *N.Z. J. Mar. Freshwater Res., 2*.

King, J. E., and T. S. Hida (1957) Zooplankton abundance in the Central Pacific. *Fish. Bull. U.S., 54*(82), 111–144.

King, J. E., and T. S. Hida (1957) Zooplankton abundance in the Central Pacific Pt. II. *Fish. Bull. U.S., 57*(118), 361–395.

King, J. E., and R. T. B. Iverson (1962) Midwater trawling for forage organisms in the Central Pacific. *Fish. Bull. U.S. 62*(210), 271–321.

Knauss, J. A. (1957) An observation of an oceanic front. *Tellus, 9*, 234–237.

Knauss, J. A. (1960) Measurements of the Cromwell Current. *Deep-Sea Res., 6*, 265–286.

Knauss, J. A. (1963) Equatorial current systems, pp. 235–252 in: *The Sea: Ideas and Observations on Progress in the Study of the Seas*, editor M. N. Hill, Vol. 2. Wiley-Interscience, New York. 554 pp.

Knox, G. A. (1957) General account of the Chatham Islands 1954 Expedition. *Bull. N.Z. Dep. Sci. Ind. Res., 122*.

Knox, G. A. (1960) Littoral ecology and biogeography of the southern oceans. *Proc. Roy. Soc. (B), 152*, 577–624.

Knox, G. A. (1963a) The biogeography and intertidal ecology of Australian coasts. *Oceanogr. Mar. Biol., 1*, 341–404.

Knox, G. A. (1963b) Problems of speciation in intertidal animals with species reference to New Zealand shores, pp. 7–30 in: *Speciation in the Sea*, editors J. P. Harding and N. Tebble, Systematics Assoc. Publ. 5.

Knox, G. A. (1968) Tides and intertidal zones. *Proc. Sym. Antarc. Oceanogr.*, Santiago, Chile, Sept. 13–16, 1966.

Koblentz-Mishke, O. I. (1965) Primary production in the Pacific. *Oceanology, 5*(2), 104–116. (Translated from *Okeanologiia, 5*(2), 325–337.)

Kort, V. G., E. S. Korotkevich, and V. G. Ledenev (1965) Boundaries of the Southern Ocean. *Sovet Antarkicheskaia Eksped., Inform. Bull., 50*, 5–7. (Translated in *Sov. Antarct. Exped. Inf. Bull., 5*(4), 261–263.)

Legand, M. (1961) Quelques résultats et problèmes de la biologie des eaux pélagiques dans le Pacifique Sud. B. L'estimation de la richesse en zooplankton. *Cahier Pac., 3*.

Mackintosh, N. A. (1946) The Antarctic Convergence and the distribution of its surface temperatures in Antarctic waters. *Discovery Rep., 23*, 177–212.

Mackintosh, N. A. (1960) The patterns of distribution of Antarctic fauna. *Proc. Roy. Soc. (B), 52*, 577–624.

Manar, T. A. (1966) Central Pacific Fisheries Resources. An introduction, pp. 1–12 in: *Proceedings of the Governor's Conference on Central Pacific Fishery Resources*, Honolulu, Hawaii, editor T. A. Manar.

Marr, J. C., and A. L. Tester (1966) Report on the working group on research program, pp. 44–50 in: *Proceedings of the Governor's Conference on Central Pacific Fishery Resources*, Honolulu, Hawaii, editor T. A. Manar.

McGowan, J. A. (1963) Geographical variation in *Limacina helicina*, in: *Speciation in the Sea*, editors J. P. Harding and N. Tebble, Systematics Assoc. Publ., No. 5, pp. 109–178.

McKenzie, M. L. (1961) A review of the present knowledge relative to a possible tuna fishery in New Zealand. *N.Z. Mar. Dept. Fish Tech. Rep., 4*, 1–49.

Menard, H. W. (1959) Geology of the Pacific sea floor. *Experimentia, 15,* 205–213.

Murphy, G. A., and I. I. Ikehara (1955) A summary of sightings of fish schools and bird flocks and of trolling in the central Pacific. *Spec. Sci. Rep. U.S. Fish Wildlife Serv., Fish., 154,* 1–19.

Murphy, G. I., and Shomura (1968) The abundance of tunas in the central equatorial Pacific in relation to the environment. Unpubl. ms. quoted in Ashmole and Ashmole (1967).

Newell, B. S. (1961) Hydrology of south-east Australian waters. *Tech. Pap. Div. Fish. Oceanogr. C.S.I.R.O., 10.*

Otsu, T., and H. O. Yoshida (1967) Distribution of albacore (*Thunnus alalunga*) in the Pacific Ocean. *Proc. Indo-Pac. Fish. Council, 12*(2), 49–64.

Ponamareva, L. A. (1966) Quantitative distribution of Pacific Euphausiids. *Oceanology, 6*(4), 564–566. (Translated from *Okeanologiia, 6*(4), 690–692.)

Ponamareva, L. A., and E. A. Lubini (1958) The quantitative distribution of plankton in the tropical waters of the western Pacific Ocean. *Dokl. Nauk SSSR, 120*(6), 1246–1248.

Radzikhovskaya, M. A. (1965) Volumes of main water masses in the South Pacific. *Oceanology, 5*(5) 29–32. (Translated from *Okeanologiia, 5*(5), 803–805.)

Reid, J. L., Jr. (1962) On the circulation, phosphate-phosphorus content and zooplankton volumes in the upper part of the Pacific Ocean. *Limnol. Oceanogr., 7,* 287–306.

Reid, J. L., Jr. (1966) Pacific Ocean. Physical oceanography, pp. 660–665 in: *Encyclopaedia of Oceanography,* editor R. W. Fairbridge, Reinhold, New York. 1021 pp.

Rotschi, H., and L. Lemasson (1967) Oceanography of the Coral and Tasman seas. *Oceanogr. Mar. Biol., 5,* 49–98.

Ryther, J. H. (1963) Geographic variations in productivity, pp. 347–388 in: *The Sea:* Ideas and Observations on Progress in the Study of the Seas, editor M. N. Hill, Vol. 2, Wiley-Interscience, New York, 554 pp.

Ryther, J. H., and C. S. Yentsch (1958) Primary production of continental shelf waters off New York. *Limnol. Oceanogr., 3,* 327–335.

Saisho, T. (1962) Notes on the early development phyllosoma of *Palinurus japonicus. Mem. Fac. Fish. Kagoshina Univ., 11*(1), 18–23.

Sandved, K. G. (1966) USNS *Eltanin:* Four years of research. *Antarc. J. U.S., 1*(4), 164–174.

Schweigger, E. (1958) Upwelling along the Coast of Peru. *J. Oceanogr. Soc. Japan, 14*(3), 87–91.

Schott, G. (1931) Der Peru-Strom und seine nordlichen Nachbargebiete in normaler und anormaler Ausbildung. *Ann. Hydrogr. Mar. Meteorol., 59,* 161–169, 200–213, 240–252.

Sette, O. E. (1955) Consideration of fish productivity in relation to oceanic circulatory systems. *J. Mar. Res., 14,* 398–414.

Sheard, K. (1953) Taxonomy distribution and development of the Euphausidacea (Crustacea). *B.A.N.Z. Antarc. Res. Exped., 1929-31. Rep. (B), Zool. Bot., 8*(1), 1–72.

Shilling, C. W., and J. W. Tyson (1964) *Aquatic Biologists of the World and Their Laboratories.* Part I. Biological Sciences Communication Project. George Washington University, Washington, D.C.

Steeman Nielsen, E., and E. A. Jensen (1957) Primary organic production: The autotrophic production of organic matter in the oceans. *Galathea Rep., 1,* 49–136.

Shokal'skiy, Yu. M. (1959) *Oceanography,* Leningrad, Gidrometeoizdat.

Stepanov, V. N. (1965) Basic types of water structure in seas and oceans. *Oceanology, 5*(5), 21–28. (Translated from *Okeanologiia, 5*(5), 793–802.)

Sverdrup, H. W., M. W. Johnson, and R. H. Fleming (1942) *The Oceans,* Prentice Hall, New York. 1087 pp.

Tebble, N. (1960) Distribution of pelagic polychaetes in the South Atlantic Ocean. *Discovery Rep., 30,* 161–300.

Tebble, N. (1962) The distribution of pelagic polychaetes across the North Pacific Ocean. *Bull. Brit. Mus. Nat. Hist., Zool., 7,* 371–492.

Thompson, J. M. (1948) *Pelagic Tunicates of Australia.* C.S.I.R.O. Aust. Handbook, J. J. Gauley, Govt. Printer, Melbourne.

Tranter, D. J. (1962) Zooplankton abundance in Australasian waters. *Aust. J. Mar. Freshwater Res., 13*(2), 106–142.

Tully, J. P. (1964) Oceanographic regions and processes in the seasonal zone of the North Pacific Ocean, pp. 68–84 in: *Studies on Oceanography,* editor Kozo Yoshida, University of Tokyo Press, 560 pp.

Vervoort, W. (1957) Copepods from Antarctic and Subantarctic plankton samples. *B.A.N.Z. Antarc. Res. Exped. 1929-31. Rep. (B), Zool. Bot., 3,* 1–60.

Viglieri, A. (1966) Oceans: Limits, definitions and dimensions, pp. 632–640 in: *Encyclopaedia of Oceanography,* editor. R. W. Fairbridge, Reinhold, New York, 1021 pp.

Voronina, N. M. (1966) Distribution of zooplankton biomass in the Southern Ocean. *Oceanology, 6*(6), 836–846. (Translated from *Okeanologiia, 6*(6), 1041–1054.)

Wooster, W. S. (1961) Yearly changes in the Peru Current. *Limnol. Oceanogr., 6*(2), 222–226.

Wooster, W. S., and T. Cromwell (1958) An oceanographic description of the eastern tropical Pacific. *Bull. Scripps Inst. Oceanogr. 7*(3), 169–282.

Wooster, W. S., and M. Gilmartin (1961) The Peru-Chile Undercurrent. *J. Mar. Res., 19*(3), 97–122.

Wooster, W. S., and J. L. Reid, Jr. (1963) Eastern boundary currents, pp. 253–280 in: *The Seas,* Observations and Ideas on Progress in the Study of the Seas, editor M. N. Hill, Vol. 2, Wiley-Interscience, New York, 554 pp.

Wyrtki, K. (1960) The Antarctic Circumpolar Current and Antarctic Polar Front. *Deut. Hydrogr. Zeitz., 13,* 153–174.

Wyrtki, K. (1962) The subsurface water masses in the Western South Pacific. *Aust. J. Mar. Freshwater Res., 13,* 18–48.

Wyrtki, K. (1966) Oceanography of the eastern Equatorial Pacific. *Oceanogr. Mar. Biol., 4,* 33–68.

Zullo, V. A., and W. A. Newton (1964) Thoracic Cirripedia from a southeast Pacific guyot. *Pac. Sci., 18*(4), 355–370.

Olga J. Koblentz-Mishke

Vadim V. Volkovinsky

Julia G. Kabanova

INSTITUTE OF OCEANOLOGY,
MOSCOW

PLANKTON PRIMARY PRODUCTION OF THE WORLD OCEAN

For humanity the most important characteristic of biological communities is their productivity. In marine communities it is possible to speak with some validity only of primary production, particularly of primary production of planktonic algae. The study of subsequent steps in marine productivity is inadequate, and to date, there are no generally accepted quantitative methods. At present the measurement of primary production is usually based on measurement of the intensity of photosynthesis of marine planktonic algae.

During the course of international expeditions of the last 15 years, extensive data have been collected concerning the primary production of different regions of the World Ocean. The data for the Pacific and Indian oceans have already been reviewed by Koblentz-Mishke (1965, 1967) and Kabanova (1968).

In this paper the primary production of the Atlantic Ocean and of the Antarctic and North Polar basins is examined in a general way for the first time. As will be shown later, not all the regions of the World Ocean are studied adequately with respect to primary production, so it is often necessary to estimate the level of production on the basis of indirect evidence.

The observations reviewed in this paper were made by most diverse modifications of the radiocarbon method. Different principles have been adopted by different authors to estimate productivity in various waters. This results in wide and systematic variations in the data from different regions, depending on the methods of collection and calculation of the raw data.

Therefore, the present attempt to summarize the available data on the production of the World Ocean is a very rough approximation of the actual situation in nature and will undoubtedly be subject to revision. To avoid making a general assessment, however, is not expedient because of the pressing demand to utilize these productivity data in evaluating the World Ocean as a potential source of food for the rapidly growing population of this planet. The intensive exploitation of commercial fisheries beyond their productivity (productivity is the increase of biomass in unit time) may bring about irrecoverable exhaustion of populations of food fish and commercially exploitable invertebrates and replacement of valuable marine organisms by others that are less valuable to mankind.

Although an immediate and direct relation between primary production of any particular oceanic region and its commercial fishery may be lacking, or not very marked, by knowing the level of primary production and the regularity of its distribution in space and time, we may be able to obtain some ideas about the regions of potential commercial fisheries as well as the relative characteristics of fish production; in combination with other data, this knowledge may considerably facilitate the investigations and general evaluation of the biological productivity of any given ocean. Finally, knowledge about the regularity of primary production helps exploitation on a scientific basis; for example, systematic marine fish farming, fish breeding and culture, and similar human activities in the marine environment can be conducted more successfully.

A summarization of the present-day knowledge of the distribution of the level of primary productivity of the oceans is only the first step in the generalization of the observed regularity of primary production. Further study involves evaluation of the effect of external environmental factors on the distribution of primary production, both in space and time.

Of course, the present work is not the first step toward evaluation of the productivity of the oceans. Such an attempt was first made by Riley (1944) and was based on a small body of data obtained by the oxygen method, which is not very sensitive and therefore is not sufficiently accurate. Riley evaluated the annual production of the World Ocean as 155×10^9 tons of carbon. This figure was critized by Steemann Nielsen and Jensen (1957) on the basis of their own measurements during the round-the-world *Galathea* expedition. The track of this expedition crossed the Pacific, Atlantic, and Indian oceans diagonally, without touching the Arctic and Antarctic waters. This was the first time that Steemann Nielsen employed the highly sensitive radiocarbon method. According to his calculations, the annual primary production is approximately 10 times lower than that estimated by Riley, or about $12-15 \times 10^9$ tons of carbon.

Later, Vinberg (1960) reviewed the results obtained by Steemann Nielsen and concluded that Steemann Nielsen had underestimated the quantity of primary production of the oceans. In a paper by Koblentz-Mishke (1965), the data for the Pacific Ocean were extrapolated to the entire World Ocean, and a value similar to that of Steemann Nielsen was obtained.

In the preparation of this paper, the authors have made use of almost all published material as well as extensive raw data. The magnitude thus obtained for the primary production of the World Ocean, and the geographical regularity of its distribution, includes much more extensive factual material than do any of the previous studies.

MATERIAL AND METHODS

The material of this paper includes data from more than 7,000 stations maintained during individual cruises and international expeditions. In the collection of these data, specialists from the Soviet Union, the United States, Japan, Australia, France, England, Denmark, Norway, the Philippines, India, Indonesia, Brazil, Argentina, South Africa, Nigeria, and the Pointe Noire Oceanographic Fisheries Center (Republic of Congo, Brazzaville) participated. Sampling of the ocean was irregular in these investigations; in some regions, especially those close to research centers, numerous measurements were made all year round, while in other regions, only occasional individual measurements were made. On the map (Figure 1) the latter areas are marked with dots. In the construction of isolines, indirect indices of primary production, e.g., distribution of phytoplankton biomass in the southern part of the Indian Ocean (Zernova, 1966), were sometimes used. Isolines for the southern part of the Pacific and Atlantic oceans were based on data from oxygen saturation of surface waters (Koblentz-Mishke, 1965; Volkovinsky and Fedosov, 1964).

The basic source of information for the Pacific and part of the Atlantic was the collection of primary production data by Doty and Capurro (1961). The paper by Kabanova (1961) serves as the source for Indian Ocean information, and, for the Atlantic and the inland seas, the list of papers that follows gives the main sources: Bessonov and Fedorov, 1965; Volkovinsky, 1966; Zaika *et al.*, 1967; Kljashtorin, 1960, 1962; Koblentz-Mishke *et al.*, 1968; Kondratieva and

Sosa, 1966; Kondratieva, 1968; Smirnova, 1959; Sorokin, 1960; Starodubzev, 1968a, b, c.; El-Sayed, 1966; Berge, 1958; Corlett, 1958; Curl, 1960; Currie, 1958; Cushing, 1958; Data report Equalant, 1964–1965; English, 1958; Fukase and El-Sayed, 1965; Hansen, 1961; Ketchum and Corwin, 1965; Mandelli and Orlando, 1966; Marshall, 1958; Menzel and Ryther, 1960; Oceanographic Research Institute, 1965; Richards, 1960; Riley, 1946; Ryther and Yentsch, 1958; Riley, 1956; Saijo and Kawashima, 1964; El-Sayed *et al.*, 1964; El-Sayed and Mandelli, 1965; Steele, 1958, 1964; Steemann Nielsen and Hansen, 1961; Thomas and Simmons, 1960. Papers on the Indian and Pacific oceans that were published later than those of Kabanova and Koblentz-Mishke are also mentioned.

The methods employed for the study of primary production were highly varied, which greatly hampered the summarization of results. The diversity of methodology peculiar to the measurement of production follows two distinct lines, concerning either the characteristic of the method or the scheme of measurement. Under the heading of method are the chemical and radiochemical treatments permitting the measurement of intensity of photosynthesis in each individual sample and the correction necessary in obtaining the value. Understanding the conditions of exposures of of the samples and the system of calculation allows us to estimate the production not only of the surface of the sea but also of the whole column. In this paper, data obtained by four direct methods were used. The majority of measurements were made with the help of the radiocarbon method suggested by Steemann Nielsen (1951). This method is now the most widely used one, and despite its shortcomings it can be considered the standard method. A few measurements in the Atlantic Ocean were made on the basis of concentration of chlorophyll *a* in the samples of water collected from different depths. This method of measurement of primary production suggested by Ryther and Yentsch (1957) is based on the use of the relation between assimilation coefficient and light penetrating to any particular depth in the sea.[*] Though such dependence is not constant, as shown by a large number of physiological experiments of many workers, the results obtained by the Ryther and Yentsch method agree well with the results obtained by the radiocarbon method.

A few productivity measurements, mainly in the northern Pacific and the continental shelf waters off the northwest African coast (Bessonov and Fedosov, 1965), were made by the oxygen bottle method. As a rule, this method gives a higher value than the radiocarbon method; in the oligotrophic waters of the open seas, this overestimation reaches the order of 1:2, while in the eutrophic regions it is

---

[*]The relationship of intensity of photosynthesis, expressed in mg of C per m^3 per hour, to concentration of chlorophyll *a*, expressed in mg per m^3 in the same sample.

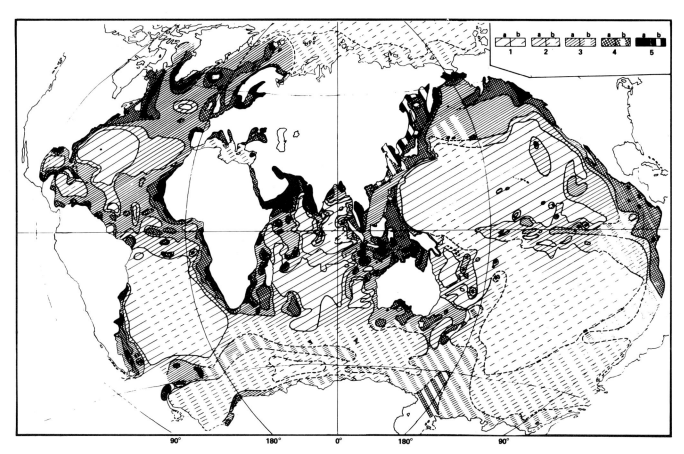

FIGURE 1   Distribution of primary production in the World Ocean. Units are in mg of C per m² per day. (1) Less than 100; (2) 100–150; (3) 150–250; (4) 250–500; (5) more than 500. *a* = data from direct ¹⁴C measurements; *b* = data from phytoplankton biomass, hydrogen, or oxygen saturation.

negligible. Therefore, in the present paper, for eutrophic waters, results obtained by the oxygen method were considered together with those taken by the ¹⁴C method. A few determinations were made in the Atlantic basin by Cushing (1958) and Kondratieva (1968) using various modifications of algological methods. According to Kondratieva, her results were closer to those of the radiocarbon method than to those of the oxygen method.

In principle, any of the above-mentioned methods may be combined with any scheme of determination for the water column. The simplest and most natural scheme is *in situ*; i.e., samples are taken from different depths and returned to the same depths for the period of exposure. Here, photosynthesis takes place under natural conditions of light and temperature. Unfortunately, during the complex conditions of oceanographic expeditions, frequent *in situ* experiments are not feasible because of their laborious and costly nature. Therefore, a number of alternate schemes have been suggested for obtaining measurements in the water column without *in situ* experiments, or with a minimum of them. The first such scheme was suggested by Steemann Nielsen and Jensen

(1957). It is based on the measurement of rate of photosynthesis in samples collected from different depths and exposed to constant conditions of artificial light during the period of experiment. Extensive application of this method has shown that it has a number of discrepancies, and the results differ widely from those obtained by *in situ* experiments. Hence, Steemann Nielsen himself gave up this scheme, but in its time the method was widespread and was used for the majority of studies of the Pacific and part of the Atlantic.

The second method of measurement of primary production in a water column under a unit area of sea surface was suggested by Sorokin (1959). This method is based on the measurement of photosynthesis of phytoplankton in samples of water taken from the surface of the sea, with subsequent correction for the change with depth. This correction depends on the unequal vertical distribution of phytoplankton and on the change with depth of the penetration of light energy. In fact, the chlorophyll method of Ryther and Yentsch does not differ from that of Sorokin. The heterogeneity of vertical distribution of phytoplankton was studied by them determining the concentration of chlorophyll, and the light correction was made in the same way as by Sorokin.

It has recently become evident that the dependence of photosynthesis of phytoplankton on light under *in situ* conditions is not as simple as thought earlier (Koblentz-Mishke *et al.*, 1966). Therefore, Sorokin's method needs further improvement.

The third scheme, which has had the greatest following during the last few years, consists of measurement of photosynthesis in samples taken from various depths and measured in natural light attenuated by light filters. The degree of attenuation of light by light filters corresponds to the extinction of light at the different depths in the sea from which the samples are taken. Almost all workers adopting this scheme, known as the "simulated *in situ* method," have used neutral filters for attenuation of light (Doty, 1962; Doty *et al.*, 1965), although Jitts (1963) uses blue glass filters for the same purpose.

In view of the difficulties of using results of measurements of primary production in the water column with the artificial illumination scheme of Steemann Nielsen when it is employed in the Pacific Ocean, one of the present authors used surface-production data for calculating the production in the entire water column. For this, she used coefficients obtained by comparing production at the surface with that in the water column as measured with the *in situ* method (Koblentz-Mishke, 1965). The other two authors, in calculating production of the Indian and Atlantic oceans, used data of production per square meter of sea surface. The major part of the data for the Atlantic Ocean (Data report Equalant, 1964–1965) and the northern part of the Indian Ocean was collected by using neutral filters, and in the southeastern part of the Indian Ocean by using blue filters. A part of the data for all the oceans was obtained using Sorokin's method.

For evaluating the differences in the values of production in the water column arising from the use of various schemes of measurement, it will be helpful to refer to Figure 2. In this diagram, values for the water column are plotted along the ordinate, and corresponding surface values are plotted along the abscissa. As is evident from the individual patterns of the figure, the scatter of points relating to each of the three techniques examined does not surpass the scatter of points from *in situ* experiments. Therefore, the use of different techniques for determining primary production in the water column apparently does not introduce any significant error in the data serving as the basis of the present paper. Of course, the diagram shows results of comparison of only three techniques, and the remaining schemes are in need of further verification.

PRIMARY PRODUCTION OF VARIOUS REGIONS OF THE WORLD OCEAN

Based on the data mentioned above, which relates to various seasons, a map of the distribution of values of primary production of the World Ocean has been drawn. The values have been divided into five classes: < 100; 100–150; 150–250; 250–500, and < 500 mg of C per m^2 per day. The statistical distribution of values of production served as the basis for these classes. Almost every class corresponds to the peak in the curve of distribution (see Figure 3).

The spatial distribution of these intervals occurs under the influence of two basic factors that determine the intensity of photosynthesis of phytoplankton in the seas: (1) quantity of light energy absorbed by algae and (2) supply of nutrient substances. The light factor is generally regarded as determining the level of primary production only in the polar regions, which are characterized by a short vegetative period and by the presence of ice, which attenuates the penetration of light into the water. Limitation of primary production by insufficient light during the vegetative period may occur in ice-packed waters. In these regions, primary production reaches the lowest limit, as shown by the results obtained by English (1958) in the drifting polar station "Alpha."

Beyond the ice-packed waters, the limit of which has been taken as the limit of lowest gradation of primary production in the North Arctic Ocean, primary production is determined by the nutrition factor alone. The relation between mineral nutrients and the level of primary production depends, as is well known, on the rate of reutilization by phytoplankton of elements such as P, N, and Si. The fastest reutilization takes place on the continental shelf and slope and in the upwelling regions of oceans. In these regions, enhanced vertical mixing in the water column occurs. Also, biogenic elements settle down from the euphotic zone in the shower of detritus; as a result of feeding, these elements can easily return to the surface. In addition, there is constant addition of these nutrients to the oceans by continental erosion and drainage. Regions particularly suitable for the growth of phytoplankton are those where upwelling and the divergence of currents occur. In the Pacific, these include regions close to coasts of Central and South America, Japan, Kamchatka, and Canada. In the Atlantic Ocean, such regions are found off West Africa, northeast Brazil, and the southeast coast of South America. In the Indian Ocean, regions of high productivity are those affected by the monsoon wind in the Arabian Sea, the Bay of Bengal, and the Indonesian Sea. The Sea of Azov, the Bering Sea, and others are also highly productive.

In the open ocean, the upwelling of deeper waters is the only significant source of transference of nutrients to the photosynthetic zone. Therefore, regions of upwelling, such

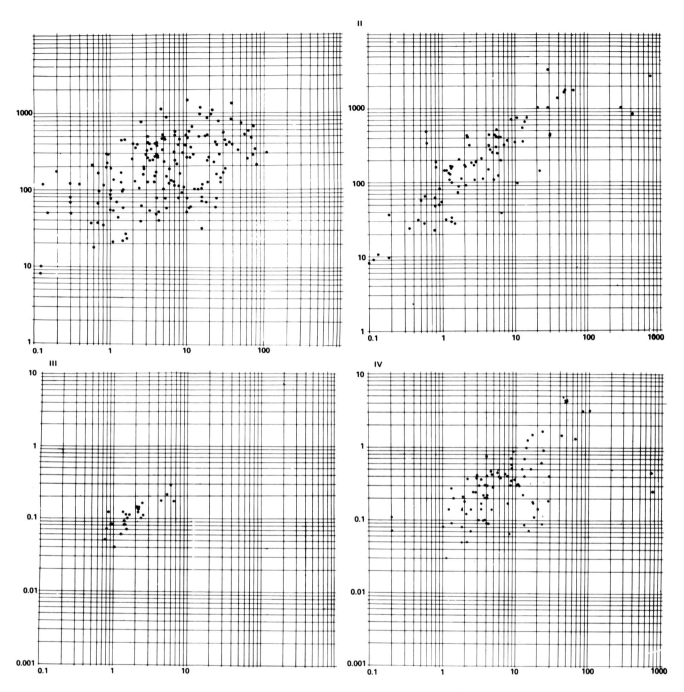

FIGURE 2   Relation between primary production in the water column (ordinate, mg of C per m^2 per day) and primary production at the surface (abscissa, mg of C per m^3 per day). Data were obtained as follows: (I) *in situ*, (II) Sorokin's scheme, (III) Jitt's scheme (daylight and blue filters), (IV) Doty's scheme (daylight and neutral filters).

as the Equatorial and Antarctic divergences, polar fronts, cyclonic halistatic regions, and regions of winter convectional mixing, are characterized by high production.

In contrast, anticyclonic halistatic regions of the oceans, where sinking of surface waters predominates, are characterized by lower levels of primary production.

Since the planetary circulation of waters of the World Ocean and the distribution of light energy penetrating the waters both have a zonal character, the geographic variation of primary production of the World Ocean is also zonal in character. The lowest levels of primary production are characteristic of polar and subtropical regions, and the highest

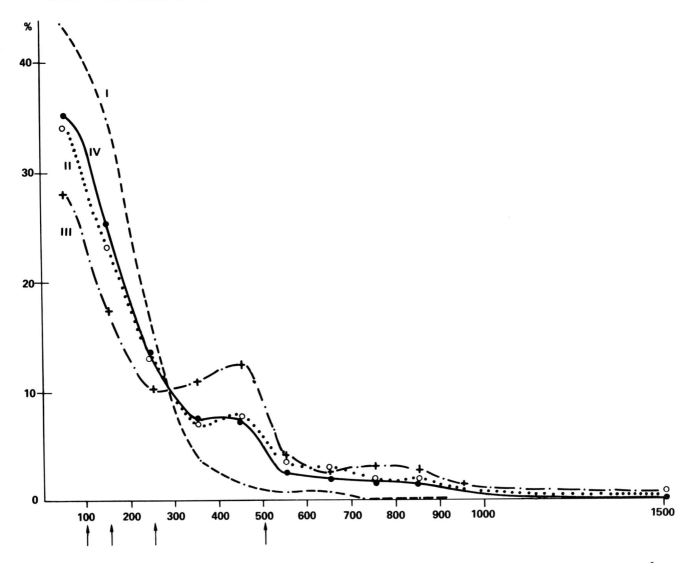

FIGURE 3 Statistical distribution of occurrence of various values of production (abscissa, production in water column in mg of C per m^2 per day; ordinate, percentage occurrence). Arrows show limits of production classes used in Figure 3. (I) Pacific Ocean; (II) Indian Ocean; (III) Atlantic Ocean; (IV) average for all oceans.

levels characterize temperate and equatorial regions. This general pattern of distribution of the level of primary production is distorted by the presence of continental projections along the meridional direction. In the wide Pacific, where the effect of land mass is very slight, the above-mentioned zonal factors of distribution of values of primary production are manifested to the highest degree. In the northern part of the Indian Ocean surrounded by continents, the isolines of levels of primary production have a more meridional orientation. But in the Indian Ocean, as well as in the Atlantic Ocean, despite the appreciable effect of continental configuration, the zonality of distribution of primary production can easily be detected, with regions of lower production occuring in the subtropical halistatic areas, and of those of higher production occurring in the equatorial and tropical regions. In Antarctic waters, where the effect of continental masses is absent, isolines of production are zonal, in agreement with the location of convergences and divergences. In preparing the chart of distribution of primary production of the World Ocean, Sverdrup (1955) studied only the zonal factors of its variation, without taking into consideration the very significant role of coastline. This scheme, through inclusion in the textbook of Gessner (1959), has become widely known and is the basis for a large number of speculations.

In looking at the map of primary production of the

World Ocean (Figure 1), it may be noted that the regions of low level of production are most widely distributed in the Pacific and least widely distributed in the Indian Ocean, with the Atlantic falling in between. As a result, these three oceans are characterized by different average values of production, as discussed in more detail below.

From the level of primary production in any particular region of the World Ocean, one can judge indirectly the relative variations in the number of phytoplankton cells (for which, unfortunately, no appropriate map is available). For the surface of the ocean (Figure 4), it has been found that despite the fact that the same production may correspond to the most varied counts of phytoplankton cells, there is a general correlation between the two. On the average, a higher level of production corresponds to a greater concentration of phytoplankton. This relationship permits one to use the map of primary production (Figure 1) as a first approximation of the desired map of phytoplankton distribution.

LEVEL OF PRIMARY PRODUCTION IN THE WORLD OCEAN

In estimating the mean annual production of the World Ocean, it would have been better to take into account the seasonal changes, as was done for the Pacific and Indian oceans in earlier publications. Taking the Pacific Ocean as an example, it has been established that the mean annual production changes only slightly from one climatic zone to another. The present paper is based on much more data than the earlier papers for individual oceans, increasing the probability that the data for primary production in each class interval has been collected more-or-less equally in the course of all the seasons. It may be supposed that the arithmetic mean of all the data relating to a given class is close to the annual mean. These values are given in Table 1, which includes data for the area represented by each class interval. Summation of these results leads to the conclusion that the total primary production of the World Ocean amounts to

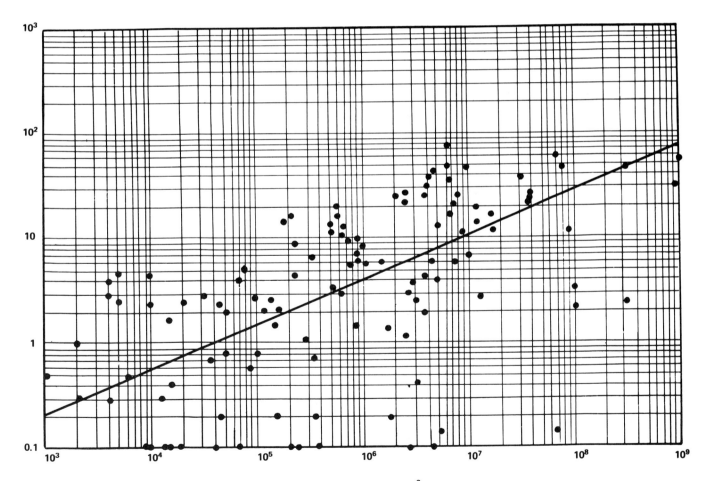

FIGURE 4   Relation between surface primary production (ordinate, mg of C per m^3 per day) and standing stock of phytoplankton (abscissa, number of cells per m^3).

TABLE 1 Primary Production Value of Various Types of Water of the World Ocean

Type of Water	Primary Production Level (mg of C per m² per day)		Areas of Each Type of Water in Different Oceans			Summary of Annual Production of the World Ocean for Each Type of Water (10⁹ tons of C)
	Mean Value	Limits of Fluctuation	Area[a]	X 10³ km²	World Ocean (%)	
Oligotrophic waters of the central parts of subtropical halistatic areas	70	<100	a	90,105	55	3.79
			b	19,599	22	
			c	30,624	31	
			d	8,000	34	
			e	148,329	41	
Transitional waters between subtropical and subpolar zones; extremity of the area of equatorial divergences	140	100–150	a	33,357	20	4.22
			b	23,750	33	
			c	22,688	22	
			d	3,051	13	
			e	82,847	23	
Waters of equatorial divergence and oceanic regions of subpolar zones	200	150–250	a	31,319	19	6.31
			b	18,886	27	
			c	32,650	32	
			d	3,642	16	
			e	86,498	24	
Inshore waters	340	250–500	a	10,422	6	4.80
			b	7,944	12	
			c	14,183	13	
			d	6,184	27	
			e	38,735	10	
Neritic waters	1,000	>500	a	243	<1	3.90
			b	5,289	<1	
			c	2,717	2	
			d	2,433	10	
			e	10,683	3	
						23 × 10⁹ tons of C per year

[a] a = Pacific Ocean; b = Indian Ocean; c = Atlantic Ocean; d = other waters (North Polar ocean; Indonesian seas; Mediterranean, Black, Asov, White, Okhotsk, Bering, Japan, China, Yellow seas); e = summary value for the whole World Ocean.

23×10^9 tons of carbon transformed from inorganic to organic form during the year. This figure is only one million higher than that obtained earlier by extrapolation from estimates of primary production of the Pacific Ocean (Koblentz-Mishke, 1965). Such a small difference was quite unexpected since the Pacific is the least productive of all the oceans (Table 2); the explanation is obviously that the Pacific and the Atlantic, with only slightly higher productivity, make up the major part of the World Ocean.

Thus, the results obtained in the present work, after correction for respiration, coincide with those of Steemann Nielsen and support the view that the productivity of the World Ocean is low in comparison to that of the land masses. It is true that the level of production calculated by Steemann Nielsen, as in the present work, was calculated from data obtained mainly by the radiocarbon method. There is some basis for believing that this method gives somewhat low results. Besides, we are certain that the area of the highly productive inshore zone was underestimated. Consequently, after making appropriate corrections, the level of primary production of the total hydrosphere may be higher than 23×10^9 tons of carbon per year gross production (or, with correction for respiration, 14×10^9 tons of carbon per year net production). We believe that a more accurate figure is $25-30 \times 10^9$ tons of carbon per year (gross production) and $15-18 \times 10^9$ tons of carbon per year (net production).

CONCLUSIONS

As noted earlier, the study of primary production is only the first stage in investigating the production of oceanic communities. In connection with the present study, several approaches to the relation between primary production and succeeding stages of its utilization can be noted.

Particular importance is attached to the effect of the magnitude of primary production.

Vinberg and Koblentz-Mishke (1966) have shown that the increased level of primary production is accompanied not only by the increase of biomass of zooplankton in the same regions of the ocean, but also by increase in the ratio of biomass of zooplankton to primary production. Judging from these indirect data, the efficiency of utilization of primary production is higher in the more productive regions than in the less productive ones. This is quite understandable, because in productive regions there is greater concentration of food, and consequently, less energy is expended in finding the prey than in the less productive regions.

The level of production calculated per surface unit area of the sea, integrated from surface to the lowest limit of photosynthesis on which we have depended so far, cannot characterize the probable efficiency of its utilization. To account for the latter, the level of primary production should be calculated as the mean production per unit of volume for the entire depth of the photosynthetic layer. This value has been established by Steemann Nielsen and Jensen (1957) as inversely proportional to the depth of the photosynthetic layer. Therefore, the concentration of production (per unit volume) changes from high-productive to low-productive waters more sharply than does the magnitude calculated per unit of surface area.

Next in importance, from the practical point of view, with respect to utilization of primary production is the demarcation of regions of high production as they relate to the location of commercial fisheries. Centers of high production are located in regions of intensive mixing of water layers where the so-called new water is formed. As water flows downstream, it progressively "ages"; the level of production also drops. Large-scale development of zooplankton—smaller in the beginning, then larger—occurs as the water flows downstream. Roughly speaking, the larger the organisms, the farther away from the centers of high primary production they accumulate. This is evident in a comparison of the map of primary production with that for distribution of zooplankton: Zones of high biomass of zooplankton are, as a rule, wider than the corresponding zones of high primary production.

In spite of the significant incompleteness of our knowledge about primary productivity of organic material in the World Ocean, it is necessary to note that increased international efforts in recent years have yielded appreciable results. The study of primary production is now at a level where it no longer impedes the thorough understanding of the problems of productivity in the oceans. In the future, the attention of scientists should be focused on the study of regularity of utilization of primary production.

REFERENCES

Berge, G. (1958) The primary production in the Norwegian Sea in June 1954, measured by an adapted ^{14}C technique. *Rapp. P.-v. Reun. Cons. Perm. Int. Explor. Mer, 144*, 85–91.

Bessonov, N. M., and M. V. Fedosov (1965) Primary production in shelf waters off the west coast of Africa. *Okeanologiia, 5(5)*, 877–883.

Corlett, J. (1958) Measurements of primary production in the western Barents Sea. *Rapp. P.-v. Reun. Cons. Perm. Int. Explor. Mer, 144*, 76–78.

TABLE 2 Mean Level of Primary Production

Production	Oceans			Seas
	Pacific	Atlantic	Indian	
mg of C per m^2 day	127	190	222	268
g of C per m^2 year	46.4	69.4	81.0	97.8

Curl, H. J. (1960) Primary production measurements in the north coastal waters of South America. *Deep-Sea Res., 7*(3), 183–189.

Currie, R. I. (1958) Some observations on organic production in the north-east Atlantic. *Rapp. P.-v. Reun. Cons. Perm. Int. Explor. Mer, 144*, 96–102.

Cushing, D. H. (1958) Some experiments using the ^{14}C technique. *Rapp. P.-v. Reun. Cons. Perm. Int. Explor. Mer, 144*, 73–75.

Data Report Equalant 1-2-3 (1964–65) International Cooperative Investigations of the Tropical Atlantic. National Oceanographic Data Center, Washington, D.C.

Doty, M. S. (1962) Data analysis on primary production during joint methodical works by scientists of different countries in Hawaii Islands region. *Okeanologiia, 2*(3), 543–553.

Doty, M. S., and L. R. A. Capurro (1961) Productivity measurements in the World Ocean. *IGY Oceanogr. Rep., 4*, 625 pp.

Doty, M. S., H. R. Jitts, O. I. Koblentz-Mishke, and Y. Saijo (1965) Intercalibration of marine plankton primary productivity techniques. *Limnol. Oceanogr., 10*(2), 282–286.

El-Sayed, S. Z. (1966) Phytoplankton production in Antarctic and Sub-Antarctic waters (Atlantic and Pacific sectors). *Second International Oceanographic Congress (abstr. of papers)*, pp. 440–441. Nauka, Moscow.

El-Sayed, S. Z., and E. Mandelli (1965) Primary production and standing crop of phytoplankton in the Weddell Sea and Drake Passage, in: *Biology of the Antarctic Seas*, 2. *Antarctic Res., 5*, 87–105.

El-Sayed, S. Z., E. F. Mandelli, and Y. Sugimura (1964) Primary organic production in the Drake Passage and Bransfield Strait. *Antarctic Res., 1*, 1–11.

English, T. S. (1958) Primary production in the central north Polar Sea, drifting station Alpha, 1957–58. Dept. Oceanogr., Univ. Washington.

Fukase, S., and S. Z. El-Sayed (1965) Studies on diatoms of the Argentine coast, the Drake Passage and the Bransfield Strait. *Oceanogr. Mag., 17*, 1–2.

Gessner, F. (1959) In *Hydrobotanic, 2 Stoffhaushalt*. VEB Deutscher Verlag der Wissenschaften, Berlin. 701 pp.

Hansen, V. K. (1961) Danish investigation on the primary production and the distribution of chlorophyll *a* at the surface of the North Atlantic during summer. *Rapp. P.-v. Reun. Cons. Perm. Int. Explor. Mer, 149*, 161–166.

Jitts, H. R. (1963) The simulation of *in situ* measurements of oceanic primary production. *Aust. J. Mar. Freshwater Res., 14*(2), 139–147.

Kabanova, J. G. (1966) Primary production in the southern Gulf of Mexico and in inshore zone of northwest Cuba. *Second International Oceanographic Congress (abstr. of papers)*, pp. 185–186. Nauka, Moscow.

Kabanova, J. G. (1968) Primary production in the north Indian Ocean. *Okeanologiia, 8*(2), 270–278.

Ketchum, B. H., and N. Corwin (1965) The cycle of phosphorus in a plankton bloom in the Gulf of Maine. *Limnol. Oceanogr., Suppl., 10*, 148–161.

Klyashtorin, L. B. (1960) Results of primary production determinations in the Atlantic Ocean. *Dokl. Akad. Nauk SSSR, 133*(4), 951–953.

Klyashtorin, L. B. (1962) Hydrobiology works. *Trud. Sov. Antarkt. Eksped., 20*, 302–312.

Koblentz-Mishke, O. J. (1965) Magnitude of primary production of the Pacific Ocean. *Okeanologiia, 5*(2), 325–337.

Koblentz-Mishke, O. J. (1967) Primary production, pp. 86–97 in: *Pacific Ocean, 7, Biology of the Pacific Ocean*, Nauka, Moscow.

Koblentz-Mishke, O. J., O. D. Bekasova, V. I. Vedernikov, B. V. Konovalov, V. V. Sapozhnicov, and V. A. Terskikh (1970) Pri-

mary production and pigments in waters to the east of the Kuril Islands during the summer of 1966. *Trud. Inst. Okeanol.*, 86

Koblentz-Mishke, O. J., A. K. Karelin, and V. A. Rutkovskaya (1966) On unusual forms of curves of change in magnitude of photosynthesis of the homogeneous phytoplankton exposed at different depths in the sea. *Second International Oceanographic Congress (abstr. of papers)*, p. 205. Nauka, Moscow.

Kondratieva, T. M. (1968) Production and diurnal changes of phytoplankton in the South Sea. Unpublished Dissertation,

Kondratieva, T. M., and A. Sosa (1966) Primary production in waters near Cuba: in *Issledovaniya Zentralnoamerikanskikh Morei, 1*, 68–80.

Mandelli, E., and A. Orlando (1966) La producción organica primaria y las características fisicoquímicas de la corriente de las Malvinas. *Boln. Serv. Hidrogr. Navy, Buenos Aires, 3*(3), 185–196.

Marshall, F. T. (1958) Primary production in the Arctic. *J. Cons. Perm. Int. Explor. Mer, 23*(2), 173–177.

Menzel, D. W., and J. H. Ryther (1960) The annual cycle of primary production in the Sargasso Sea off Bermuda. *Deep-Sea Res., 6*(4), 351–367.

Oceanographic Research Institute (1965) Measurement of primary production by the R/V *Lady Theresa* at the fixed station = 29°54'S = 31°07'E. Oceanographic Research Institute, Durban, South Africa.

Richards, F. A. (1960) Some chemical and hydrographic observations along the north coast of South America. *Deep-Sea Res., 7*(3), 163–182.

Riley, G. A. (1944) The carbon metabolism and photosynthetic efficiency of the Earth as a whole. *Amer. Sci., 32*, 129.

Riley, G. A. (1946) Factors controlling phytoplankton populations on Georges Bank. *J. Mar. Res., 6*, 54–73.

Riley, G. A. (1956) Oceanography of Long Island Sound, 1952–54. IX. Production and utilization of organic matter. *Bull. Bingham Oceanogr. Coll., 15*, 324–344.

Ryther, J. H., and C. S. Yentsch (1957) The estimation of phytoplankton production in the ocean from chlorophyll and light data. *Limnol. Oceanogr., 2*(3), 281–286.

Ryther, J. H., and C. S. Yentsch (1958) Primary production of the continental shelf waters off New York. *Limnol. Oceanogr., 3*(3), 327–335.

Saijo, Y., and T. Kawashima (1964) Primary production in the Atlantic Ocean. *J. Oceanogr. Soc. Japan, 19*(4), 190–196.

Smirnova, L. I. (1959) Phytoplankton of Okhotsk Sea and of the waters near Kuril region. *Trud. Inst. Okeanol., 30*, 3–51.

Sorokin, Yu. I. (1959) Determination of the photosynthetic productivity of phytoplankton in water using the ^{14}C method. *Fiziologiia Rast., 6*, 125.

Sorokin, Yu. I. (1960) Determination of phytoplankton primary production in the Atlantic Ocean by isotopic method. *Dokl. Akad. Nauk SSSR, 131*(4), 941–944.

Starodubtsev, E. G. (1968a) Primary production in the area of the Kuroshio Current in February, 1966. *Trud. Inst. Okeanol.*

Starodubtsev, E. G. (1968b) The report on the results of oceanographic surveys on board R/V *Shokalsky* in the winter season: Seasonal variation of primary production in the area of the Kuroshio Current (ms).

Starodubtsev, E. G. (1968c) Technical report of the results of oceanographic surveys in the summer season. On the subject: Seasonal variations of primary production in the area of the Kuroshio Current (ms).

Steele, J. H. (1958) Production studies in the northern North Sea. *Rapp. P.-v. Reun. Cons. Perm. Int. Explor. Mer, 144*, 79–84.

Steele, J. H. (1964) A study of production in the Gulf of Mexico. *J. Mar. Res., 22*(3), 211–222.

Steemann Nielsen, E. (1951) Measurement of the production of organic matter in the sea by means of carbon-14. *Nature*, 167, 4252.

Steemann Nielsen, E. (1952) The use of radioactive carbon (^{14}C) for measuring organic production in the sea. *J. Cons. Perm. Int. Explor. Mer, 18*(2), 117–140.

Steemann Nielsen, E. (1958) A survey of recent Danish measurements of the organic productivity in the sea. *Rapp. P.-v. Reun. Cons. Perm. Int. Explor. Mer, 144*, 92–95.

Steemann Nielsen, E., and V. K. Hansen (1961) The primary production in the waters west of Greenland during July 1958. *Rapp. P.-v. Reun. Cons. Perm. Int. Explor. Mer, 149*, 158–159.

Steemann Nielsen, E., and E. A. Jensen (1957) Primary oceanic production. The autotrophic production of organic matter in the oceans. *Galathea Rep., 1*, 49–135.

Sverdrup, H. U. (1955) The place of physical oceanography in oceanographic research. *J. Mar. Res., 14*(4), 287–294.

Thomas, W. H., and E. G. Simmons (1960) Phytoplankton production in the Mississippi delta, pp. 103–116 in: *Recent Sediments, Northwest Gulf of Mexico, 1951–1958*, editors F. P. Shepard,

F. B. Phleger and T. H. Van Andel, American Association of Petroleum Geologists.

Vinberg, G. G. (1960) *The Primary Production of Bodies of Water.* Institute of Biology of the Byelorussian Academy of Sciences, Minsk. 329 pp.

Vinberg, G. G., and O. I. Koblentz-Mishke (1966) Problems of primary production of water basins, pp. 50–62 in: *Ekologiya Vodnych Organizmov*. Nauka, Moscow.

Volkovinsky, V. V. (1966) The study of primary production in the South Atlantic waters. *Second International Oceanographic Congress (abstr. of papers)*, pp. 99–100. Nauka, Moscow.

Volkovinsky, V. V., and M. V. Fedosov (1964) On primary production formation in Antarctic waters. *Okeanol. Issled., 13*, 115–122.

Zaika, V. E., A. D. Gordina, T. M. Kovaleva, and L. V. Kuzmenko (1967) Preliminary results of biological investigations during the 19th cruise of R/V M. *Lomonsov*. In *Issledovaniya Severo-Zapadnoy Chasti Indiyskogo Okeana*, pp. 83–89.

Zernova, V. V. (1966) Distribution of phytoplankton biomass in layer 0–100 m, pp. 127–128 in: *Atlas Antarktiki, 1*, Moscow.

Sayed Z. El-Sayed

TEXAS A & M UNIVERSITY

PHYTOPLANKTON PRODUCTION OF THE SOUTH PACIFIC AND THE PACIFIC SECTOR OF THE ANTARCTIC

INTRODUCTION

The first measurements of primary organic production in the Pacific Ocean, using the radioactive ^{14}C uptake method, were made by Steemann Nielsen (1952) during the Danish *Galathea* Deep-Sea Expedition (1950–1952). These measurements were made in the Tasman Sea, off New Zealand, and in a diagonal crossing of the Pacific Ocean between northern New Zealand and California.

The first extensive survey of primary production and phytoplankton standing crop in the eastern tropical Pacific was made during the EASTROPIC Expedition in 1955 (King *et al.*, 1957; Holmes, 1958); few measurements, however, were made south of 10°S. Subsequent expeditions investigated the regions north of the equator, and a few extended to 35°S (Holmes, 1961; Blackburn, 1966).

Thanks to the extensive investigations of the Inter-American Tropical Tuna Commission during the past two decades and, more recently, the investigations carried out by the research vessel *Anton Bruun* and by the EASTRO-PAC Expedition (1967), the biological oceanography of the eastern tropical Pacific (to 30°S) is among the best studied regions in the eastern South Pacific Ocean.

On the other side of the Pacific, the productivity of the Tasman Sea, the Coral Sea, and the inshore regions off Australia and New Zealand are comparatively better studied than any other area in the western South Pacific, due mainly to the efforts of Australia's Commonwealth Scientific and Industrial Research Organization (C.S.I.R.O.) (Humphrey, 1960; Jitts, 1965).

The productivity of the waters in the western tropical Pacific from the Solomon Islands to New Caledonia and New Hebrides were studied by the Institut Français

d'Océanie (Angot, 1961). The data collected by the French, Philippine, and Russian investigators in these tropical waters made the southwestern Pacific Ocean also one of the better known regions.

Our knowledge of primary productivity in the vast expanses of the Pacific Ocean, south of 35°S, until very recently, was almost nonexistent. The first published estimates of the productivity of this enormous region were given by El-Sayed (1967) and Burkholder and Burkholder (1967). Figures 1 and 2 summarize the existing knowledge, published to date, of the distribution of surface phytoplankton standing crop (in terms of chlorophyll *a*) and primary production (in terms of ^{14}C uptake) between 20°N and the Antarctic Continent.

During the past 3 years, the Department of Oceanography at Texas A & M University, has been carrying out an investigation of the biological productivity of the South Pacific Ocean and of the Pacific sector of the Antarctic aboard the USNS *Eltanin*. In this investigation we are interested primarily in assessing the standing crop of phytoplankton and in measuring primary productivity in the various areas studied. We are also interested in correlating the distribution and abundance of the phytoplankton with the hydrographic conditions in the areas investigated.

The results of this investigation, to be reported herein, are based on data collected from 274 stations and 812 surface-water samples during nine cruises of the *Eltanin* between May 1965 and May 1967 (see Figure 3). The author wishes to emphasize the preliminary nature of these results, which are part of a still-continuous investigation of the South Pacific. However, enough material has accumulated on the geographic and seasonal variations in the productivity parameters to warrant reporting at this stage.

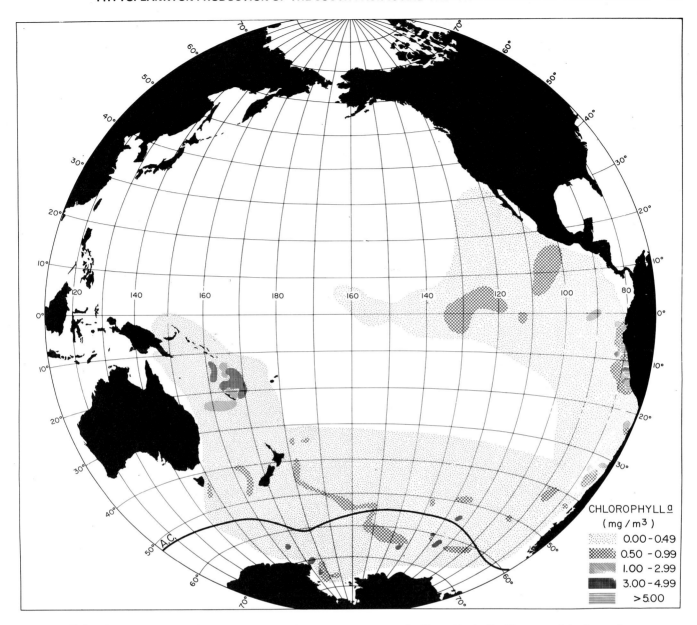

FIGURE 1   Distribution of chlorophyll *a* in surface water in the South Pacific and in the Pacific sector of the Antarctic.

RESULTS

In this paper we will limit our discussion to the spatial and temporal distribution of chlorophyll *a* and primary production collected south of 35°S during the nine cruises of the *Eltanin*. The distribution and concentration of the nutrient salts (i.e. phosphates, silicates, nitrates, and nitrites) and the particulate and dissolved organic carbon collected during these cruises has been published by the American Geographical Society and will not be discussed in this paper. (Antarctic Map Folio Series, Folio 10, 1968.) The pertinent data collected during the *Eltanin* cruises are summarized in Table 1.

GEOGRAPHIC VARIATIONS IN PRODUCTIVITY PARAMETERS IN SOUTH PACIFIC DISTRIBUTION OF SURFACE CHLOROPHYLL *a* AND ^{14}C UPTAKE

In the South Pacific the phytoplankton standing crop in surface-water samples was generally poor; chlorophyll *a* values are less than 0.50 mg per m^3, except for isolated pockets ranging between 0.50 and 0.99 mg per m^3 (Figure 1). The coastal regions off Chile, New Zealand, and Tasmania and the Ross Sea showed higher chlorophyll *a* values than the oceanic areas.

The distribution of ^{14}C uptake in surface-water samples in the areas studied shows a more-or-less similar picture to

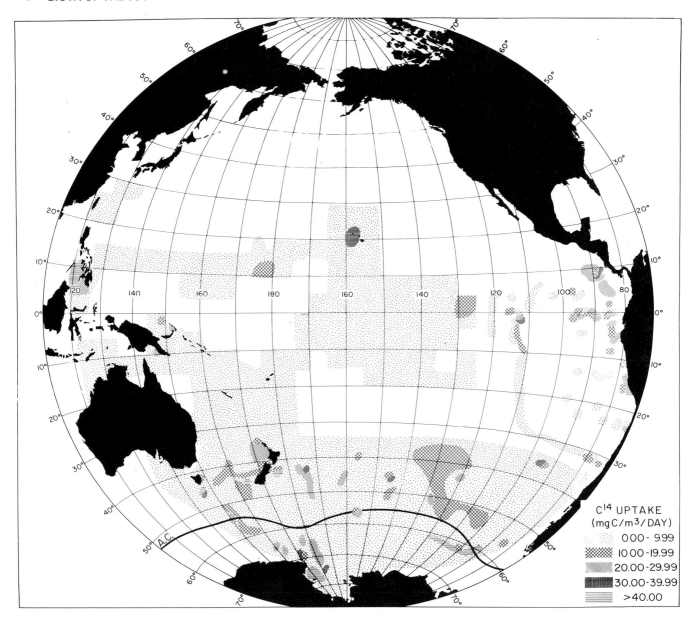

FIGURE 2   Distribution of ^{14}C uptake in surface water in the South Pacific and in the Pacific sector of the Antarctic.

that of phytoplankton standing crop. Most of the stations occupied in the South Pacific gave low rates of photosynthesis (< 5.0 mg of C per m^3 per hr); however, patches of relatively high photosynthetic rates were recorded between 170° and 180°W (Figure 2).

LATITUDINAL VARIATIONS IN PHYTOPLANKTON STANDING CROP AND ^{14}C UPTAKE

Since the areas investigated in the South Pacific encompass a large geographical region that is bound to the north by the ill-defined Subtropical Convergence (roughly at 40°S), and to the south by the coast of Antarctica (about 78°S), it is instructive to study the latitudinal variations in the productivity parameters in this vast region. For this reason, the data collected during *Eltanin* cruises 18–28 (less cruise 22, which was made in the Scotia Sea) were grouped by 5° latitude, and were plotted as shown in Figure 4. It is clear from this figure that the standing crop of phytoplankton in surface-water samples taken between 30° and 60°S did not show great variations. The average value is slightly less than the overall average for the South Pacific (0.25 mg per m^3). A slight increase in chlorophyll *a* concentration is found be-

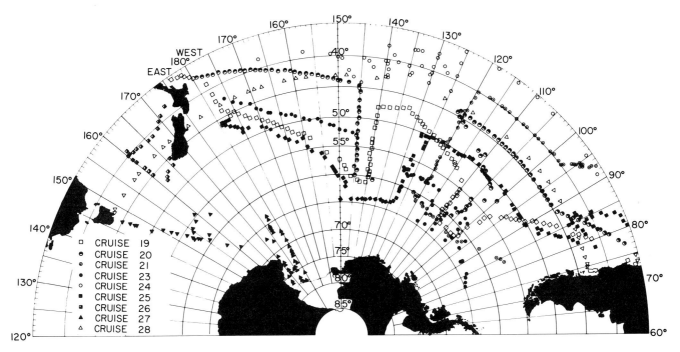

FIGURE 3   Stations occupied by USNS *Eltanin* in the Pacific sector during Cruises 19-28, less Cruise 22.

TABLE 1   Observations Made of the Standing Crop of Phytoplankton, Primary Production, Nutrient Salts and Particulate and Dissolved Organic Carbon during *Eltanin* Cruises 18-28 (less Cruise 22) in the South Pacific and in the Pacific Sector of the Antarctic

		Minimum	Maximum	Average	Standard Deviation	Number of Observations
Chl	a (mg/m^3)	0.01	.5.80	0.26	0.34	723
Chl	a (g/m^2)^a	0.23	41.32	12.62	6.32	217
^{14}C	(mg C/m^3/hr)	0.03	22.50	1.22	1.69	656
^{14}C	(g C/m^2/hr)^a	3.54	194.73	32.01	23.93	213
PO$_4$	(μg-at./1)	0.01	4.66	1.09	0.46	350
PO$_4$	(g-at./m^2)^a	0.84	457.00	42.98	37.01	146
SiO$_3$	(μg-at./1)	0.1	79.9	13.5	13.5	368
SiO$_3$	(g-at./m^2)^a	0.01	5.41	0.57	0.59	154
NO$_3$	(μg-at./1)	0.1	30.2	12.9	5.5	313
NO$_3$	(g-at./m^2)^a	0.01	2.93	0.63	0.41	161
NO$_2$	(μg-at./1)	0.01	1.23	0.18	0.11	296
NO$_2$	(g-at./m^2)^a	0.6	33.8	10.7	6.7	146
POC	(mg C/1)	0.003	0.520	0.058	0.051	149
POC	(g C/m^2)^a	2.46	18.84	5.52	2.27	153
DOC	(mg C/1)	0.25	2.66	0.95	0.27	156
DOC	(g C/m^2)^a	14.1	243.1	58.3	23.4	157

^{a} Integrated values in euphotic zone.

tween 60° and 70°S, followed by a substantial increase between 70° and 80°S, due mainly to the high values encountered off the Ross Ice Shelf.

In terms of carbon fixation, the picture is more-or-less similar to that of the distribution of chlorophyll a; i.e., low photosynthetic values were generally found between 30°

and 70°S, and conspicuously high values were observed between 75° and 80°S.

The integrated values (i.e., values integrated through a column of water extending from sea surface to the depth of the euphotic zone) of chlorophyll a and ^{14}C uptake are shown in Figure 5. In this figure, relatively high values are

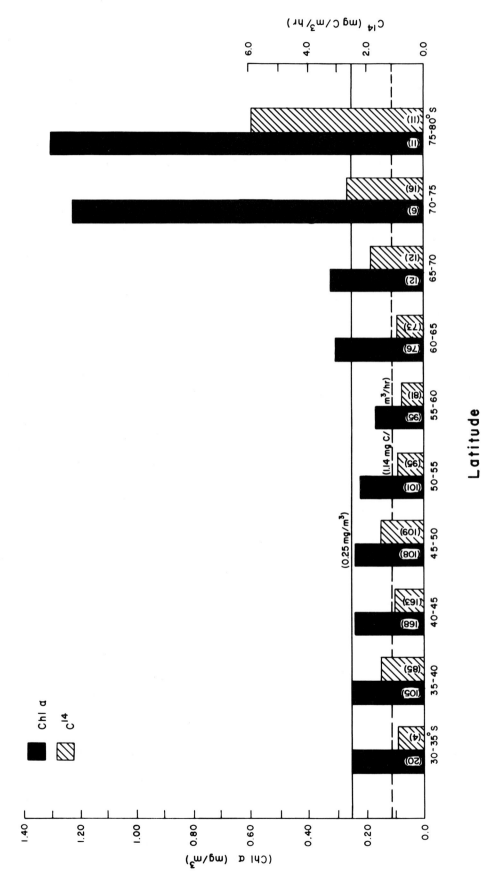

FIGURE 4 Latitudinal distribution of chlorophyll *a* and ^{14}C uptake in surface-water samples collected during *Eltanin* Cruises 18–28 (less Cruise 22). Dashed lines indicate average values; numbers in parentheses refer to number of observations.

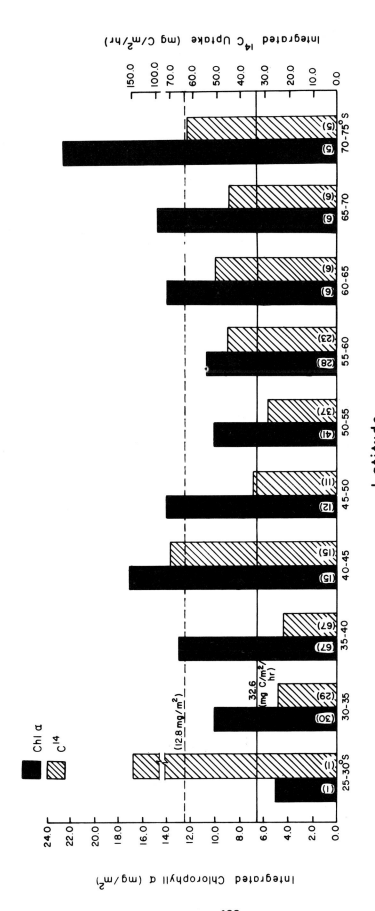

FIGURE 5 Latitudinal distribution of chlorophyll *a* and ^{14}C uptake in the euphotic zones during *Eltanin* Cruises 18–28 (less Cruise 22). Dashed lines indicate average values; numbers in parentheses refer to number of observations.

found between 45° and 50°S, and between 75° and 80°S, except for one observation made between 30° and 35°S, where high ^{14}C uptake was recorded.

LONGITUDINAL VARIATIONS IN PHYTOPLANKTON STANDING CROP AND PHOTOSYNTHESIS

The vast expanses of the South Pacific made it imperative to group the productivity data by 10° longitude in order to discern any longitudinal variations in productivity parameters. The *Eltanin* data collected in that region are plotted in Figure 6. It is clear from this figure that the western section of the South Pacific, between 140°E and 170°W (less the region between 150° and 160°E), is conspicuously richer in surface chlorophyll *a* concentration than the poverty-stricken central and eastern South Pacific. It should be remembered, however, that the substantial increase in phytoplankton in the western regions of the South Pacific could be attributed to two factors: (1) the proximity of the observations made to the coasts of New Zealand, Australia, and Tasmania (the so-called land-mass effect of Doty and Oguri, 1956); and (2) the location of several of the stations in the western regions in the Pacific sector of the Antarctic, which are several times more productive than the subantarctic regions (El-Sayed, 1968). The low chlorophyll *a* values encountered between 150° and 160°E (in the Tasman Sea) reflect the poor phytoplankton conditions in these predominantly subtropical and nutrient-impoverished regions.

The regions between 80° and 170°W, by and large, have chlorophyll values that are lower than the overall average for the South Pacific. Hart (1942), based on pigment values obtained by net hauls, also found the eastern South Pacific to be markedly poor in phytoplankton. Yet, as the Chilean coast is approached, the rise in pigment concentration between 70° and 80°W is clearly manifested in Figure 6.

In terms of the photosynthetic activity of surface phytoplankton between Chile and Australia, the picture is very much similar to that of the phytoplankton standing crop. Again, it is evident that high photosynthetic assimilation is encountered in the coastal regions off Chile, New Zealand, and Australia. The integrated values of chlorophyll *a* and ^{14}C uptake (Figure 7) showed a rather similar distribution to that of surface values, except for the low carbon assimilation between 70° and 80°W.

SEASONAL VARIATIONS IN THE DISTRIBUTION OF CHLOROPHYLL *a* AND ^{14}C UPTAKE

Figure 8 shows the seasonal variations of surface productivity parameters during the nine cruises of the *Eltanin* in the South Pacific. Initial increase in surface chlorophyll *a* and ^{14}C uptake values is noted between Cruise 19 (July–September 1965), Cruise 20 (September–November 1965), and Cruise 21 (November 1965–January 1966). This is followed by a marked decline during the two subsequent cruises: Cruise 23 (April–May 1966) and Cruise 24 (July–September 1966). During the austral spring and summer, as expected, significant increases in phytoplankton standing crop and photosynthetic activity of the primary producers were noted during Cruise 26 (November–December 1966) and Cruise 27 (January–February 1967). Cruise 28, made during the austral fall (March–May 1967), showed appreciable decline in the productivity values collected.

The seasonal variations in the integrated productivity parameters, also shown in Figure 8, follow closely those of surface values.

YEAR-TO-YEAR VARIATIONS IN THE PRODUCTIVITY PARAMETERS

The *Eltanin* data from the South Pacific, also enabled us to study the year-to-year variations in the productivity parameters. For this purpose, the data of Cruises 19, 20, 24, and 25 were chosen. These cruises were selected primarily for their coverage of the South Pacific from coast to coast, and for the coincidence of their schedules, which fell on about the same dates in 1965 and 1966, as noted below.

Cruise Number	Dates	Areas Investigated
19	July 7–September 2, 1965	Chile to New Zealand
20	September 15–November 10, 1965	New Zealand to Chile
24	July 13–September 3, 1966	Chile to New Zealand
25	September 26–November 18, 1966	New Zealand to Chile

The integrated values of chlorophyll *a* and primary production in the euphotic zone for the above-given cruises are plotted in Figure 9. It is clear from this figure that the standing crop of phytoplankton during the midaustral winter in 1966 was significantly larger than that found in the same period a year earlier (average values: 8.59 mg per m^2 for Cruise 24, and 5.6 mg per m^2 for Cruise 19). Carbon assimilation in the euphotic zone also showed higher values during the former cruise compared to the latter (average values: 25.53 mg of C per m^2 per hr and 19.2 mg of C per m^2 per hr, respectively).

Comparison between austral fall and early spring (Cruises 20 and 25), on the other hand, showed about the same amount of pigment concentration in the euphotic zone in

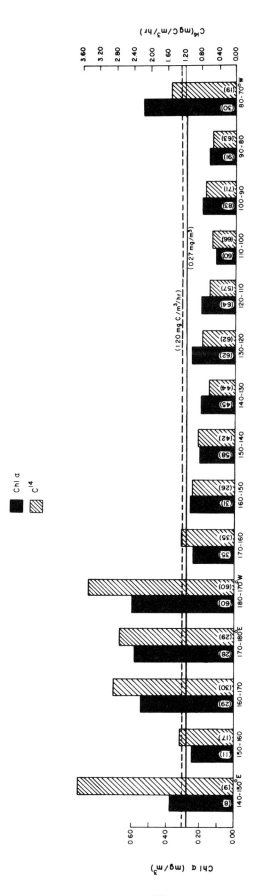

FIGURE 6 Longitudinal distribution of chlorophyll *a* and ^{14}C uptake in surface-water samples collected during *Eltanin* Cruises 18–28 (less Cruise 22). Dashed lines indicate average values; numbers in parentheses refer to number of observations.

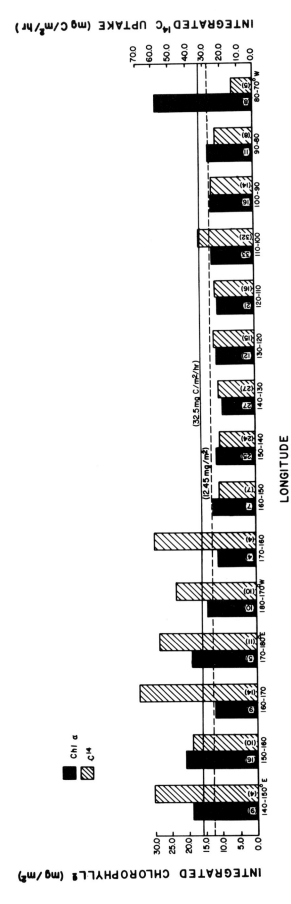

FIGURE 7 Longitudinal distribution of chlorophyll *a* and ^{14}C uptake in the euphotic zones during *Eltanin* Cruises 18–28 (less Cruise 22). Dashed lines indicate average values; numbers in parentheses refer to number of observations.

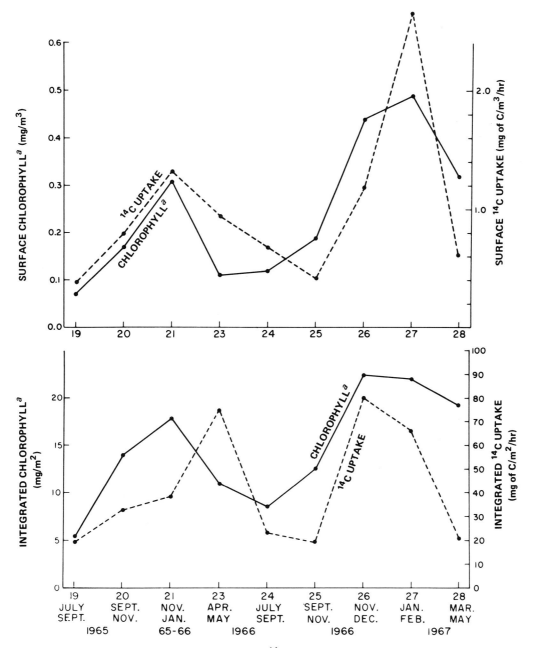

FIGURE 8   Seasonal variations in chlorophyll *a* and ^{14}C uptake (surface and integrated values) taken during *Eltanin* Cruises 19–28 (less Cruise 22).

both cruises (average 13.53 and 11.88 mg per m^2); however, the photosynthetic values were higher during Cruise 20 (32.67 mg of C per m^2 per hr) compared with Cruise 25 (19.56 mg of C per m^2 per hr).

VERTICAL DISTRIBUTION OF CHLOROPHYLL *a* AND PRIMARY PRODUCTION IN THE SOUTH PACIFIC

Although the vertical distribution of chlorophyll *a* and ^{14}C uptake in the euphotic zone was studied at all the Local Ap-

parent Noon (LAN) stations occupied by the *Eltanin*, no attempt will be made to show here the vertical productivity profiles for each station. Instead, the distribution of the average chlorophyll *a*, primary production, and light penetration for the combined stations in each cruise will be discussed. These are plotted in Figure 10. Noticeable variations in the vertical distribution of chlorophyll *a*, ^{14}C uptake, and light penetration seem to characterize the various cruises. Except for Cruises 27 and 28, where the chlorophyll values tended to increase with depth, the vertical distribu-

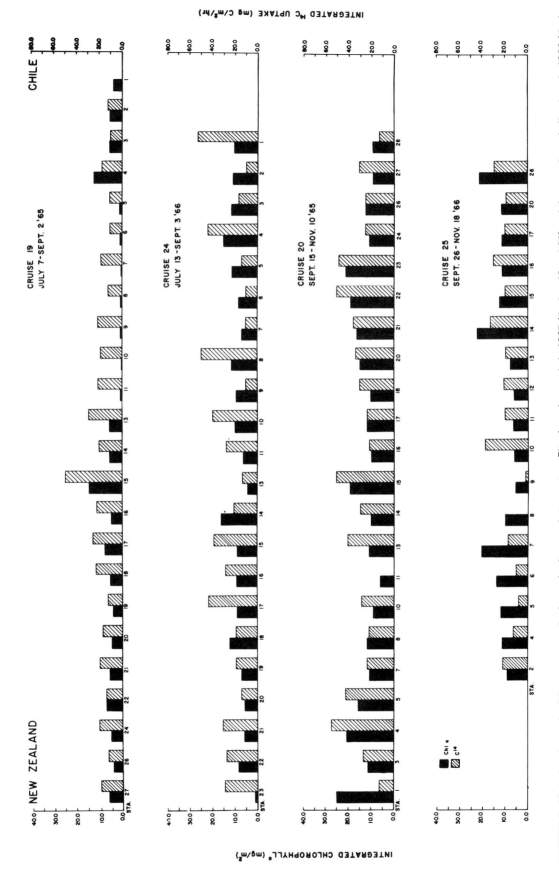

FIGURE 9 Variations in productivity parameters in the euphotic zones between two *Eltanin* cruises made in 1965 (Cruises 19 and 20) and those made at similar dates in 1966 (Cruises 24 and 25).

204

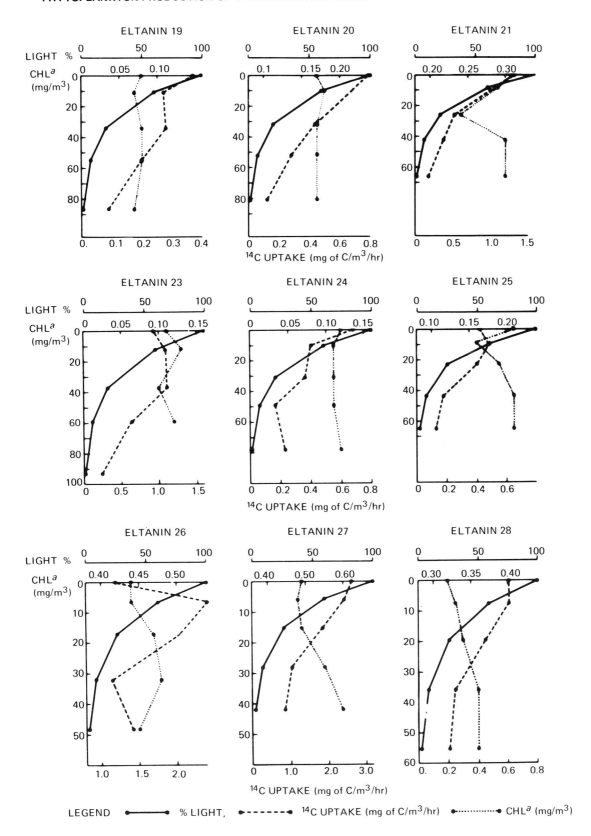

FIGURE 10 Vertical distribution of average chlorophyll *a* and ^{14}C uptake and percent light penetration at different depths in the euphotic zones during *Eltanin* Cruises 19–28 (less Cruise 22).

tion of the phytoplankton standing crop showed, in general, a more-or-less homogeneous distribution within the euphotic zone. Although not shown in Figure 10, on several occasions we have noted that chlorophyll *a* concentrations below the euphotic zone were higher than those found in the lighted zone. However, below the depth of 150 m the size of the phytoplankton populations is drastically reduced.

Carbon assimilation, on the other hand, showed high values at the surface; these decreased gradually to minimal values at depths ranging between 42 m (Cruise 27) and 94 m (Cruise 23). During Cruise 26, the highest photosynthetic activity was recorded at a subsurface level (6 m), rather than at the surface. This subsurface increase could be partly attributed to the photoinhibition of surface phytoplankton due to the early austral summer in the Tasman Sea.

The depths of the euphotic zone also showed considerable variation between the nine cruises. The average depths ranged between 50.7 m for Cruise 27 (between Tasmania and the Ross Sea) and 105 m for Cruise 23 (west of the Drake Passage, between 55° and 65°S). It is noteworthy that the average depth of the euphotic zones for all the combined cruises (based on 213 LAN observations) was 79.6 m. Thus, if the transparency of seawater can be used as an indication of the productivity in the water column, as suggested by Hart (1962) and others, there is little doubt that the clear waters of the South Pacific do reflect the paucity of their phytoplankton populations.

CORRELATION BETWEEN DIATOMS, DINOFLAGELLATE, AND *Trichodesmium thiebauti* POPULATIONS WITH PRODUCTIVITY PARAMETERS[*]

It is generally recognized that quantitative distribution of the individual species of phytoplankton has been neglected in most of the biological investigations carried out in the South Pacific. In this respect, the quantitative estimates of the diatoms, dinoflagellates, and the blue-green alga, *Trichodesmium thiebauti*, populations that were made during *Eltanin* Cruise 28 (Figure 11) are of special significance. In Figure 12, we have plotted the quantitative distributions of chlorophyll *a*, ^{14}C uptake, diatoms, dinoflagellates, and the blue-green algae. Close similarity in the quantitative distribution of the diatom and dinoflagellate populations at the stations occupied in this cruise are noticeable in Figure 12. *T. thiebauti*, on the other hand, shows a pattern of distribution quite different from that of the diatoms and dinoflagellates. Further, chlorophyll *a* and ^{14}C uptake data showed marked discrepancies from those of diatom and dinoflagellate populations, especially in the region between stations 25 and 35. However, good correlation was found between stations 5 and 10. It is possible, however, that the above-mentioned discrepancies could be attributed partly to the differences in the vertical range of sampling, as the phytoplankton populations were taken in 0–200-m net hauls, whereas chlorophyll *a* and ^{14}C uptake data were

[*]The author would like to acknowledge the contribution made by Dr. Ryuzo Marumo to this section (also see Marumo, 1968).

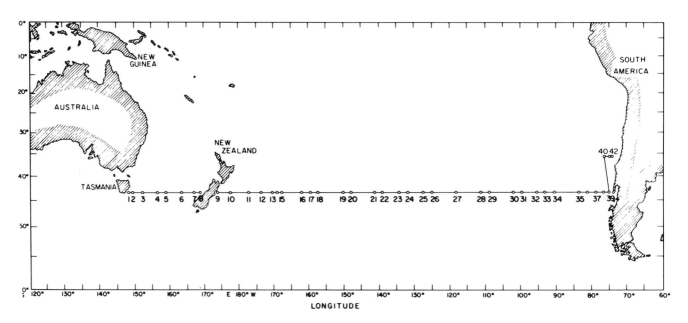

FIGURE 11 Positions of stations occupied during *Eltanin* Cruise 28.

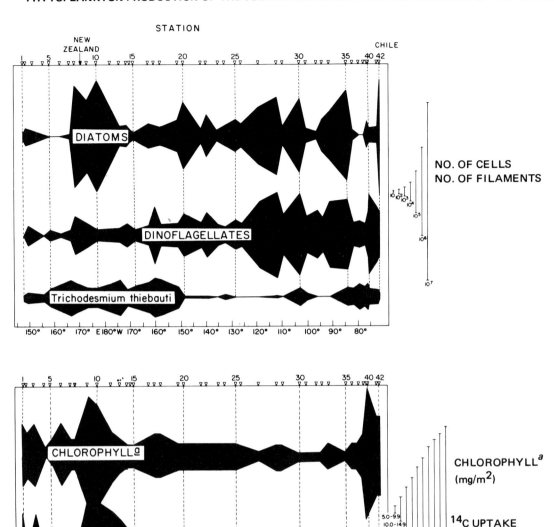

FIGURE 12   Cell numbers of diatoms, dinoflagellates, and number of filaments of *Trichodesmium thiebauti* per m^2 of water column compared with integrated chlorophyll *a* and ^{14}C uptake (in euphotic zone) in a transect across the Pacific Ocean (43°S).

limited to the depth of the euphotic zones, which seldom exceeded 100 m during this cruise.

Aside from the obvious correlation (or lack of it at some stations) in the quantitative distribution of the parameters used in Figure 12, it was, nonetheless, interesting to corre-late the distribution of diatom, dinoflagellate, and *T. thiebauti* populations with the hydrographic data collected during Cruise 28. From Figure 13 it is possible to distinguish at least five bodies of water on the basis of the character of the phytoplankton community that could be correlated with the existing hydrographic conditions in these regions, as discussed below.

1. In the region between stations 1 and 19 (in predominantly subtropical waters and under the influence of the warm East Australian Current) *Trichodesmium thiebauti* is abundant. The diatom and dinoflagellate populations, on the other hand, are scarce or moderate, except for the large diatom population found off New Zealand and for the extreme abundance of the neritic *Chaetoceros* sp. at stations 8, 9, and 10.

2. In the region between stations 27 and 35 temperature and salinity suggest subantarctic conditions. Here the diatom and dinoflagellate populations are abundant, whereas *Trichodesmium thiebauti* is scarce or totally absent in some stations. In this region, *Chaetoceros atlanticus*, a typical antarctic diatom, was frequently found.

3. The region between stations 20 and 26 could be considered hydrographically as a transition region between the subtropical and subantarctic waters. This was reflected in the mixed nature of its pelagic flora.

4. The region between stations 36 and 38 is characterized by a floral community that resembles that found between stations 1 and 7—i.e., by scarce diatom and dinoflagellate populations, and an abundance of *T. thiebauti*. One would suspect that this water mass is of subtropical character.

5. The region between stations 39 and 42 is coastal in character; this is manifested by the presence of the neritic dinoflagellates *Ceratium furca* and *C. gibberum*, which predominate in this region.

A LOOK TOWARD THE FUTURE

1. It is evident from Figures 1 and 2 that, in terms of phytoplankton standing crop and primary production, the areas off South America to the east, off Australia and New Zealand and New Guinea to the west, and the western tropical Pacific to the north are among the better known areas in the South Pacific. The areas between the equator and 35°S, on the other hand, still can be considered, biologically, as *terra incognita*.

Qualitative and quantitative studies of phytoplankton similar to those made by Hasle (1959) in the tropical Pacific, and more recently in the Antarctic Pacific (Hasle, 1968a), and those made by Kozlova (1966) in the Pacific and Indian sectors of the Antarctic are badly needed if the productivity of the South Pacific is to be properly understood. Further, maps showing the distribution of the main species of diatoms and dinoflagellates, similar to those prepared by Hasle (in press) and Balech (in press), would contribute greatly to our knowledge of the biology of the South Pacific.

In this connection, it would be good if a map showing the quantitative distribution of the phytoplankton similar to that of primary production prepared by Koblentz-Mishke and her colleagues (see p. 185 in this volume), were prepared for the South Pacific.

2. The role of hydrography and its effect on primary production is of cardinal importance to our understanding of the variability of organic production from region to region. Except for isolated studies in which the distribution and abundance of phytoplankton were correlated with hydrographic phenomena, the regions of the South Pacific have been neglected in this respect.

In this connection we should be mindful of the contributions the marine biologists are capable of rendering to the physical oceanographers in the quest for delineating water masses in such vast and complicated regions as the South Pacific. A number of phytoplankters have been successfully used as "indicator species" by Hart (1937), Braarud (1962), Smayda (1958), Balech (1962), and many others. The example given in this paper based on Marumo's study (Marumo, 1968), where he used *Trichodesmium thiebauti* and *Chaetoceros atlanticus* to delineate water masses, is, indeed, most promising.

3. The seasonal and year-to-year fluctuations in the productivity parameters discussed in this paper are based on observations made by the author in the region south of 35°S. Whether similar fluctuations exist in the virtually unknown regions between the equator and 35°S is subject to speculation and should be pursued with great interest by future investigators.

4. The vast expanse of the South Pacific presents a formidable problem to any investigator or to any one nation. Now that the *Eltanin* is shifting its center of operation to the Indian sector of the Antarctic and the Australian, Chilean, and Peruvian investigators are preoccupied with problems related to their countries' contiguous coastal regions, I am afraid that the biological exploration of this vast region of the World Ocean will be left in abeyance for quite some time. Unless other contributors to this symposium will tell us of some yet-untapped fisheries resources in the South Pacific, it is highly unlikely that an all-out attack on the biological oceanography of that region will be seriously considered in the near future. This, together with the sparse human population scattered in the islands of the South Pacific, would not make it politically or economically expedient to suggest an ambitious undertaking similar to that of the International Indian Ocean Expedition.

We should be reminded in this respect that the scientific investigations in the better known regions in the eastern South Pacific, e.g., off Peru, Chile, and Equador, were prompted by fisheries interests, and we should be eternally grateful to the whale fishery in the Antarctic for it has resulted in the valuable contributions made by the *Discovery* Expedition to our knowledge of hydrography, chemistry, and biology of the Southern Ocean.

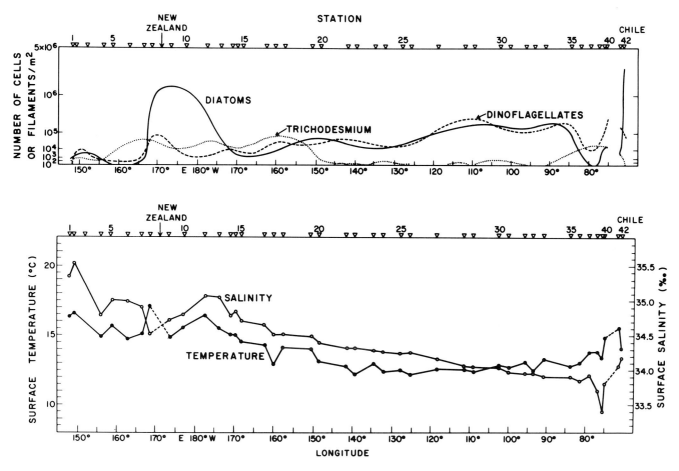

FIGURE 13   Smoothed curves showing the distribution of diatoms, dinoflagellates, and *Trichodesmium thiebauti* per m² of water column (upper) along 43°S during *Eltanin* Cruise 28. Surface-water temperature and salinity at the stations occupied are also shown (lower).

It would be helpful, then, to explore other avenues and other approaches whereby a good coverage of the South Pacific could be achieved in the not-too-distant future. In the first place, the methodology and techniques employed in quantitative study of phytoplankton and primary productivity should be improved upon. For instance, we should capitalize on the new techniques that are now being developed to monitor phytoplankton concentration by use of manned or unmanned satellites. One would look forward to the day when an automated system (similar to the STD system) could be used in estimating primary productivity *in situ.* Further, fluorometric methods for continuous measurements of chlorophyll *a in vivo*, which gave very encouraging results during the EASTROPAC Expedition, should be further explored and improved in order to give us the widest coverage with the least expenditure of effort.

5. Although only small beginnings have been made in the study of the biological productivity of the South Pacific, there are many exciting problems awaiting future investigators. We have very little knowledge of the role played by the nannoplankton in the productivity of the South Pacific. During the International Indian Ocean Expedition, Norris (1966) showed that most of the small-sized (5-90-μ) photosynthetic unarmored dinoflagellates contribute between 55 and 72 percent of the photosynthetic activity in the Indian Ocean. Similar investigations are needed in the South Pacific to assess the contribution of the nannoplankton to the productivity of these waters. Another interesting problem is to assess the role played by the dissolved and particulate organic carbon in the ecosystem of the South Pacific.

Another problem pertains to phytoplankton heterotrophy. The findings of Vinogradov (1962) and Wolff (1960) that copepods occur at great depths throughout the Pacific Ocean, which implies an indigenous food source for these animals, are of intrinsic interest and of far-reaching importance in our understanding of the food cycle in the Pacific Ocean.

The research reported in this paper was supported by the Office of Antarctic Programs, National Science Foundation, under Grant GA-915.

REFERENCES

Angot, M. (1961) A summary of productivity measurement in the southwestern Pacific Ocean, in: *Proceedings of the Conference on Primary Productivity Measurement, Marine and Freshwater* held at University of Hawaii, August 21–September 6, 1961, 1–9.

Balech, E. (1962) The changes in the phytoplankton population off the California Coast. *Rep. Calif. Coop. Oceanogr. Fish. Invest.*, 7, 127–132.

Balech, E. (1968) Dinoflagellates in: V. C. Bushnell (ed.) Antarctic Map Folio Series, Folio *10*, 8–9, plates 11 and 12.

Blackburn, M. (1966) Biological oceanography of the eastern tropical Pacific: Summary of existing information. *Spec. Sci. Rep. U.S. Fish Wildl. Serv. Fish.* No. 540: 1–18.

Braarud, T. (1962) Species distribution in marine phytoplankton. *J. Oceanogr. Soc., Japan, 20th Anniv. Vol.*, 628–649.

Burkholder, P. R., and L. M. Burkholder (1967) Primary productivity in surface waters of the South Pacific Ocean. *Limnol. Oceanogr.*, *12*, 606–617.

Doty, M. S., and M. Oguri (1956) The island mass effect. *J. Cons. Perm. Int. Explor. Mer*, *22*, 33–37.

El-Sayed, S. Z. (1967) Biological productivity investigations of the Pacific sector of Antarctica. *Antarct. J. U.S.*, 2(5), 200–201.

El-Sayed, S. Z. (1968) On the productivity of the southwest Atlantic Ocean and the waters west of the Antarctic Peninsula, in: *Biology of the Antarctic Sea, III. Antarct. Res. Ser.*, editors, W. L. Schmitt and G. A. Llano, Amer. Geophys. Union, *11*, 15–47.

Hart, T. J. (1937) *Rhizosolenia curvata* Zacharias, an indicator species in the Southern Ocean. *Discovery Rep.*, *16*, 413–446.

Hart, T. J. (1942) Phytoplankton periodicity in Antarctic surface waters. *Discovery Rep.*, *21*, 261–356.

Hart, T. J. (1962) Notes on the relation between transparency and plankton content of the surface waters of the Southern Ocean. *Deep-Sea Res.*, *9*, 109–114.

Hasle, G. R. (1959) A quantitative study of phytoplankton from the equatorial Pacific. *Deep-Sea Res.*, *6*, 38–59.

Hasle, G. R. (1968) An analysis of the phytoplankton of the Pacific Southern Ocean: abundance, composition and distribution during the "Brategg" Expedition, 1947–48. Hvalrådets Skrifter Nr. 52:1–168.

Hasle, G. R. (1968) Marine Diatoms, in: V. C. Bushnell (ed.) Antarctic Map Folio Series, Folio *10*:6–8, plates 9 and 10.

Holmes, R. W. (1958) Surface chlorophyll *a*, primary production, and zooplankton volumes in the eastern Pacific Ocean. *Rapp. P.-v. Cons. Perm. Int. Explor. Mer*, *144*, 109–116.

Holmes, R. W. (1961) A summary of productivity measurements in the southeastern Pacific Ocean, in *Proceedings of the Conference on Primary Productivity Measurement, Marine and Freshwater* held at University of Hawaii, August 21–September 6, 1961, 18–57.

Humphrey, G. F. (1960) The concentration of plankton pigments in Australian waters. *Tech. Pap. Div. Fish C.S.I.R.O.*, *9*.

Jitts, H. R. (1965) The summer characteristics of primary productivity in the Tasman and Coral seas. *Aust. J. Mar. Freshwat. Res.*, *16*, 151–162.

King, J. E., T. S. Austin, and M. S. Doty (1957) Preliminary report on Expedition EASTROPIC. *Spec. Sci. Rep. U.S. Fish Wildlife Serv., Fish*, *201*, 1–155.

Kozlova, O. G. (1966) Diatoms of the Indian and Pacific sectors of the Antarctic, Israel Program for Scientific Translations, 191 pp.

Marumo, R. (1968) The study of phytoplankton in the section along the 43°S between Australia and South America. *Tex. A & M Res. Found., Tech. Rep.* (Unpublished)

Norris, R. (1966) Unarmoured marine dinoflagellates. *Endeavour*, *25*, 124–128.

Smayda, T. J. (1958) Biogeographical studies of marine phytoplankton. *Oikos*, 9, 158–191.

Steemann Nielsen, E. (1952) The use of radioactive carbon (C^{14}) for measuring organic production in the sea. *J. Cons. Perm. Int. Explor. Mer*, *18*, 117–140.

Vinogradov, M. E. (1962) Feeding of the deep-sea zooplankton. *Rapp. P.-v. Cons. Perm. Int. Explor. Mer*, *153*, 114–120.

Wolff, T. (1960) The hadal community, an introduction. *Deep-Sea Res.*, *6*, 95–124.

Brian McK. Bary

INSTITUTE OF OCEANOGRAPHY,

UNIVERSITY OF BRITISH COLUMBIA

BIOGEOGRAPHY AND ECOLOGY OF PLANKTON IN THE SOUTH PACIFIC

INTRODUCTION

It is a pecularity of the South Pacific Ocean that investigations to date have been predominantly of the peripheral areas. Thus, equatorial and tropical waters have received much attention, chiefly from laboratories in Hawaii and from the Scripps Institution of Oceanography, and to an increasing amount from New Caledonia. The Southern Ocean has had a long history of investigations, primarily during the Falkland Islands Dependencies Surveys and the Discovery series, beginning in the 1920's and continued more recently by the Soviet Union and the United States. In the west, New Zealand and Australian laboratories are studying the near-shore and oceanic waters in their vicinity, and the reports of several expeditions include these waters. On the west coast of South America, chiefly toward the northern half, data from indigenous laboratories are increasing and are being added to by extensive surveys, for example, by the Inter-American Tropical Tuna Commission. In the central South Pacific waters, observations have been made by the *Carnegie* (1928–1929), Downwind (1958), and Monsoon (1962) expeditions; the transect made by the *Dana* (1928–1929) was in the north and west, and more recently, Russian work has been in equatorial, western, and southern waters.

In a broad view of the ecology and distribution patterns in the South Pacific, much can be added by analogy with the South Atlantic, North Pacific, or Indian Oceans. Such analogies may seem plausible, but an essential factor to be incorporated into future investigations is their substantiation.

BIOGEOGRAPHY AND ECOLOGY

It is convenient to the understanding of the biogeography and ecology of the South Pacific—or any other region—that some basis be available to provide a cohesive interpretation of the scattered, diverse data. One such basis is the relating of occurrences of pelagic organisms to, and interpreting certain ecological features in terms of, one or another oceanographic feature, such as water masses, currents, convergences and divergences, and upwelling (see the discussion, page 221).

Reid (1961, 1965) discusses the currents of the Pacific, and Knauss (1963) reports the recent interpretations of the equatorial system of water movements. (See also Forsbergh and Joseph, 1964, whose diagram is incorporated in Figure 1). Gunther (1936) and Wooster and Reid (1963) describe conditions in the eastern boundary current, and Burling (1961) and Rochford (1957) cover currents and the water masses in the western South Pacific. For the Southern Ocean, accounts of the circulation and water masses have been given by Sverdrup (1933), Deacon (1933, 1963, 1964), and Crease (1964), among others. Figure 1 is a composite diagram that includes the water masses of Sverdrup *et al.* (1942).

A dominant feature is the large gyre in the southeast to which water is contributed from the West Wind Drift, the Peru (Humboldt) Current, the South Equatorial Current, and the water flowing southwest toward Australia and New Zealand. The East Wind Drift is Antarctic Water adjacent to the land; the West Wind Drift incorporates both Antarctic and Subantarctic Waters, but is predominant to the north of the Antarctic Convergence (Crease, 1964). The Antarctic and Subantarctic convergences include transition zones of variable width, in more-or-less stable positions (Deacon 1964), at which waters of differing temperatures and salinities meet. A divergence and upwelling have been observed to the south of the Antarctic Convergence, and they are attributable in part to "stationary cyclones" (Beklemishev, 1958, 1959) which are located more-or-less at the same position for long periods. Beklemishev subdivides the Antarctic Water into two circumpolar zones—

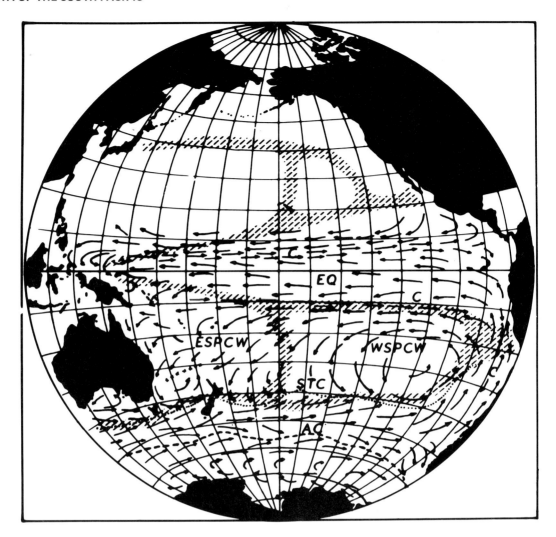

FIGURE 1 Generalized diagram of currents (after various authors) and water masses in the South Pacific. C = countercurrents; EQ = Equatorial Water; ESPCW = Eastern South Pacific Central Water; WSPCW = Western South Pacific Central Water; STC = Subtropical Convergence; AC = Antarctic Convergence. (Reprinted with permission from Sverdrup *et al.*, 1942; copyright © 1942, Prentice-Hall, Inc.)

"Low" Antarctic Water to the north, and "High" Antarctic Water to the south.

Of the other water masses, Pacific Equatorial Water includes the equatorial system of currents; the Eastern South Pacific Central Water (ESPCW), the bulk of the South Pacific gyre; the Western South Pacific Central Water (WSPCW), the southwest flow of water toward Australia and New Zealand; and the Subtropical Convergence, which demarcates the northern edge of the Southern Ocean waters. On present knowledge, these water masses and boundaries appear to be important in the distribution and ecology of at least some of the biota.

In the equatorial system (Knauss, 1963), upwelling is associated with divergence, and sinking is associated with convergence of waters. The processes are significant to the biology of this zone. Off western South America, the northward-flowing Peru Current is a contributory factor to upwelling, with very high primary and secondary productivity as a consequence (Wooster and Reid, 1963; Forsbergh and Joseph, 1964). On the other hand, biologically, disastrous effects ensue on the southward flow of surface water from the north (El Niño) (Bjerknes, 1961).

Data on the horizontal and vertical distribution of inorganic phosphate phosphorus in the Pacific Ocean have been compiled by Reid (1962, 1965). The pattern of distribution at 100 m (Figure 2) parallels that of the water masses (Figure 1), although there is no distinction between the ESPCW and WSPCW. Peripheral waters of the South Pacific have high, and central waters have low, phosphate phosphorus, and the influence of the concentrations reflects in the general planktonic distributions, as Reid (1962) has pointed out (Figure 5).

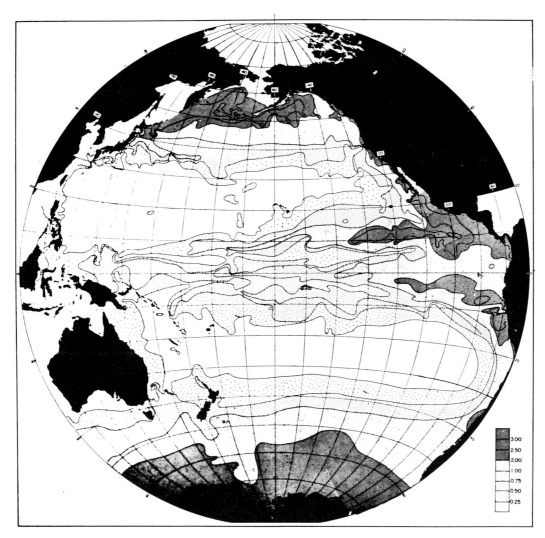

FIGURE 2   Distribution of phosphate phosphorus in the Pacific Ocean at 100 m. (Reprinted with permission from Reid, 1962.)

The peripheral waters are "new" (Sverdrup and Allen, quoted by Graham, 1941a). Along the equator, off the coast of South America and in the Antarctic (and probably around New Zealand), divergences cause upwelling, with the result that nutrient-rich water reaches the surface (Figure 2). This water supports higher phytoplankton production, accompanied at an early stage by a few species of herbivorous zooplankton, usually reaching high numbers of specimens (Longhurst, 1967). This stage persists in areas of upwelling in high latitudes and temperate coastal waters where prolonged stability of the upper waters does not obtain; very high biomasses are typical of these waters. In equatorial waters, where a strong thermocline, and a resultant stability, develops, phosphate becomes reduced, and species diversity increases, but the number of specimens of each species (i.e., the dominance) is reduced. The biomass is much

lower than in temperate or high latitude waters, but is maintained in excess of that in the "poor" waters of the gyres to north and south, or associated with the sinking at the Equatorial Convergence (King and Demond, 1953; King and Hida, 1954). The South Pacific gyre consists of "old," oligotrophic water. A permanent thermocline prevents all but a minimal transfer of nutrients into the upper waters, and consequently, there is a low primary productivity and amount of zooplankton.

In Figure 3 the productivity determined during a transect of the Pacific by the *Galathea* is illustrated (Nielsen, 1954; Nielsen and Jensen, 1957), together with data on numbers of species and specimens collected by other workers; Figures 4 and 5 show the quantities of zooplankton collected by various expeditions in the South Pacific. There is an accord to the extent that where primary productivity

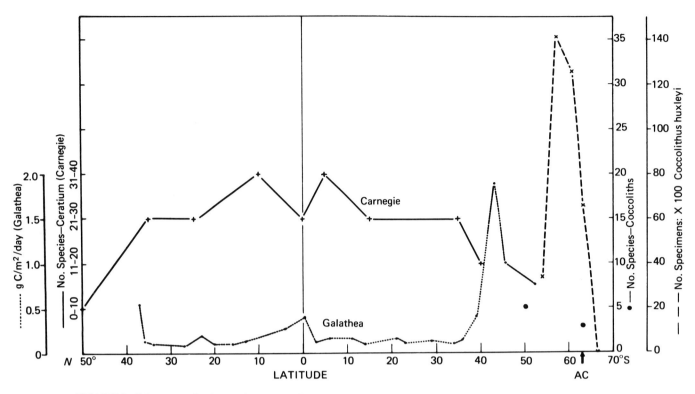

FIGURE 3   Primary production and numbers of species and specimens of phytoplanktonic organisms in the Pacific.
Carnegie—Nos. of species of *Ceratium*. (Reprinted with permission from Graham and Bronikovsky, 1944.)
Galathea—Carbon fixation in transect of Pacific. (Reprinted with permission from Nielsen and Jensen, 1957.)
Coccoliths—(Reprinted with permission from Hasle, 1960.)

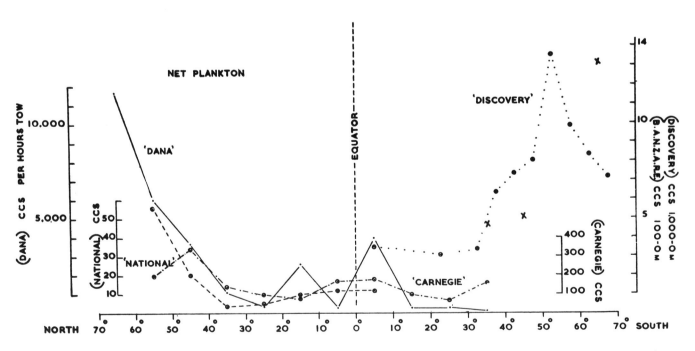

FIGURE 4   Quantities of zooplankton collected by several expeditions in the Pacific. (Reprinted with permission from Foxton, 1956).

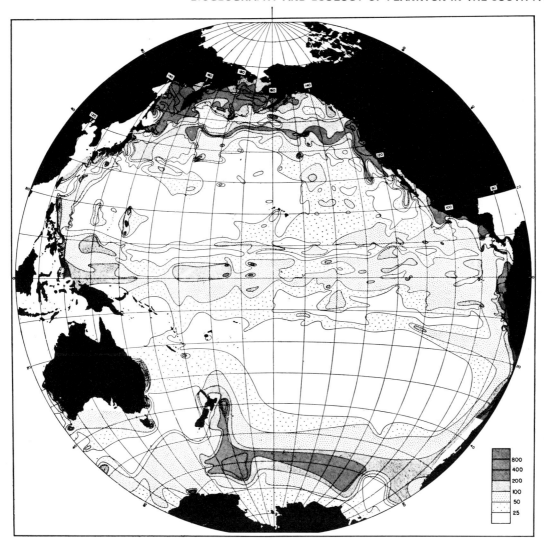

FIGURE 5   Volumes of zooplankton (cm³ per 1,000 m³ of water) of the Pacific Ocean (cf. with Figure 2). (Reprinted with permission from Reid, 1962.)

is high or low, quantities of zooplankton are also high or low. For example, there are very high values in both primary productivity and zooplankton off the North American coast and New Zealand (Jespersen, 1935; Nielsen and Jensen, 1957); and in a north–south direction (Figure 5) there are high values in high latitudes, low values in middle and low latitudes, and variable, moderate values when crossing the equator.

In the Southern Ocean there is little east–west variation in volumes collected in either Antarctic or Subantarctic Waters (Foxton, 1956). The largest volumes occur southward of 50°S, in Antarctic Water, and there is an increase in summer over winter for all depths sampled (down to 1,000 or 1,500 m) throughout the Southern Ocean. Foxton estimates that the standing crop in Antarctic Water is at least four times that of the tropics, but if larger macroplankton

were included (e.g., *Euphausia superba*), the disparity would be "considerably increased."

King and Demond (1953) have shown that in crossing the equator in 150°W and 170°W there were no significant differences in the mean volumes of series of collections, but that there were significant differences among samples within each series, dependent on latitude and season. In the upwelling water of the Equatorial Divergence, volumes (and numbers of species) were low; numbers increased with distance from the upwelling, and toward the convergence when this was strongly developed, but somewhat farther south when there was no convergence. Planktonic Foraminifera are numerous in specimens and species (Bradshaw, 1959) between about 4°S and 4°N; Hasle (1960) lists 32 species of coccolithophores near the equator, and Graham (1941b) showed that the highest numbers of species of

Ceratium (31–40) were centered at about 10°N and 10°S of the equator (Figure 5). Thus, relatively high volumes as well as numerous species are present in the plankton of the Equatorial Water. Within this, however, there is fluctuation that apparently is dependent on the hydrographic conditions, a feature emphasized by Vinogradov and Voronina (1963).

There is a strong tendency toward faunal homogeneity (Vinogradov and Voronina, 1963) in the Equatorial Water, but differentiation has been demonstrated between the eastern and western parts. Thus Brinton (1962) and Johnson and Brinton (1963) depict a group of 6 species of euphausiids (typified by *Euphausia diomediae; Nematoscelis gracilis*) that occurs in a band between about 15°–20° N and S, from west to east; a second group of 5 species (e.g., *Euphausia distinguenda*) occurs toward the east and extends southward to the Peruvian coast, and a third group of 7 is present in western equatorial waters, and southward to eastern Australia and northward to Japan. Among the chaetognaths (Beri, 1959), *Sagitta pacifica*, although occurring in a broad north-south band, is concentrated along the equator, but *Sagitta*

sp. (*S. bieri*) (Alvariño, 1961) occurs in the eastern equatorial waters and north and south along the coasts of the Americas. Three others, *Sagitta bedoti, S. neglecta*, and *S. pulchra* "appear to be disjunct equatorial species that are absent from the central equatorial Pacific" (Bieri, 1959); neither euphausiids nor Foraminifera has equivalent disjunct distributions. *Ceratium lunula* is an equatorial species (Graham, 1941b), and *C. tripos atlanticum* (Graham and Bronikovsky, 1944) is associated with upwelling in eastern tropical areas; and there are parallel distributions among the Peridiniales (Graham, 1942). It seems reasonable to enquire not only as to how many other species and groups of species are distributed similarly to the euphausiids and chaetognaths and what maintains these distributions in equatorial waters, but also as to the amount, locations, and timing of fluctuations in the plankton and the causes behind them.

The central South Pacific waters appear to have one euphausiid (*Euphausia gibba*, Figure 6 and Table 1) confined to them; but there are several species (e.g., *E. Brevis* and *Stylocheiron affine*) that are common to central waters in

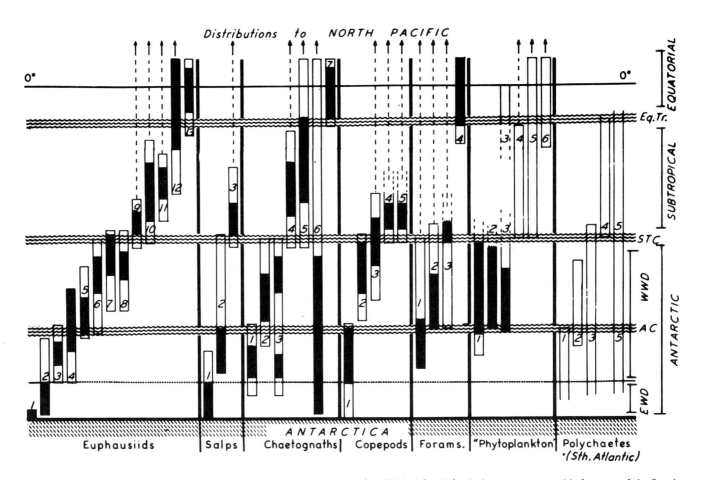

FIGURE 6 Generalized diagram of distributions of planktonic organisms (see Table 1 for list) relative to oceanographic features of the South Pacific (various authors). EWD = East Wind Drift; WWD = West Wind Drift; AC = Antarctic Convergence; STC = Subtropical Convergence; Eq. Tr. = Equatorial Transitional.

TABLE 1 List of Species Included in the Generalized Distributions Shown in Figure 6

Euphausiids	Salps	Chaetognaths	Copepods	Forams	Phytoplankton	Polychaetes (South Atlantic)
E. crystallorophias	S. gerlachei	Sagitta marri	Calanoides acutus	Globigerina pachyderma (sin. and dex.)	Rhizosolenia simplex	Tomopteris carpenteri
E. superba	S. thompsoni	Sagitta maxima	Eucalanus acus	Globoratalia inflata	Nitzschia spp.	Maupasia caeca
E. frigida	S. fusiformis	Sagitta gazellae	Paracalanus parvus	Globigerina bulloides	Coccolithus huxleyi	Travisiopsis levinseni
E. triacantha		Sagitta californica	Sapphirina spp.	Globigerinoides conglobata	Ceratium tripos	Tomopteris nisseni
E. vallentini		Sagitta serratodentata	Pleuromamma spp.		Ceratium macroceros	Pelagobia longicirrata
E. longirostris		Eukrohnia hamata			Ceratium breve	
E. lucens		Sagitta robusta				
E. similis						
E. recurva						
E. brevis						
E. mutica						
E. "gibba" group						
E. diomediae						

both the North and South Pacific (Brinton, 1962). There are no chaetognaths (Figure 6) peculiar to the central South Pacific (Bieri, 1959); *Sagitta californica*, however, is widely distributed in central waters north and south of the equator, and the similar distribution of *Ceratium cephalotum* typifies several species of this genus and of the Peridiniales (Graham, 1941; Graham and Bronikovsky, 1944).

The southeastern South Pacific appears to have distinctive faunistic features. The area is one of transitional water to which ESPC and Subantarctic waters contribute (Figure 1). Although there appear to be few indigenous species, numerous others either have not been recorded there or occur less abundantly there than in neighboring areas. Among chaetognaths (Bieri, 1959), *Sagitta friderici* is abundant (and also in transitional waters of California), but *Pterosagitta draco*, which is widespread in central waters to the west, is absent. The only euphausiid confined to the area is *Euphausia mucranata*, but *Nyctiphanes simplex* occurs there and again off the coast of California; *Thysanoessa gregaria* (Figure 8) is present across the Pacific and is numerous in the transitional water off South America, whereas *Nematoscelis megalops*, with a similar oceanic distribution, is absent. Among species of *Ceratium* (Graham and Bronikovsky, 1944) and the Peridiniales (Graham, 1942), some are absent from the area and are common elsewhere (e.g., *Ceratium gravidum, C. falcatum, Peridinium oceanicum*, and *Ceratocorys* spp.); others occur frequently there, but less so elsewhere (*Ceratium cephalotum* and *Peridinium crassipes*), and others again are noticeably reduced in frequency of occurrence (*C. breve* and *Gonyaulax pacifica*). Additionally, species that are present there are also widespread throughout the South Pacific (*Ceratium massiliense, C. macroceros, C. furca*, and *Goniodoma polyedricum*).

It would seem then that some species are entering the transitional water in the eastern South Pacific along with the contributing waters and are able to survive. On the other hand, there are species that are either reduced in abundance or eliminated; in addition, there are those few that are present in the transitional water and have not been found in the contributing waters, but may occur in transitional water elsewhere. Clearly, the species are demonstrating a range of tolerances to a variable environment. Whether those that do occur in transitional water are there as expatriates or breeding populations is unknown. The transitional waters cover a reasonably limited area in the southeast Pacific into which waters of distinctive properties and faunas are entering. Interactions there between species and the environment would make an interesting study for comparison with transitional waters elsewhere.

Even though, in one or other of the sectors of the Southern Ocean, the distribution and ecology of the plankton have received a considerable amount of attention, Mackintosh (1964a and b) warns that, in the main, only broad

distributional features are known; other ecological aspects require more definitive study. Probably the fact that the circumpolar distribution of species, on the whole, is continuous (Mackintosh, 1937, 1964a; David, 1958; Foxton, 1961; Baker, 1959) will help in that biological situations described for one sector can be expected to apply also to another (David, 1964). Such an accord, however, needs establishing: for example, *Euphausia lucens* occurs mainly south of the Subtropical Convergence in the Pacific (John, 1936; Bary, 1956; and see Figure 6 and Table 1) but north of it in the Indian Ocean (Baker, 1965); and *Salpa gerlachei* is limited to the Pacific sector of the Antarctic (Foxton, 1961).

Much attention has been given to the effects of the Antarctic and Subtropical convergences on distributions (Figure 6 and Table 1). Convergences are essentially fronts in which the transition between the properties of the waters is rapid. Because of this change, a convergence can (and often does) act as a faunal boundary. From its structure, the Subtropical Convergence probably affects occurrences mainly of those species inhabiting depths to about 100 m. The few subtropical species recorded to the south of this convergence presumably have been transported in Subtropical Water moving southward below Subantarctic Water. For example, *Thysanoessa gregaria* was collected at 51°S (south of New Zealand) (Bary, 1956), although it is regarded more usually as an inhabitant of southern (and northern) transitional waters (Brinton, 1962) or "temperate and subtropical waters" (Hansen, 1911; Tattersall, 1924; Boden, 1954). David (1964) regards occurrences of *Sagitta serratodentata* in Subantarctic Water as originating northward of the convergence.

The sinking water at the Antarctic Convergence contributes to the Antarctic Intermediate Water, which flows northward at around 800–1,000 m. Species may remain in this water, at least seasonally (Mackintosh, 1937, 1964a; David, 1958, 1964), and extend their range northward at the deeper levels. Examples are "yeasts" (Fell, 1967), *Rhincalanus gigas, Calanoides acutus* (Mackintosh), and *Sagitta maxima, S. marri*, and *Eukrohnia hamata* (David). How many species are concerned and how far to the north they may be distributed is known in only a general way (Figure 6). However, *Eukrohnia hamata* may be continuous with the Northern Hemisphere distribution of the species; *Primno macropa* (Amphipoda) has been collected from New Zealand waters in areas of upwelling; but *Sagitta marri* and *Calanoides acutus* extend barely beyond the Antarctic Convergence.

The Subtropical and Antarctic convergences have been shown as effective barriers in distributions of some organisms; on the other hand, it is true also that species "cross" these barriers (Figure 6). The distribution of *Coccolithus huxleyi* extends from Equatorial into Subantarctic waters, and it appears to multiply in the latter (Hasle, 1960), but it has been collected only very rarely south of the Antarctic

Convergence (Figure 3). There are distinct species groups of Foraminifera (Figure 6) in the Subtropical and Subantarctic waters of the Atlantic (Boltovskoy, 1966a); some species from each group are contributed to the transitional water separating these, and the abundant *Globoratalia inflata* is almost confined there. In the southern Pacific, *Globigerina* ex. gr. *pachyderma* (sin. and dex. forms) occurs in high numbers from Subantarctic into Antarctic Water (Boltovskoy, 1966b), but a number of species (typified by *Globoratalia bulloides* and *G. truncatulinoides*), while common in Subantarctic Water, barely extend into the transitional water associated with the Subantarctic Convergence, and do not extend into Antarctic Water at all.

Foxton (1961) has clarified the systematics and distributions of some Southern Hemisphere salps. *Salpa fusiformis* and *S. aspera* do not penetrate southward of the Subtropical Convergence (Figure 6); a new species, *Salpa thompsoni*, has a circumpolar distribution in both Subantarctic and Antarctic waters, but, again, does not cross the Subtropical Convergence; a second new species, *S. gerlachei*, extends westward from the Ross Sea almost to Bransfield Strait, but only in the highest latitudes.

The distributions of some pelagic polychaetes in the South Atlantic (Tebble, 1960) are limited by convergences, but others are not (Figure 6). Treadwell (1943) does not show the equivalent for the South Pacific, but presumably a more complete survey than was possible from *Carnegie* material may do so. The Antarctic Convergence clearly separates two distinct populations of the predominantly near-surface *Sagitta gazellae* (David, 1955, 1964).

John (1936) indicates that occurrences of some euphausiids in the Pacific sector (Figure 6) are limited by either the Antarctic Convergence (e.g., *Euphausia frigida*) or the Subtropical Convergence (*E. longirostris*), but others are not (Mackintosh, 1964a). These cross one or another of the convergences (e.g., *E. triacantha*) or occur apparently without relation to any known oceanographic situations (e.g., *E. crystallorophias; E. similis*).

Species occurring across a convergence may be able to tolerate and occupy a wide range of conditions (Figure 8A, B, and C), or they may be less tolerant and not able to survive outside the confines of the water mass they inhabit and within which they are being transported (Figure 9A). The resulting distribution, viewed geographically, may be similar in the two cases, but the "causes" differ fundamentally. It seems important that the distinctions be recognized and investigated because they reflect on other ecological considerations.

The distributions of *Euphausia superba*, *E. triacantha*, *Salpa thompsoni*, and *S. gerlachei* may bear on another situation. *Euphausia superba* occurs predominantly in the most southern Antarctic Water; *E. triacantha*, in the more northern Antarctic Water (Figure 6; Table 1). The distributions, while not exclusive, approach this state (Marr, 1962; Baker, 1965). *Salpa gerlachei* occurs in the southernmost waters of the Pacific sector; *S. thompsoni*, northward of it. Beklemishev (1958, 1959) and Beklemishev and Korotkevitch (1958), in discussing distributions of zooplankton and phytoplankton in the Antarctic with reference to upwelling associated with the Antarctic Divergence, cite instances when *Euphausia superba* has concentrated in the upwelling water. Perhaps the zone of divergence is persistent enough a feature at times to constitute a boundary limiting the northward distribution of *E. superba* and *S. gerlachei* and the southern penetration of *E. triacantha* and *S. thompsoni*.

Despite the fact that the Southern Ocean is relatively well known, much remains to be investigated. Mackintosh (1964a) discusses the problems in some detail, together with a hypothesis linking distributions and other ecological considerations with speciation in which an explanation for species that "cross" convergences, or other supposed barriers, and those that do not, is expounded. To try to elucidate such features from field data will be time-consuming and difficult, but at present no alternative presents itself, so it ought to be tried. Detailed sampling of the biota down to great depths at all seasons and then relating occurrences of carefully identified species and all their life-history stages to the water masses (and mixtures) would provide a considerable amount of relevant information.

OTHER ECOLOGICAL CONSIDERATIONS

There are other ecological features for which more information is needed. Mackintosh (1937), David (1955), and Baker (1959), using collections from vertical hauls over standardized discrete depth ranges, have shown, for example, that there is a seasonal change in vertical distribution in several species (e.g., *Sagitta gazellae*, Figure 9; *Euphausia superba;* and *Calanoides acutus*) in Antarctic waters; specimens occur in the upper levels in summer and descend to deep water in winter. Breeding appears to be associated with this change. There is also the probability that the species that move northward (and eastward) during the summer are contained within the Antarctic circumpolar orbit by descending to the deep water and returning southwards during winter, either as adults, or at an earlier stage in life-history. Such movements suggest that stages of the life cycle may inhabit alternative environments (water masses) (Figures 9A, 10). Implicit are several questions: Are all, or only some species affected; and if the latter, then what is the fate of those individuals carried northward and destined not to return; do they breed or do they become expatriates? How regularly do particular life-history stages become inhabitants of a particular water? Is the stimulus initiating the descent the same as that initiating ascent? Is the *correct* water recognized and if so, how; or is entry into the correct water a matter of chance and therefore beneficial only to the fraction of the population that happens to enter it?

Other and different ecological relationships have been described for equatorial waters, and, in particular, between phytoplankton and zooplankton (Vinogradov and Voronina, 1962) in the Indian Ocean. Nutrient-rich, upwelled water provides the basis for high phytoplankton productivity. Associated with this is an increase of herbivorous copepods, which remain in close proximity to the phytoplankton. The more slowly increasing carnivores (other copepods and larger zooplankton) are displaced horizontally by water movements (to the north in the Indian Ocean). King and Hida (1954) indicate that a similar situation exists in Pacific equatorial waters. These sorts of relationships in the equatorial waters could be compared with the equivalent in areas of upwelling in high and low latitudes for the other peripheral South Pacific waters (Longhurst, 1967). Possible causes for species diversity and the "balanced" and "unbalanced" cycles in relations between phytoplankton and zooplankton

(Heinrich, 1962) are associated ecological features for which information would be obtained.

Data are needed on the food and feeding of species undertaking seasonal and diel vertical migrations, including in the great depths, in order to determine the general nature of the downward transfer of energy (Figure 7). Vertical distributions, their permanency (Vinogradov, 1962b), or changes in the distribution of the species and their life-history stages within and between areas (Mackintosh, 1937) would be relevant. Comparison of data from areas rich and poor in phytoplankton would indicate not only differences and similarities in near-surface trophic relations (e.g., Heinrich, 1962), but the nature of changes in the relations with depth and their effects on distribution, abundance, and diversity.

The influence of an island on ecology may be local, but it can also become important in aggregate where islands are

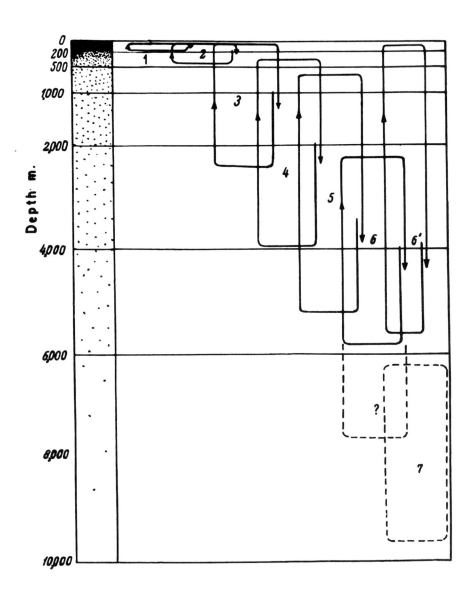

FIGURE 7   Schematic presentation of the overlapping ranges of depths of 7 groups of zooplanktonic organisms and the vertical migration of each, to indicate one way in which energy could be transferred downwards from the surface.

(Left panel) The range of biomass of plankton with depth; frequency of points in each layer is proportional to the biomass. (Reprinted with permission from Vinogradov, 1962a.)

numerous. The South Pacific is peculiarly suited to studying planktonic assemblages and ecology for areas free of islands or encumbered by them. The islands may be isolated or in groups, with or without encircling reefs, and affected by a variety of oceanographic influences. Higher primary productivity east of the Galápagos Islands (Forsbergh and Joseph, 1964) probably results from the effect that these islands have on circulation and distribution of nutrients. How the neritic pelagic faunas are maintained in the diverse conditions encountered about islands (Boden, 1952; Boden and Kampa, 1953) is a study in physical oceanography as well as in biology. Similar considerations should be extended to the shoal waters in the vicinity of seamounts.

DISCUSSION

The better known areas of the South Pacific appear to be the Southern Ocean, the equatorial system (more especially in the eastern half), the northeastern waters off South America, and the southwest Pacific, in that order. The less well known are the central waters, including the southeastern gyre, and a poorly defined, oceanwide zone lying between midlatitudes and south of the Subtropical Convergence. For example, no faunal distinctions have been drawn between eastern and western South Pacific Central waters. Possibly Reid's suggestion (1962) that the gyre in the central waters is the basis of the faunal entity applies. If so, distinctions may not be significant; but the differing histories of the WSPCW and ESPCW indicate that the faunas should also differ.

It is likely that the better known areas will continue to be investigated, partly because they have economic associations and partly because there are enough data on which to base further studies. However, it is important that the whole region be more fully explored rather than concentrating efforts on the undoubtedly interesting peripheral areas so far studied in some detail.

The general dearth of data on an oceanwide scale in the South Pacific stems from spotty collecting and also because methods used and the analyses accorded collections have been such that only the broadest evaluations or comparisons can be made (e.g., Foxton, 1956; and see Figure 4). Collecting techniques in future do not have to be identical for everyone, although it would help if they were similar. The least requirement is that one set of data may be related to another with more telling effect than has been possible to date. To this end, all collections should first be as quantitative as available equipment allows. Second, investigations for the foreseeable future should be concerned primarily with elucidating some broadly acceptable, unifying concept.

Quantitative procedures involve volumetric sampling using equipment such as pumps, closing containers, or metered closing samplers ("nets"), or techniques such as

the use of ^{14}C for determining productivity. All samples should be monitored for depth and duration (where appropriate) and be carefully documented. Preferably, zooplankton samples should be from stratified or divided hauls and be uncontaminated by organisms from above or below the depth or range of depth sampled. Each sample should be accompanied by appropriate physical and chemical data; at the very least, temperature and salinity data should be provided.

There is evidence supporting the contention that occurrences (and therefore distributions) of species are related to water masses in the South Pacific (Figures 8, 9, and 10). However, there is other evidence indicating that species are not so controlled (Figure 6; Table 1). Contradictions of this sort are to be expected, but they should be scrutinized, which would be possible if occurrences were studied in relation to the vertical and horizontal extent and movements of the water masses, instead of emphasizing the geographic location of samples. Nor would this approach jeopardize a considerable range of investigations (see below). Identical sampling procedures among several programs would not be essential, but if each were quantitatively based and if the analyses produced data with some compatibility, then results could be compared or combined to provide a basis for elucidating or investigating principles underlying distributions and ecology. Detailed studies of a part of an ecological or an oceanographic system are important, but data from them often can be interpreted adequately only in the light of such principles.

To further such an aim it would be necessary to associate all occurrences with concurrent hydrographic observations, whether it is intended to correlate the occurrences with water masses as defined by Sverdrup et al. (1942; viz., McGowan, 1960; Johnson and Brinton, 1963), or by using Hida's technique (1957), or the method for near-surface waters used by Bary (1963; Figures 8, 9, and 10). In addition to revealing the relationships of the species to water masses, or the included water "bodies" (Bary, 1963), these techniques, and the statistical approach of Fager and McGowan (1963) can point up differences in tolerances among species; provide information on the geographical and vertical extent and composition of faunal entities, and on interactions between adjacent entities; demonstrate the "home water" of species and their breeding and succession within it or possible instances of expatriation from it; and assist in interpreting species diversity, whether covering broad distances (e.g., oceanwide traverses) and/or depths, or areas associated with currents, divergences, upwelling, and convergences.

The more useful data discussed above are those that have been preceded by exhaustive systematics applied to whole groups of organisms within recognized oceanographic units, e.g., the Antarctic and/or Subantarctic waters (John, 1936; David, 1955, 1958; Foxton, 1961, or within a system, such

FIGURE 8   The geographic portrayal (A) of the distribution of *Thysanoessa gregaria* and its occurrences relative to water masses of the North Pacific (B), and South Pacific (C). (Reprinted with permission from Brinton, 1962.)

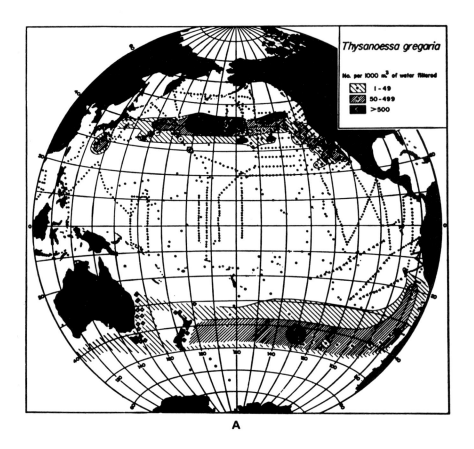

A

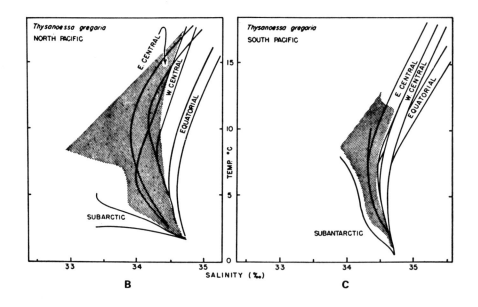

B

C

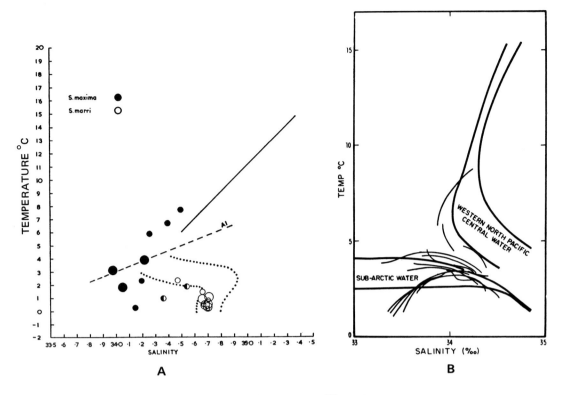

FIGURE 9   *A*, Occurrences of *Sagitta maxima* (●) and *S. marri* (○) relative to water masses.
AI = Antarctic Intermediate Water; (.....) = Envelope of the Warm Deep Water; (——) = Core of South Atlantic
Central Water. (Reprinted with permission from David, 1958.) *B*, Occurrences of *Poeobius meseres* relative to
water masses of North and South Pacific. Thin lines show temperature and salinity over the length of each haul
in which specimens were present. (Reprinted with permission from J. A. McGowan, The relationship of the distribution of the plankton worm, *Poeobius meseres* Heath, to the water masses of the North Pacific. *Deep-Sea
Res., b*; copyright © 1960, Pergamon Press, Inc.)

as a large part of the Pacific Ocean (Graham, 1941a and b;
Bieri, 1959; Tebble, 1960; Brinton, 1962; Boltovskoy,
1966a and b). It is essential to future investigations that
such systematics be expanded. Unless identifiable species
and populations, or both, are being considered in relation
to their occurrences and ecology, information will be lost
or neglected. Thus, as David (1958) points out for *Eukrohnia hamata*, specimens from Antarctic Water differ morphologically from those in Subantarctic Water; and specimens
submerged below tropical waters may differ from the southern and northern representatives. He has also shown that of
two morphologically differing "races" of *Sagitta gazellae*,
one inhabits Antarctic Water and the other, Subantarctic

Water (Figure 6; Table 1). In recognizing *Salpa gerlachei*,
Foxton (1961) has opened up the problem of why only this
species has a limited distribution in high-latitude Antarctic
Water. Brodsky (1961) has shown that *Calanus finmarchicus*
(S.L.) is in reality a complex of species (and subspecies),
several of which inhabit the South Pacific. Clarification of
the distributions and environmental relations of such taxa
has already provided information of greater particular use
than was available when considering them, for example, as
"*Salpa fusiformis*" or as "*C. finmarchicus.*" Before definitive and meaningful distributional or other ecological data
can be obtained, the systematics of organisms must be
secure.

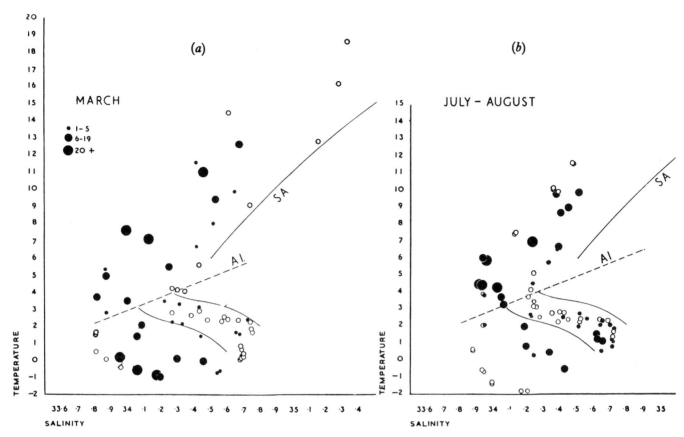

FIGURE 10 Summer (a) and winter (b) occurrences of *Sagitta gazellae* in the Atlantic sector (0° meridian) of the Southern Ocean. For water masses, see caption to Fig. 8 a; o = specimens absent. (Reprinted with permission from David, 1955.)

REFERENCES

Alvariño, A. (1961) Two new chaetognaths from the Pacific. *Pac. Sci., 15* (1), 67–77.

Baker, A. de C. (1959) Distribution and life history of *Euphausia triacantha* Holt and Tattersall. *Discovery Rep., 29,* 309–340.

Baker, A. de C. (1965) The latitudinal distribution of *Euphausia* species in the surface waters of the Indian Ocean. *Discovery Reps., 33,* 309–334.

Bary, B. McK. (1956) Notes on ecology, systematics and development of some Mysidacea and Euphausiacea (Crustacea) from New Zealand. *Pac. Sci., 10*(4), 331–367.

Bary, B. McK. (1963) Distribution of Atlantic pelagic organisms in relation to surface water bodies. *Proc. Roy. Soc., Canada, Spec. Publ., No. 5,* 51–67, University of Toronto Press.

Beklemishev, K. V. (1958) The dependence of the phytoplankton distribution on the hydrological conditions in the Indian sector of the Antarctic. *Dokl. Acad. Nauk SSSR, 119* (4), 494–497.

Beklemishev, K. V. (1959) Antarctic Divergence and whaling grounds. *Isvestia Akad. Nauk SSSR, 6,* 90–93.

Beklemishev, K. V., and V. S. Korotkevitch (1958) Zooplankton of the Indian sector of the Antarctic. *Dokl. Akad. Nauk SSR, 121* (6), 1009–1011.

Bieri, R. (1959) The distribution of the planktonic Chaetognatha in the Pacific and their relationship to the water masses. *Limnol. Oceanogr.,* 4(1), 1–28.

Bjerknes, J. (1961) "El Niño" study based on analysis of ocean surface temperatures 1935–57. *Inter-Amer. Trop. Tuna Comm., Bull., 5* (3), 219–303.

Boden, B. P. (1952) Natural conservation of insular plankton. *Nature, 169* (4304), 697–699.

Boden, B. P. (1954) The euphausiid crustaceans of southern African waters. *Trans. Roy. Soc. S. Africa, 34* (1), 181–243.

Boden, B. P., and E. M. Kampa (1953) Winter cascading from an oceanic island and its biological implications. *Nature, 171* (4349), 426–427.

Boltovskoy, E. (1966a) La zona de Convergencia Subtropical/Subantarctica en el oceano Atlantico (Parte Occidental) (Un estudio en base a la investigacion de Foraminiferos-indicadores). *Serv. Hidrogr. Naval, Argentina, H. 640,* 1–69.

Boltovskoy, E. (1966b) Zonacion en las latitudes altas del Pacifico sur segun los Foraminiferos planctonicos vivos. *Revta. Mus. Argent. Cienc. Nat. Bernardino Rivadavia Inst. Nac. Invest. Cienc. Nat. (Hydrobiol.),* 2(1), 1–56.

Bradshaw, J. W. (1959) Ecology of living planktonic Foraminifera in the North and Equatorial Pacific Ocean. *Contrib. Cushman Found. Foraminifera Res., 10* (2), 25–64.

Brinton, E. (1962) The distribution of Pacific euphausiids. *Bull. Scripps Inst. Oceanogr., 8* (2), 51–270.

Brodsky, K. (1961) Comparison of *Calanus* species (Copepoda) from the Southern and Northern Hemisphere. *N.Z. Dep. Scient. Ind. Res., Info. Ser. No. 33,* 22 pp.

Burling, R. W. (1961) Hydrology of circumpolar waters south of New Zealand. *N.Z. Dep. Scient. Ind. Res., Bull., 143,* 66 pp., maps.

Crease, J. (1964) The Antarctic circumpolar current and Convergence. *Proc. Roy. Soc., (A), 281,* 14–21.

David, P. M. (1955) The distribution of *Sagitta gazellae* Ritter Zahony. *Discovery Rep., 27,* 235–278.

David, P. M. (1958) The distribution of the Chaetognatha of the Southern Ocean. *Discovery Rep., 29,* 199–228.

David, P. M. (1964) The distribution of Antarctic Chaetognaths. *Biologie Antarctique: 1st. Sym. of SCAR, Paris*, pp. 253–256.

Deacon, G. E. R. (1933) A general account of the hydrology of the South Atlantic Ocean. *Discovery Rep., 7*, 171–238, Plates viii–x.

Deacon, G. E. R. (1963) The Southern Ocean. Vol. 2, Ch. 12 in: *The Sea: Ideas and Observations on Progress in the Study of the Seas*, editor M. Hill, Wiley-Interscience, New York.

Deacon, G. E. R. (1964) Introduction to "A Discussion on the physical and biological changes across the Antarctic Convergence." *Proc. Roy. Soc., (A), 281*, 1–6.

Fager, E. W., and J. A. McGowan (1963) Zooplankton species groups in the North Pacific. *Science, 140* (3566), 453–460.

Fell, J. W. (1967) Distribution of yeasts in the Indian Ocean. *Bull. Mar. Sci., 17* (2), 454–470.

Forsbergh, E. D., and J. Joseph (1964) Phytoplankton production in the eastern Pacific Ocean. *Inter-Amer. Trop. Tuna Comm., Bull., 8* (9), 479–527.

Foxton, P. (1956) The distribution of the standing crop of zooplankton in the Southern Ocean. *Discovery Rep., 28*, 191–236.

Foxton, P. (1961) *Salpa fusiformis* Cuvier and related species. *Discovery Rep., 32*, 1–32.

Graham, H. W. (1941a) Plankton production in relation to character of water in the open Pacific. *J. Mar. Res., 4* (3), 189–197.

Graham, H. W. (1941b) An oceanographic consideration of the dinoflagellate genus *Ceratium*. *Ecol. Monogr., Durham, 11* (1), 99–116.

Graham, H. W. (1942) Studies in the morphology, taxonomy, and ecology of the Peridiniales. *Sci. Results, Cruise VII of the Carnegie, 1928–29, Biol. 3, Carnegie Inst. Washington Publ., 542*, i–v, 129 pp.

Graham, H. W., and N. Bronikovsky (1944) The genus *Ceratium* in the Pacific and North Atlantic Oceans. *Sci. Results, Cruise of the Carnegie, 1928–29, Biol. 5, Carnegie Inst. Washington Publ., 565*, 1–187.

Gunther, E. R. (1936) A report on oceanographical investigations in the Peru Coastal Current. *Discovery Rep., 13*, 107–276.

Hansen, H. J. (1911) The genera and species of the order Euphausiacea with account of remarkable variation. *Bull. Inst. Oceanogr., Monaco, 210*, 1–54.

Hasle, G. R. (1960) Plankton coccolithophorids from the Subantarctic and Equatorial Pacific. *Nytt. Mag. for Botanikk, 8*, 77–88, 2 pl.

Heinrich, A. K. (1962) The life histories of plankton animals and seasonal cycles of plankton communities in the oceans. *J. Cons. Int. Explor. Mer, 27* (1), 15–24.

Hida, T. S. (1957) Chaetognaths and pteropods as biological indicators in the North Pacific. *Spec. Sci. Rep. U.S. Fish Wildlife Serv., Fish, 215*, 13 pp.

Jesperson, P. (1935) Quantitative investigations on the distribution of macroplankton in different oceanic regions. *Rep. "Dana" Exped., 2*(7), 1–44.

John, D. (1936) The southern species of the genus *Euphausia*. *Discovery Rep., 14*, 193–324.

Johnson, M. W., and E. Brinton (1963) Biological species, watermasses and currents. Vol. 2, Ch. 18 in: *The Sea: Ideas and Observations on Progress in the Study of the Seas*, editor M. N. Hill, Wiley-Interscience, New York.

King, J. E., and J. Demond (1953) Zooplankton abundance in the central Pacific. *Fish. Bull., U.S., 82*, 111–144.

King, J. E., and T. S. Hida (1954) Variation in zooplankton abundance in Hawaiian waters, 1950–52. *Spec. Sci. Rep. U.S. Fish Wildlife Serv., Fish, 118*, 66 pp.

Knauss, J. A. (1963) Equatorial current systems. Vol. 2, Ch. 10 in: *The Sea: Ideas and Observations on Progress in the Study of the Seas*, editor M. N. Hill, Wiley-Interscience, New York.

Longhurst, A. R. (1967) Diversity and trophic structure of zooplankton communities in the California Current. *Deep-Sea Res., 14*, 393–408.

McGowan, J. A. (1960) The relationship of the distribution of the planktonic worm, *Poeobius meseres* Heath, to the water masses of the North Pacific. *Deep-Sea Res., 6*, 125–139.

Mackintosh, N. A. (1937) The seasonal circulation of the Antarctic macroplankton. *Discovery Rep., 16*, 365–412.

Mackintosh, N. A. (1964a) Distribution of the plankton in relation to the Antarctic Convergence. *Proc. Roy. Soc., (A), 281*, 21–38.

Mackintosh, N. A. (1964b) A survey of Antarctic biology up to 1945. *Biologie Antarctique: 1st. Sym. of SCAR, Paris*, pp. 29–38.

Marr, J. W. S. (1962) The natural history and geography of the Antarctic krill (*Euphausia superba* Dana). *Discovery Rep., 32*, 33–464.

Nielsen, E. S. (1954) On organic production in the oceans. *J. Cons. Int. Explor. Mer, 19*(3), 309–328.

Nielsen, E. S., and A. A. Jensen (1957) Primary oceanic production. The autotrophic production of organic matter in the oceans. *Galathea Rep., 1*, 49–136. (Sci. Res., Danish Deep-Sea Exped. Round the World, 1950–52.)

Reid, J. L. (1961) On the geostrophic flow at the surface of the Pacific Ocean with respect to the 1,000 decibar surface. *Tellus, 13*(4), 489–502.

Reid, J. L. (1962) On circulation, phosphate–phosphorus content, and zooplankton volumes in the upper part of the Pacific Ocean. *Limnol. Oceanogr., 7*(3), 287–306.

Reid, J. L. (1965) Intermediate waters of the Pacific Ocean. *Johns Hopk. Oceanogr. Stud., 2*, 85 pp.

Rochford, D. J. (1957) The identification and nomenclature of the surface water masses in the Tasman Sea (Data to the end of 1954). *Aust. J. Mar. Freshwater Res., 8*(4), 369–413.

Sund, P. N. (1961) Some features of the autecology and distribution of Chaetognatha in the eastern tropical Pacific. *Inter-Amer. Trop. Tuna Comm., Bull., 5*(4), 307–340.

Sverdrup, H. U. (1933) On vertical circulation in the ocean due to the action of the wind with application to conditions within the Antarctic Circumpolar Current. *Discovery Rep., 7*, 141–169.

Sverdrup, H. U., M. W. Johnson, and R. H. Fleming (1942) *The Oceans, Their Physics, Chemistry and General Biology*. Prentice-Hall, Inc., New York, i–x, 1060 pp.

Tattersall, W. M. (1924) Euphausiacea. *Bull. Brit. Mus. Nat. Hist., Zool., 8*(1), 1–36.

Tebble, N. (1960) The distribution of pelagic polychaetes in the South Atlantic Ocean. *Discovery Rep., 30*, 161–300.

Treadwell, A. L. (1943) Polychaetous annelids. *Sci. Res. Cruise VII of the Carnegie, 1928–29, Biol. 4, Carnegie Inst., Washington, Publ. 555*, 29–60.

Vinogradov, M. E. (1962a) Feeding of the deep-sea zooplankton. *Rapp. P.-v. Cons. Perm. Int. Explor. Mer, 153*(18), 114–120.

Vinogradov, M. E. (1962b) Quantitative distribution of deep-sea plankton in the western Pacific and its relation to deep-water circulation. *Deep-Sea Res., 8*, 251–258.

Vinogradov, M. E., and N. Voronina (1962) The distribution of different groups of plankton in accordance with their trophic level in the Indian Equatorial Current area. *Rapp. P.-v. Cons. Perm. Int. Explor. Mer, 153*(33), 200–204.

Vinogradov, M. E., and N. Voronina (1963) Quantitative distribution of plankton in the upper layers of the Pacific Equatorial Currents. *Trud. Inst. Okeanol., 71*, 22–59.

Wooster, W. S., and J. L. Reid (1963) Eastern boundary currents. Vol. 2, Ch. 11 in: *The Sea: Ideas and Observations on Progress in the Study of the Seas*, editor M. N. Hill, Wiley-Interscience, New York.

Michel Legand

Philippe Bourret

René Grandperrin

OCEANOGRAPHES BIOLOGISTES AU
CENTRE O.R.S.T.O.M. DE NOUMÉA

Jacques Rivaton

ASSISTANT DE RECHERCHE

A PRELIMINARY STUDY OF SOME MICRONEKTONIC FISHES IN THE EQUATORIAL AND TROPICAL WESTERN PACIFIC

Micronekton studies, at least those dealing with concepts of quantitative distribution, are relatively new. They were developed mainly with the increased use of the Isaacs-Kidd midwater trawl. This gear provides biologists a relatively easy method for sampling of the larger organisms of the food chain in the open ocean. Because the study of these organisms is important to understanding the biocenosis in which tunas live, a relatively large effort on this subject has been made by the O.R.S.T.O.M. Oceanographic Department of Nouméa. By now, after 3 years of cruises by the research vessel *Coriolis*, some general features of the composition of the samples have been established; to some extent, they give a new composite view of the ichthyology of the equatorial currents and surrounding area of the western Pacific and a confirmation of hitherto sparse indications of the great abundance of some micronektonic fishes. We do not know to what extent a generalization from these features to the whole tropical South Pacific will be valid, although we think that some can be justified. As the results considered are only those of cruises preparatory to more extensive work, we have restricted our aim here to a synthesis of those results and have not tried to relate them to other data.

ORIGIN OF DATA

The data are derived from three cycles of cruises of the research vessel *Coriolis* (Figure 1).

1. November 1964 to March 1965: Alize cruises from 92°W to 162°E along the equator; 33 stations made with a 5-ft Isaacs-Kidd midwater trawl (IKMT) towed obliquely to 300 m between 2000 to 2200 hours. (References to Alize cruises are only incidental.)
2. December 1965 to October 1966: Bora I–IV cruises, mainly along 170°E from 20° to 5°N; 96 stations made with a 10-ft IKMT towed obliquely down to various depths according to the cruises. The detail of the stations is: 26 (0–300 m), 14 (0–650 m), 21 (0–900 m or 0–1200 m) night hauls; 4 (0–300 m), 8 (0–650 m), 23 (0–900 m or 0–1,200 m) day hauls.
3. December 1966 to September 1967: Cyclone I to VI cruises along 170°E from 0° to 5°S, 10-ft IKMT oblique hauls were made; Cyclone I: 39 stations at various times of the day generally to 150 m (among them 5–75 m and 2–1,200 m). (Only incidental references to this cruise are made here.) Cyclone II to VI (March to September 1967): 89 oblique hauls, 0–1,200 m, made successively at 0000, 0400, 0800, 1200, 1600, and 2000 hours.

COMMENTS ON THE CATCHING EFFICIENCY OF THE IKMT

Restrictions

Several restrictions have to be kept in mind. Graded coarse meshes were used in the gear, and the escapement through the meshes is dependent not only on the size, but also on the shape of the species of fishes considered. The percentage of the stock sampled will vary as a function of these factors. Because of these considerations we have referred to observed numbers of organisms on standard courses (distance towed) of the net in the water, usually 5,000 m for shallow hauls and 10,000 m for deep hauls, rather than to standard volumes of water filtered; otherwise, we have explicitly called the results "observed." Table 1 gives examples of the size distributions for five characteristic fishes. The extreme two species in Table 1 are also those with extreme shapes: *Sternoptyx diaphana* Hermann is very high, and *Serrivomer* sp. is very long and narrow; *Ceratoscopelus townsendi* (Eigenman and Eigenman) and *Cyclothone pallida* Brauer are good examples of the average size and shape of the fishes collected with the IKMT. We can expect

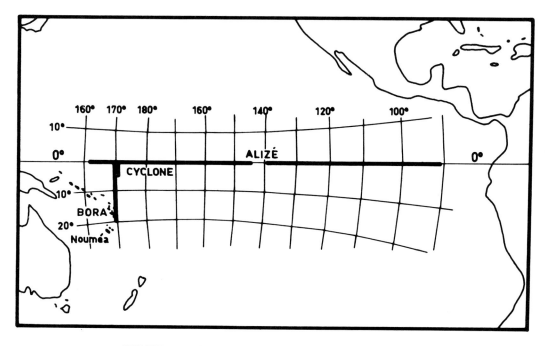

FIGURE 1 N. O. *Coriolis*: Cruises Alize, Bora, Cyclone tracks.

TABLE 1 Size Groups Frequencies for IKMT Fishes at 96 Stations of the Bora Cruises (170°E, 5°N to 20°S, 4 Cruises) (Modal Size Groups Underlined; First 50 Percent of Total Observed Number Bordered)

L (total length in mm)	Sternoptyx diaphana (in % of the total number)	L	Vinciguerria nimbaria	L	Ceratoscopelus towsendi	L	Cyclothone pallida	L	Serrivomer sp.
5	23								
10	27	10	33	10	1	10	2		
15	13	15	28	15	18	15	13		
20	15	20	20	20	42	20	25		
25	7	25	11	25	15	25	28		
30	10	30	5	30	14	30	23		
		35	2						
40	5	40	1	40	8	40	6		
				60	2	50	2		
								50	0.3
								75	9
								100	30
								150	33
								200	14
								250	9
								400	5
	N = 998		N = 1,035		N = 369		N = 6,388 (No. of measured specimens)		N = 289

that small species of "normal" shape, as well as small specimens of species of "normal" size and an important fraction of the elongated species, are undersampled by the trawl, while higher species and larger specimens are relatively oversampled, at least until they again become undersampled because of their size-related ability to avoid the IKMT.

Conditions of Sampling

It should also be noted that the conditions of sampling are not really constant during the whole haul. Figure 2 gives a reproduction of bathykymograph traces for seven night stations made to 1,200 m on the Bora IV cruise. Usually after lowering and before starting up, the trawl follows a more-or-less horizontal course; this could result in oversampling of the fauna of the deepest level. Conversely, shallower species could be undersampled even if they were very densely distributed, but in a thin layer. Their density in a short vertical section would be divided by the depth of the whole column sampled; so we can get an idea of their importance relative to the whole column, but not of their eventual predominance at a particular level. Irregularities in the sampling might also arise from variations in speed and direction of the currents crossed by the net on its way down and up, as is the case in equatorial waters.

Replicability of IKMT Hauls

The replicability of the IKMT hauls should also be considered. During the Cyclone I cruise, six 150-m stations were made at 0200 hours during 6 consecutive nights (November 23–28, 1966) at a constant position (00°36'S, 169°32'E). These results allow an approach to the study of the variability of sampling in this area for a 10-ft IKMT. One can even consider that sampling variability is overestimated because of an evident continuous evolution during these 6 days of the hydrology as well as of the populations of the various organisms sampled. The following coefficients of variation were calculated from the six original counts, corrected for an average tow distance of 5,000 m: Fishes, 14 percent; Cephalopods, 21 percent; Fish larvae, 30 percent; Crustacea larvae, 36 percent; Carids, 39 percent; Amphipods, 41 percent; Stomatopod larvae, 45 percent; Euphausiids, 118 percent.

So if we consider that adult and larval fishes and juvenile cephalopods were the most "micronektonic" organisms caught by the IKMT, they were also the organisms best sampled by this gear.

The detail by species for the fishes collected in the six shallow hauls follows (average M and coefficient of variation V, in percentage of the average):

Fishes	$M = 346, V = 14\%$
Diaphus fulgens	$M = 38, V = 85\%$
Diaphus schmidti	$M = 38, V = 38\%$
Diaphus malayanus	$M = 8, V = 44\%$
Vinciguerria nimbaria	$M = 146, V = 21\%$
Diaphus regani	$M = 29, V = 28\%$
Bregmaceros	$M = 13, V = 47\%$

From these probably overestimated variations, we can conclude that a ratio of 1 to 2 between numerations of two different stations for the most common species corresponds to a difference significant at the 5 percent level.

In the following text, we shall not consider variations from station to station but compare means calculated on various numbers of stations, so the significance level between these means should be reached for a ratio smaller than 1 to 2.

The Nature of the Cruises Considered

Finally, referring to the cruises considered, it must be pointed out that they were preliminary in nature and were devoted to the study of sampling conditions or of the particular features of the distribution of some organisms.

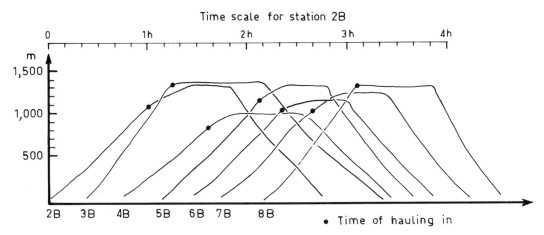

FIGURE 2   Bora IV: bathykymograph tracks for seven night stations.

Alize stations were made with a 5-ft trawl and only reached shallow depths.

The seasonal coverage of the Bora cruises was complete, but the cruises were too infrequent and the methodology was not consistent.

On Cyclone II to VI, the methodology was consistent, and the density of the stations was sufficient to conveniently describe diurnal and seasonal variations, but the cruises covered only 7 months and were too restricted geographically.

Nevertheless, to summarize, we shall consider here the results from 185 Bora and Cyclone stations made during 22 months, and for the most part, reaching depths greater than 600 m. We can thus expect to have a sufficient description of the usual composition of the ichthyofauna sampled by the 10-ft IKMT, except for a certain overestimation, relative to other species, for the smaller specimens of *S. diaphana* and for a large underestimation of the smaller specimens for the Apodes, and generally for the very elongated fishes. One must also point out that the juveniles of most species are much more abundant in the sampled population than appears in the results. An empiric rule could be that for "normally" shaped fishes, specimens and species of sizes between 2 and 10 cm are the bulk of the ichthyofauna sampled by the 10-ft IKMT.

ICHTHYOFAUNA IN THE WESTERN EQUATORIAL PACIFIC

Only the most common species were generally determined in the western equatorial Pacific.

Table 2 shows the composition of the ichthyofauna caught by the 10-ft IKMT during Bora and Cyclone cruises.

There are some differences in the results of these cruises, which could derive to some extent from the differences in sampling methods or differences in the areas sampled; there are also some clear and important common points.

1. The three dominant species are identical in the two series of cruises; *C. pallida** is largely the main species, constituting more than one half the total number; the two others, *V. nimbaria* and *S. diaphana*, represent together from 11 to 14 percent of the total number.

2. We can notice the poor influence of the Myctophids in the whole 1,200-m column, i.e., they represent only 13 to 15 percent of the total number.

*Some preliminary studies were made to check the homogeneity of the *Cyclothone* designated here under *C. pallida*. Their number is so high, and in many cases these delicate fishes are in such bad condition that further consideration is needed. Nevertheless, in the first examination, the *Cyclothone* considered here appear to constitute a monospecific group with exceptions limited to a few individuals, and generally conform to the *C. pallida* description according to V. A. Mukhacheva (1964, On the genus *Cyclothone (Gonostomidae pisces)* of the Pacific Ocean. *Trud. Inst. Okeanol.*, 73, 93–108).

3. In the two sets of cruises, 18 to 22 species composed approximately 90 percent of the total number.

4. Despite some differences in their order, of the eight species following the three main ones, by order of abundance, six were common to the two sets of cruises.

SEASONAL VARIATIONS IN THE EQUATORIAL AREA AND SOME DOMINANT SPECIES

The average number and percent of the dominant species of fishes for the equatorial area are presented in Tables 3a and 3b, respectively, for each cruise.

Considering the coefficient of variation of the averages per cruise as an index of the variability in time, the contrast between the great stability of *Cyclothone* and *S. diaphana* and the great variability of *V. nimbaria* is quite noticeable. *V. nimbaria* show two very high peaks during the 2 years of the cruises, and consequently, have varied each year from less than 1 percent to 18 percent of the total catch.

SOUTH–NORTH VARIATIONS IN THE TROPICAL AND EQUATORIAL AREA FOR SOME DOMINANT SPECIES

Table 4 summarizes the variations in abundance along the track of some Bora cruises.

Again we can notice a strong differentiation between the main species.

● As a general rule there are two marked peaks at the two ends and a minimum between 14° to 4°S.
● *C. pallida* is very stable from south to north, with only slight increases at the ends of the legs.
● *S. diaphana* shows a more marked increase northward but remains well distributed elsewhere.
● *V. nimbaria* shows a very high variability along the leg, with a small peak at the south and a strong development in the equatorial area.

When examining in detail the latitudinal variations of the number per haul of 10,000-m tow distance for *V. nimbaria*, this species shows that the northward increase is never observed at more than two or three consecutive stations between 0° and 2°N. This suggests a relation between this increase and the equatorial circulation. For example, its abundance in number per haul in the equatorial area for two Bora cruises is as follows:

Position Cruises	4°S	3°S	2°S	1°S	0°	1°N	2°N	3°N	4°N
Bora III	6	30	10	30	<u>258</u>	<u>222</u>	<u>194</u>	0	0
Bora IV	0	1	1	2	<u>55</u>	<u>14</u>	9	5	–

TABLE 2   Fishes Collected by Midwater Trawling in the Equatorial Southwestern Pacific (170°E) on Bora and Cyclone Cruises

Species[a]	Bora Cruises, Various Depths, Stations between 4°N and 5°S[b] (% of total)	Cyclone Cruises 0–1,200 m 0° to 5°S[c] (% of total)
Cyclothone pallida Brauer 1902	59.6	56.1
Vinciguerria nimbaria (Jordan & Williams 1895)	5.8	8.6
Sternoptyx diaphana Hermann 1781	5.1	5.1
	70.5	69.8
Lampanyctus festivus Tåning 1928	1.4	2.6
Diaphus schimdti Tåning 1932	1.3	1.9
Serrivomer sp.	1.5	1.8
Ceratoscopelus townsendi (Eigenman and Eigenman 1889)	1.0	1.8
Diaphus regani Tåning 1932	1.8	1.7
Diaphus fulgens Brauer 1904	3.3	1.4
Myctophum rufinum Tåning 1928	0.4	1.3
Chaulodius sloanei Block & Schneider 1801	1.2	1.2
	11.9	13.6
Bregmaceros sp.	not sorted	0.9
Gonostoma elongatum Günter 1878	1.7	0.8
Lampanyctus pyrsobolus Alcock 1890	0.8	0.8
Diaphus malayanus Weber 1913	0.2	0.6
Lampanyctus niger Günter 1887	0.9	0.5
Valencienellus tripunctulatus (Esmark 1871)	0.4	0.5
Hygophum benoiti (Cocco 1838)	1.0	0.4
Diaphus lutkeni Brauer 1904	not sorted	0.4
Gonostoma bathyphilum (Vaillant 1888)	not sorted	0.4
Sternoptychid undetermin	1.1	0.3
Myctophum asperum Richardson 1844–1848	not sorted	0.3
Diogenichthys laternatus (Garman 1899)	0.4	0.2
	6.5	6.1
Various sorted species	*4 species*	*15 species*
	1.1	1.2
Total sorted species	*23 species*	*38 species*
	90.0	90.8
Total percent of Myctophids	*13 species sorted*	*all species sorted (25)*
	13.2	14.7

[a]By order of importance for the Cyclone cruises.
[b]Number of stations selected = 62; number of fishes collected per station = 262 (corrected for a course of 10,000 m of the trawl).
[c]Number of stations selected = 84; number of fishes collected per station = 377 (corrected for a course of 10,000 m of the trawl).

INDICATIONS ON THE NIGHTTIME VERTICAL DISTRIBUTION OF THE DOMINANT SPECIES. RELATIONSHIP WITH THE SHALLOWER FAUNA ALONG THE EQUATORIAL CURRENT (ALIZE CRUISES)

Vertical Distribution of *C. pallida* and *S. diaphana*

Considering that *C. pallida* and *S. diaphana* were more stable in number during the 2-year period of sampling than the other species, we can examine in Table 5 their variation of abundance on the Bora cruises as a function of the depths of the haul.

To complete Table 5, we note that on Cyclone I, during which night hauls to 150 m were made, no specimens of the two species were collected. Also, on the Bora cruises, south of 5°S, *S. diaphana* was never observed shallower than 300 m at night, and only 6 percent of the total of the *Cyclothone* were taken in the shallow tows. During the Alize cruise (see Table 6), *Cyclothone* were no more than 0.8 to 3 per haul in the night 0–300-m hauls, and *S. diaphana* in their peak region were no more than 0.4 per

TABLE 3a Number per Haul (for a 10,000-m Course) for the Main Species in Equatorial Area

Species	Bora Cruises (1966)				Cyclone Cruises (1967)					Gen. Average	$V = \dfrac{100\sigma}{m}$
	B I (Dec.)	B II (March)	B III (June)	B IV (Sept.–Oct.)	C II (March–Apr.)	C III (May)	C IV (June)	C V (July)	C VI (Sept.)		
C. pallida	192	155	222	205	225	198	173	211	238	202	13 %
V. nimbaria	2	5	84	10	79	27	29	13	12	29	107 %
S. diaphana	13	15	24	17	18	16	18	22	21	18	17 %
L. festivus	1.0	3.1	7.6	5.0	8.5	12.1	5.4	10.0	10.9	7.1	52 %
D. schmidti	0.4	3.8	6.2	7.6	13.2	4.9	5.3	5.0	7.0	5.9	58 %
Serrivomer sp.	4.0	4.0	4.6	3.4	5.8	5.9	4.3	8.8	9.0	5.5	38 %
C. townsendi	5.7	1.3	1.8	2.4	4.7	6.7	1.7	6.7	11.8	4.8	71 %
Total number of fishes	282	238	458	317	447	343	298	356	415	350	22 %
Number of stations selected[a]	9	6	9	9	18	18	12	18	18	tot. = 117	σ calculated on 9 values
Types of hauls selected	900-m/day and night	900-m/day night	650-m/ night	1,200-m/night	1,200-m/six times a day	1,200-m/ six times a day	1,200-m/ six times a day	1,200-m/ six times a day	1,200-m/ six times a day	—	—

[a] For the Bora Cruises, the deeper stations only were selected.

TABLE 3b Percentage of the Total Number of Fishes, Corresponding to the Values Presented in Table 3a

Species	Bora Cruises (1966)				Cyclone Cruises (1967)					Gen. Average
	B I (Dec.)	B II (March)	B III (June)	B IV (Sept.–Oct.)	C II (March–Apr.)	C III (May)	C IV (June)	C V (July)	C VI (Sept.)	
C. pallida	69.1	65.1	48.8	63.6	50.2	57.7	58.1	59.2	57.4	57.7
V. nimbaria	0.7	2.1	18.5	3.1	17.7	8.0	9.6	3.7	2.9	8.3
S. diaphana	4.7	6.3	5.3	5.3	4.1	4.8	5.9	6.2	5.0	5.1
L. festivus	0.4	1.2	1.7	1.6	1.9	3.5	1.8	2.8	2.6	2.0
D. schmidti	0.1	1.6	1.3	2.3	2.9	1.4	1.8	1.4	1.6	1.7
Serrivomer sp.	1.4	1.7	1.0	1.0	1.3	1.7	1.4	2.5	2.2	1.6
C. townsendi	2.0	0.5	0.4	0.7	1.0	2.0	0.6	1.9	2.8	1.4

231

TABLE 4 South–North Repartition of the Dominant Species during Bora Cruises (Number per Haul)

Area of the 170°E	20°–15°S	14°–10°S	9°–5°S	4°–0°	1°–4°N	Ratio, max. average / min. average
C. pallida	198[a]	153	164	148	209	1.4
S. diaphana	7.5	6.4	10.4	11.7	22.9	3.6
No. of stations selected[b]	(6)	(5)	(7)	(15)	(9)	–
V. nimbaria	8.6	2.9	1.2	13.6	28.3	23.4
L. festivus	1.8	1.3	5.0	2.7	4.8	3.8
D. schmidti	2.9	1.7	0.3	4.1	4.8	16.0
Serrivomer sp.	0.7	1.8	2.8	4.7	3.0	6.7
C. townsendi	7.4	7.3	5.9	3.8	1.2	6.2
Total no. of fishes	290	151	166	240	286	1.9
No. of stations selected[b]	(7)	(7)	(9)	(20)	(11)	–

[a]Maxima and minima underlined.

[b]This table has been established by using for each Bora cruise only the hauls of the same type made all along the leg. C. pallida and S. diaphana being absent in tropical area and rare in equatorial area above 300 m, the Bora I 300-m hauls were not considered in their case.

haul (but *Argyropelecus lynchus*, another Sternoptychid, was important in the eastern part of the cruise).

So, for these two species there is a clear trend for the bigger adults and also the youngest juveniles to remain in the middle and deeper levels. In equatorial waters, only medium-sized *C. pallida* and *S. diaphana* were found in appreciable quantities shallower than 300 m; but, compared with *C. pallida*, the shallow water maximum of *S. diaphana* was composed of younger and less numerous classes. Figure 3 shows this situation as a hypothetical vertical distribution of the main species according to their size development.

Vertical Distribution of *V. nimbaria*

Because of their large seasonal variations it is impossible to apply a similar treatment to the *Coriolis* data on *V. nimbaria*. But there are other possible sources for evaluation of its depth distribution. First, Alize results should be considered (Table 6).

Vinciguerria sp. are always important along the equator. The percentages of *Vinciguerria* in the Alize 300-m hauls, relative to the absence of *Cyclothone* and *Sternoptyx*, are quite comparable with what they were in some of the Bora or Cyclone cruises, in which they were at their maximum. Thus, the Alize sampling to 300 m gives better results than the Bora and Cyclone deep sampling for *V. nimbaria*. It is interesting to note that east of 110°W all the *Vinciguerria* were *V. lucetia; V. nimbaria* replaces this species quite sharply at 115°W, only one or two individuals of each species being found in the area of the other one.

On Cyclone I, hauls to 150 m were made at the same hour on six successive nights at 00°36′S, 169°32′E; on the

seventh night, one haul to 75 m was made at the same position. The observed number of *V. nimbaria* in the 75-m haul was only slightly smaller than the average number observed in the six 150-m hauls, and the size distribution for the shallower station was different. In the 75-m haul, only 9 percent of the fishes were longer than 20 mm, and none was longer than 29 mm, while in the 150-m hauls 40 percent of the observed fishes were longer than 20 mm and 11 percent were longer than 29 mm, reaching 45 mm.

The average number of *V. nimbaria* in the 0–150-m hauls, corrected for a 10,000-m course, is 292; assuming that the entire population was concentrated in the 0–150-m layer, a 0–1,200-m haul would collect 37 individuals. On the last day of the above observations, two 0–1,200-m hauls were made, yielding a corrected average of 23 *Vinciguerria*. This is not in contradiction of the hypothesis that all the *Vinciguerria* were shallower than 150 m (Figure 3).

Relationship with the System of Equatorial Currents

Finally, all this being considered, it is quite probable that *V. nimbaria* is mainly concentrated in the upper 300 m at night and probably in a much shallower layer. The relative abundance of the species appears to be about the same all along the equator west of 115°W. The specimens smaller than 20 mm caught at night could be related with the westward South Equatorial Current, and the older ones with the eastward Cromwell Undercurrent.

A part of the medium-sized *Cyclothone* can also be influenced by the Cromwell Current, but younger and older ones, and practically all the *S. diaphana*, seem to be independent of both currents (see Figure 3).

Different patterns in the horizontal distribution of the

TABLE 5 Schematic Vertical Distribution of *Cyclothone pallida* and *Sternoptyx diaphana* at Night, According to Their Size (Bora cruises, equatorial area only)

Cyclothone pallida (size groups)

Number of Hauls Averaged	Considered Layers	Theoretical Percent of the Population[a]	10mm	15mm	20mm	25mm	30mm	35mm	Over 40mm	Total Population
4 (0–300m)	0–300m	25	11	30	33	[51]	31	18	12	31 %
5 (0–650m)	300–650m	29	11	[28]	21	[28]	27	8	22	21 %
4 (0–1,200m)	650–1,200m	46	[78]	42	46	21	42	[74]	66	48 %
Observed number per 0–1,200-m haul for the whole section			100 = 4.5	26	60	54	26	34	22	227

Sternoptyx diaphana (size groups)

			5mm	10mm	15mm	20mm	25mm	30mm	Over 40mm	Total Population
11 (0–300m)	0–300m	25	8	[28]	18	7	0	5	0	13 %
9 (0–650m)	300–650m	29	[55]	40	45	[93]	56	5	5	44 %
9 (0–1,200m)	650–1,200m	46	37	32	37	0	44	90	[95]	43 %
Observed number per 0–1,200-m haul for the whole section			100 = 4.8	4.7	2.7	1.5	1.6	2.1	1.8	19.2 %

NOTE: Numbers are expressed as percentage of the observed average number of the deepest night hauls. The shallower hauls were successively subtracted from the next deeper. For instance, the 300–600-m result is:

$$\left(\frac{0-600 \text{ result} \times 100}{0-1\,200 \text{ result}}\right) - \left(\frac{0-300 \text{ result} \times 100}{0-1\,200 \text{ result}}\right).$$

[a] Assuming a homogeneous distribution of the population and homogeneous conditions of hauling on the whole 1,200-m section.

233

TABLE 6 Some Data on the Ichthyofauna of the Upper 300 m along the Equator (Alize cruise).

Limit of the Considered Area Along 0°	A 92°–110°W	B 115°–138°W	C 145°–168°W	D 170°W–162°E
Number of Stations (IKMT 5-foot)	7	9	8	9
Number of Fishes per Haul[a]	97.8	35.5	89.8	51.0
Percent of the Number of Fishes for:				
Cyclothone sp.	0.8	8.4	1.0	0.4
V. lucetia (Garman 1899)	60.2	1.6	0	0
V. nimbaria	0	23.7	48.9	59.2
S. diaphana	0	0	0	0
A. lynchus Garman 1899	13.0	0.3	0.4	0
Total of Myctophids	23.3	61.6	46.2	36.9

[a]Corrected for a 5,000-m course

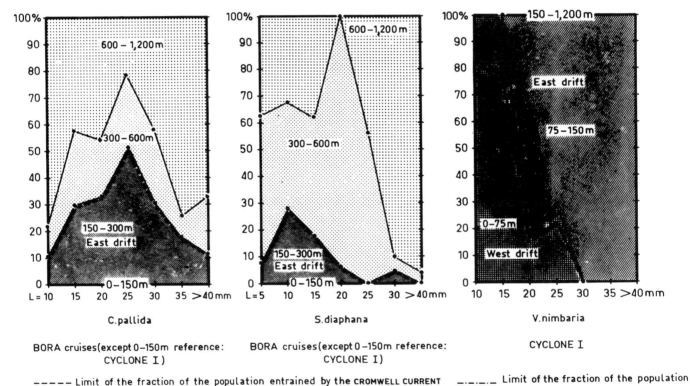

FIGURE 3 Assumed vertical distribution of the three main species of fishes in the West Equatorial Pacific as a function of their sizes, referred to the Equatorial Current and the Equatorial Undercurrent (*Coriolis* cruise data).

biomass of these three species could derive from differences in their vertical distributions. For instance, from 150° to 155°W enormous concentrations of postlarval *V. nimbaria* were collected during Alize cruises, causing an increase from one to four of the biomass of the total IKMT plankton evaluated at the peak station. So we can assume that *V.*

nimbaria could be aggregated in large patches in the equatorial currents in the central Pacific; the movement of these patches to the west is a possible explanation for the marked seasonal peaks of juveniles observed for this species along 170°E during the Bora and Cyclone cruises.

CONCLUSION

These results constitute a description of the ichthyofauna of the upper 1,200 m in the western equatorial and southern tropical adjacent area and give more composite ideas for the study of the so-called micronektonic fishes in the equatorial area of the western Pacific.

1. Three dominant species constitute the biggest part of the fauna, the *Cyclothone* being more than one half in number of the total.

2. *Cyclothone* and *Sternoptyx* were observed to be very stable in abundance during the 2 years of sampling, and the density of *Cyclothone* did not change greatly from 0° to 20°S.

V. nimbaria was also present everywhere; however, it appears to exhibit marked seasonal peaks in equatorial water*; this could be related to enormous concentrations of larval stages further east.

3. *V. nimbaria* in the equatorial region appears to be shallower and restricted to the westward South Equatorial Current and eastward Cromwell Undercurrent. This distribution could influence successively the juveniles and the maturing adults. Some of the medium-sized *Cyclothone* are possibly influenced by the undercurrent; oldest and juvenile *Cyclothone* and most of the *Sternoptyx* remain below the active part of the Equatorial Current system.

Questions of interest concerning these organisms include the following:

1. At present we have no results to report on the food of the *Cyclothone* and the *Sternoptyx*. *V. nimbaria* food is under study in our laboratory; results for the eastern tropical Indian Ocean gave an average food weight of 4.5 percent of their body weight on night stations, consisting mainly of Copepods.

2. Around New Caledonia, *Sternoptyx diaphana* was, by order of occurrence, one of the most important prey species of the *Alepisaurus ferox* (Lancet fish) and was also important in the stomach contents of the Albacore tuna (*Alepisaurus* being also an important food for this tuna in the same area). In the same area *Vinciguerria* were found in the stomachs of Yellowfin more often than any other micronektonic fishes. The enormous concentration of post-larval *Vinciguerria* could be an important source of food for the bigger epipelagic species but could also have an important predatory effect on the available plankton.

3. Relation of tropical stocks of these species with their equatorial stocks have to be studied mainly for *V. nimbaria*.

The abundance of these species and their relationships with different parts of the meso and epipelagic zones,

mainly in the Equatorial Current system, suggest that it could be very interesting to confirm the above remarks and to understand better their biology and their position in the oceanic food chain.

BIBLIOGRAPHY

Backus, R. H., G. W. Mead, R. L. Haedrich, and A. W. Ebeling (1965) The mesopelagic fishes collected during cruise 17 of the R/V *Chain* with a method for analyzing faunal transects. *Bull. Mus. Comp. Zool. Harv.*, *134*(5), 139–198.

Brauer, A. (1906) Die Tiefsee Fisch. I Systematischer Teil. *Wiss. Ergebn. 'Valdiva'.* Bd. XV.

Fraser-Brunner, A. (1949) A classification of the fishes of the family Myctophidae. *Proc. Zool. Soc. Lond.*, *118*(4), 1019–1106.

Gilbert, C. H. (1905) The deep sea fishes of the Hawaiian Islands. In *The aquatic resources of the Hawaiian Islands, II.* D. S. Jordan and B. W. Evermann. *Bull. U.S. Fish Comm.*, *23*(2), 575–713, 45 Plates, 44 Illus.

Grandperrin, R. (1967) Etude comparative d'échantillons de macro-plancton et de micronecton récoltes par trois filets différents. *Cahier O.R.S.T.O.M. Océanogr.*, *5*(4), 14–29.

Grandperrin, R., and M. Legand (1967) Influence possible du système des courants équatoriaux du Pacifique sur la répartition et la biologie de deux poissons bathypélagiques. *Cahier O.R.S.T.O.M. Océanogr.*, *5*(2), 69–77.

Grandperrin, R., and J. Rivaton (1966) *Coriolis* Croisière Alize. Individualisation de plusieurs ichtyofaunes le long de l'équateur. *Cahier O.R.S.T.O.M. Océanogr.*, *4*(4), 35–49.

Jespersen, P., and A. V. Taning (1926) Mediterranean Sternoptychidae. *Rep. Danish Oceanogr. Exped. Med.*, *2*(9), 1–59.

King, J. E., and R. I. B. Iversen (1962) Midwater trawling for forage organisms in the central Pacific. *Fish Bull. U.S.*, *62*(210), 271–321.

Koefoed, E. (1958) Isospondyli 2. Heterophotodermi -1-. *Rep. Sars. N. Atl. Deep Sea Exped.*, 4, part 2(6).

Legand, M., and J. Rivaton (1967) Cycles biologiques des poissons mésopélagiques dans l'est de l'Océan Indien. Deuxième note: Distribution moyenne des principales espèces de l'ichtyofaune. *Cahier O.R.S.T.O.M. Océanogr.*, *5*(4), 73–98.

Norman, J. R. (1930) Oceanic fishes and flat fishes collected in 1925–27. *Discovery Rep.*, *2*, 261–370.

Parr, A. E. (1927) The Stomiatoid fishes of the suborder Gymnophotodermi. *Bull. Bingham Oceanogr. Coll.*, *3*(2).

Parr, A. E. (1928) Deep sea fishes of the order Iniomi from the waters around the Bahama and Bermuda Islands. *Bull. Bingham Oceanogr. Coll.*, *3*(3).

Parr, A. E. (1928) Deep sea Berycomorphi and Percomorphi from the waters around the Bahama and Bermuda Islands. *Bull. Bingham Oceanogr. Coll.*, *3*(6).

Pearcy, W. G., and R. M. Laurs (1966) Vertical migration and distribution of mesopelagic fishes off Oregon. *Deep Sea Res.*, *13*, 153–165.

Regan, T., and E. Trewavas (1929) The fishes of the families Asthonestidae and Chauliodontidae. *Rep. Danish Dana Exped.* (5).

Regan, T., and E. Trewavas (1930) The fishes of the families Stomiatodae and Malacosteidae. *Rep. Danish Dana Exped.* (6).

Roule, L., and L. Bertin (1929) Les poissons apodes appartenant au sous-ordre des Nemichthydiformes. *Rep. Danish Dana Exped.*, (4).

Schultz, L. P. (1961) Revision of the marine silver hatchet fishes (family Sternoptychidae). *Proc. U.S. Nat. Mus.*, *112*(3449), 587–649.

---

*Work in progress in the O.R.S.T.O.M. laboratory also indicates seasonal increases for this species in the tropical waters of the Indian Ocean.

A HISTORY OF SOUTH PACIFIC FISHES

Giles W. Mead

MUSEUM OF COMPARATIVE ZOOLOGY,
HARVARD UNIVERSITY

INTRODUCTION

In one very real sense there is no such thing as a distinctive South Pacific fish fauna. There are South Pacific fishes, to be sure, and some of these live only in the South Pacific Ocean. The fauna as such, however, is not a South Pacific fauna but a composite one of diverse origins. It contains the richest tropical coastal fauna in the world or, more correctly, it houses parts of two such faunas, neither of which is its own. The greater of these faunas, the western South Pacific, is continuous with the fauna of the Indian Ocean west to South Africa and the Red Sea and the warmer parts of the North Pacific. It diminishes eastward, but little of it survives off the Americas. The lesser fauna is a tropical American one shared with parts of the Atlantic. There are peculiar island fish faunas in the South Pacific, as there are elsewhere, but the biological features are not South Pacific but island features. There is a very characteristic polar fish fauna, but it is shared with all other oceans that wash Antarctica. The large oceanic fishes belong to groups common to all comparable seas, and the little fishes that live below them are of mixed ancestry; some are not now known from elsewhere, but many are common to waters as distant as the North Atlantic. Nor has the South Pacific served in any important way as a source fauna from which other parts of the world received their fishes. Indeed, the tropical Indo-Pacific is clearly such a source area, but it is not South Pacific *per se* but a tropical evolutionary spawning ground that extends two thirds of the way around the world. Although some groups are now confined to the South Pacific and may have arisen there, they are surprisingly few and relatively unimportant in an evolutionary sense. The subject that I was asked to review, the ecology and zoogeography of South Pacific fishes, thus becomes complex and must be restated as the history of those fishes that now live in the South Pacific.

Contemporary zoogeographers tend toward one of several directions: the delineation of faunal boundaries (e.g., Ekman, 1953); the dynamic histories of groups—their dispersal patterns, radiations, colonizations, etc. (e.g., Darlington, 1957, 1965); and more recently the theoretical approach of MacArthur and Wilson (1967) together with associated experimental studies such as the creation, through extermination, of artificial Surtseys or Krakatoas and the subsequent analysis of the patterns of repopulation. The last of these approaches will not be considered here, and the first will be discussed but briefly.

The total distributional pattern and faunal boundaries of a complex fauna, be it in the South Pacific or elsewhere, is a mosaic reflecting the end points of a series of interacting origins, dispersals, and colonizations as well as extinctions working in consort with a geologically and ecologically changing sea. This mosaic can never be fully understood. It is often argued that a characterization of such a pattern without a fuller knowledge of the countless individual pieces is unwarranted—that distributional generalities should not be attempted until classification at the species level is nearly complete. If accepted, that argument is painfully applicable to the temperate South Pacific, for which source material for distributional analysis is both scarce and scattered.

Nonetheless, contributions are regularly being made, especially with respect to the coastal fauna. Of particular pertinence are the excellent reviews of south-temperate littoral zones, based on the distributions of both plants and invertebrates, by Knox (1960) and other contributors to the Royal Society symposium chaired by Pantin (1960), the papers on Pacific basin biogeography (Gressitt, 1963), Andriashev's (1965) excellent review of the distribution of the Antarctic fish fauna and its zoogeographic subdivisions, the papers by Myers (1941) and Rosenblatt (1967) with their cautious proposals of faunal discontinuities within the Peruvian subtropics, those cited by Whitley (1965), which consider certain problems of Australasian zoogeography, and, of course, the classic by Ekman (1953).

Comments on the South Pacific oceanic fish fauna are necessarily few. The area is vast, the sampling framentary, and the solid systematics but recently emergent (e.g., Ebeling, 1962). The larger members of this fauna, such as the scombroids, are considered in connection with the fisheries (Kasahara, this volume, page 252); and the distribution of the smaller oceanic forms such as the myctophids and gonostomatids, with respect to the physical and biological factors that influence their occurrence, are appropriately best considered as part of the South Pacific plankton (Bary, this volume, page 211). The only mesopelagic fish survey made in the South Pacific and studied in its entirety is that from a transect, at about 34°S, from coastal Chile westward for about 2,000 km during January and February, 1966 (*Anton Bruun* Cr. XIII; Mead, 1966; Craddock and Mead, in press), although Bussing (1965) reported on midwater collections taken at several *Eltanin* stations in the area. As in other mixed eastern oceanic environments, such as that in the Pacific off North America (Aron, 1962; Lavenberg and Ebeling, 1967; Ebeling, 1967; among many others), the oceanic fish fauna off central Chile is a mixed one containing principally subantarctic, panoceanic, South Pacific Central, and a few endemic elements. Notable, however, was the complete absence from the *Bruun* collection, about 16,000 specimens of 133 species, of any representative of the Antarctic fish fauna.

The results of this *Bruun* cruise serve to emphasize the need for further intensive study of the southern "transitional" fauna, which may be characterized in part by the occurrence of certain elements specifically associated with other hydrological discontinuities such as the Subtropical Convergence; and of the relationship of this fauna to the eastern temperate seas that have been termed "transitional." (Applied to the fauna, "transitional" is an unfortunate term for this area, hundreds of thousands of cubic miles of lebensraum, which, while it does house species characteristic of each of the contributory water masses, has its own mosaic of species associates and peculiar faunal elements.) An intensive international effort directed toward an understanding of the waters off Chile and the Subtropical Convergence to the west, together with the plants and animals living there, is in order.

In general, however, the dearth of information about the fauna throughout the South Pacific is obvious. With the possible exception of Peruvian fishery surveys, the only study of the oceanic nekton conducted by a nation bordering that ocean is that reported in this volume by Legand, page 226. With respect to the fauna, the number of localities collected from expeditionary vessels using gear larger than a standard meter net is almost inconsequential. Knowledge of the coastal temperate fish fauna is also poor. The Australian fish fauna was examined as a whole 40 years ago (McCulloch, 1929–1930), and was simply listed, without distributional commentary, by Whitley (1965). McDowall (1964) has discussed the nature of the New Zealand freshwater fishes, a complementary fauna and thus of some significance to marine zoogeography, but Graham's simple listing of the New Zealand marine fish fauna (1956) is of limited scientific use. Work toward a monograph on the fishes that live along the 3,000 important miles of Chilean coast was interrupted by the tragic death in 1961 of Fernando de Buen, which left Chile with few scientists capable of undertaking such a survey and none either able or, apparently, so inclined. There are de Buen papers, a literature review by Fowler (1945), and the popular compilation, worthy of little scientific praise, by Mann (1954). The South Pacific thus harbors the least well-known fish fauna in the oceans. Any analysis, such as that presented here, is admittedly suspect and will be of limited longevity, unless current neglect by ichthyologists persists.

THE PROBLEM

I will be concerned here with the following question: Are there common elements or trends in the apparent origins, dispersals, and colonization patterns of several groups of fishes of diverse ancestry now living in the South Pacific Ocean; and, if so, what do they reveal about the fish fauna of that ocean? Prerequisite to any consideration of this question, however, is a partial dismantling of this fauna into subunits that are more-or-less zoogeographically compatible.

Initially, a bisection of the South Pacific fish mosaic along ecological lines is required. This is the separation of the primary deep-water fish fauna, in the sense of Andriashev (1953) from the surrounding coastal fauna. Stripped of secondary invaders into the midwater realm, such as apogonids, scombroids and bramids, and of the deep-sea benthic and engybenthic* forms, the deep-water and pelagic fauna and associated zoogeographical problems are distinct, despite the occasional pseudopelagic species—those of off-shore phylogenetic affinities but more often caught close to, rather than far from, the continental margins. This primary deep-sea and oceanic fauna is distributionally as distinct from the remaining teleosts as are the latter from the primary freshwater fishes. The separation is far greater than that between the faunas of other marine environments, such

---

*The term "engybenthic," derived from the Greek and meaning to approach or draw close to the bottom, was coined in discussions at Harvard University some years ago for that part of the deep-sea fish fauna that lives characteristically not directly on, but within ten or twenty meters of, the bottom. It was introduced by Haedrich (1967) but has been generally suppressed in the interest of minimizing the number of such terms in the distributional literature. However, the concept is useful, and the apparent unwillingness of many to use terms such as "near-bottom" while introducing such barbarisms as "bentho-pelagic" suggest that the term "engybenthic," which is both euphonious and correct, may have a place in the distributional and ecological vocabulary.

as mainland versus island or coral versus sand bottom.

Second-level dismemberment of the distributional mosaic formed by the remaining shore and shelf fishes is both more complex and more critical. It is associated with higher taxa, and it involves the phylogenetic history of the teleosts and the problem of identifying for closer examination those groups that have been most important in shaping the present fauna of the South Pacific. It requires a reading of a classification of fishes, such as that recently proposed by Greenwood *et al.* (1966) from an ecological as well as a morphological point of view. Such a reading is most enlightening, certain of the rearrangements formally proposed by these authors, although not necessarily originating with them, are fitting distributionally as well as phylogenetically. An example is the inclusion of the deep-sea ceratioid fishes within an order that also contains the cods rather than in advanced grade parallel to, or derived from, the more recent percoids. In degree of modification and adaptation to the bathypelagic environment, these fishes belong with the primary deep-sea fishes, for which an earlier rather than a later origin is more readily understandable. Similarly relocated downward in the evolutionary scale are the cling fishes (Gobiesocidae), a group the present distribution of which appears to reflect a secondary radiation, with strength in temperate waters but with advanced forms in the Indo-Pacific seas that were once the point of origin. Also helpful with respect to distribution is the characterization of the scorpaenoid fishes as a group parallel to, rather than derived from, the other perch-like fishes, thus opening the possibility that these may have evolved and radiated before or concurrently with, rather than after, the percoids proper.

I will confine my discussion to certain major groups of teleost fishes, some conservative and others not, and a generalized and perhaps oversimplified account of these. Data have been selected, and comments on criteria for selection thus are in order.

I do not believe that there are "two kinds of evolution" (Briggs, 1966), but I do believe, as do Briggs and many others, that evolutionary descendants differ greatly in their potential for the production of further significant evolutionary lines. I am interested here more in lineages that are of such significance than in others which are not.

That major groups of fishes, as of other animals, arose through the invention of distinctly new adaptive types (Simpson, 1944) is clear. I do not believe, however, that among fishes major lineages of lasting significance were produced by phenomena similar to those that have produced the specialized forms common to island groups or those evolved through explosive evolution in the ancient lakes. That explosive evolution in such places has occurred, that morphological divergence to generic or family level has taken place, and that such situations are of immense interest in any study of the roles of isolation, time, and other relevant factors in evolution is apparent. But that such evolution, which has generally taken place in a new and biologically noncompetitive environment and has resulted in specialization of the most detailed kind, has been an important feature in the mainstream of fish evolution, as Myers (1960) suggested with respect primarily to lake fish faunas but to the bathypelagic groups as well, seems doubtful.

I do not consider as distinctly new or important types of lasting significance lineages that have had to become narrowly and excessively specialized to survive. For example, the carapids have developed the physiological, morphological, and, I presume, psychological adaptations that enable them to make a home in the south end of a northbound sea cucumber, and this ability certainly can be called unique and new. But this ability can hardly be expected to lead to new and diverse groups of important fishes. Other minor groups in which adaptation and modification have been so irreversibly complete that evolution of another *major* lineage from them is unlikely are not hard to find. It is difficult, for example, to derive from a sea horse anything but another sea horse, or but another angler fish from an existing one.

While the minor highly specialized dead-ends will not be further considered here, not all of these are minor. One such group is the Apodes, the true eels eulogized by Gosline (1960). In these, adaptation to a burrowing environment has involved virtually every structural component. Adaptation and modification have been so irreversibly complete that the evolution of another major lineage from the eels, regardless of their success as eels, is unlikely. Eels can give rise to other eels, as indeed have the burrowing eels, by throwing successful descendants into the deep sea. It is difficult, however, to envision eels evolving into distinctly new evolutionary grades. Further along in the evolutionary hierarchy appears another such group—adaptive, successful, speciose, and widespread from pole to pole and into fresh waters as well, but unlikely candidates for the next major evolutionary development. These are the flatfishes, distinguished percoid derivatives with a brilliant present but not an expansive future as the ancestors of a higher and distinctly new structural grade unless evolution should proceed from the larval rather than the adult form. Again, it is difficult to derive anything from a flatfish other than another flatfish.

Each such group contributes to the faunal complexity of an area such as the South Pacific, and each has its point of origin and zoogeographical history. Each, however, is generally limited by irreversible specialization to a specific environment, and the extent of its radiation is controlled accordingly.

The changes that produced the important new adaptive types and thus the new and lasting lineages of primary importance were not such as these. The changes of importance have been more subtle advances that enable the lineages to explode in an evolutionary sense and to disperse unimpeded by extreme specialization and its corollary, ecological and environmental restriction. Changes such as the development of the ability to eat coral, the ability to hover, the ability to digest plant material, or the ability to thrive at polar temperatures are subtle, conservative, and immensely important, but seldom bizarre. Such changes require a relatively unspecialized ancestor. The theory is consistent with the observation of Gosline (1960) who, on morphological grounds, noted that the long and major lineages within the teleosts have been essentially conservative ones, and that from this conservative mainstream have evolved the many specialized offshoots that, although sometimes speciose, can be considered evolutionary dead-ends.

The successful evolution of the new and lasting lineages requires competition. Evolution with relaxed selection permits the survival of almost anything. With evolution in the presence of close competition, the champion competitor will hold his ground and gradually improve his competitive position. The loser will become extinct, and the runners up can either become highly specialized in the original habitat or move along into another, both alternatives being precarious for survival. It is thus understandable that the Indo-Pacific has been the primary marine source of fish fauna in the world (Myers, 1941; Briggs, 1966, 1967). I do not believe, however, that it is location or long-term physical stability, historical and seasonal, or diversity of so-called niche that led to the creation of superior descendant species in that area, species capable of giving rise to new and lasting lineages and of colonizing outlying areas. The primary reason is that these species were slowly, surely, and conservatively selected for in a speciose and thus brutally competitive environment.

THE PATTERNS

Excluded from consideration here, though not arbitrarily, are not only the deep-sea fishes and their few inshore derivatives (e.g., the Holocentridae, large-eyed, red, and largely nocturnal reef fishes probably descendant from an early offshore berycoid; see Greenfield, 1968) but also many of the lesser highly specialized groups that lend complexity to fish classification but whose distributions should be superimposed on the more general patterns. The number of major groups to be considered are relatively few. Following the classification of Greenwood *et al.* (1966), they are the following:

DIVISION I   (or Cohort Taeniopaedia of Greenwood *et al.*, 1967)
Superorder Elopomorpha
    Order Elopiformes (bonefish, tarpons, etc.)
    Order Anguilliformes (true eels)

Superorder Clupeomorpha (herrings and anchovies but not the smelts)

DIVISION III   (Euteleostei)
Superorder Protacanthopterygii
    Order Salmoniformes (omitting the few inshore representatives of the primarily deep-sea myctophoids)
    Order Gonorynchiformes (Gonorynchidae and Chanidae)
Superorder Paracanthopterygii
    Order Batrachoidiformes
    Order Gobiesociformes
    Order Lophiiformes (of limited concern)
    Order Gadiformes (including zoarcids and ophidioids as well as codlike forms)
Superorder Atherinomorpha
    Order Atheriniformes (atherinoids and exocoetoids)
Superorder Acanthopterygii
    Order Scorpaeniformes (including Dactylopteridae)
    Order Perciformes (the immense array of spiny-rayed fishes) and its principal derivatives, the Pleuronectiformes (flatfishes), and Tetraodontiformes (puffers)

Elopomorpha

Elopiformes and Anguilliformes   Of those groups examined, few reveal as little of their origins as do the tarpons and their allies, and the true eels. The Elopiformes, few in number but successful, primitive in appearance, and old (dating at least from the Cretaceous; Romer, 1966, p. 354), are broadly distributed in tropical seas. A few, such as *Dixonina*, are not presently circumtropical; a few, such as *Pterothrissus*, are engybenthic forms; and all are probably survivors of a once more prominent group that never became cold-tolerant and have thus contributed nothing to the fish fauna of the South Pacific south of the tropics. Far more speciose are the true eels. These, having dominated the infaunal environment, have become more highly adapted to that environment and apparently precluded large-scale occupation of it by other, possibly competitive, forms (Gosline, 1960). Some such as *Myroconger* are extremely limited in distribution (St. Helena); others (*Gymnothorax*) are as broadly distributed as any shelf genus; and a few (*Anguilla*) live in fresh water, although only as adults. The progenitor, and hence its habitat and point of origin, is unknown. None has become totally cold-adapted. It is true that the bathypelagic descendants are taken from deep waters of polar temperatures, but there is not yet evidence that spawning occurs at these depths. Indeed, there is some evidence to the contrary (see Mead *et al.*, 1964, p. 576 and also the discussion there of the reproductive habits of other eel derivatives in the deep sea). All eels, as far as is known, are migratory. Also indicative of warm-water reproductive requirements is the abundance of leptocephalid eel larvae in the warmer central water masses and the relative scarcity of both young and adults in both the California and the Humboldt currents, cold-water eastern-boundary currents that separate the littoral environment from the warmer waters offshore. The coast inshore to these currents is also devoid of true eels, or nearly so. The elongate Chilean *Genypterus* and the northern eel-blennies may be ecological replacements for the true coastal eels along these shores. Similarly, *Anguilla*, able to live in fresh waters off Scandinavia and Labrador as adults, must repair to the Sargasso Sea to reproduce.

Eels are thus essentially tropical or subtropical and undoubtedly have radiated in a complex way closely related to stringent reproductive requirements—requirements that remain almost completely unknown.

Clupeomorpha

The history of the herrings has been admirably discussed by Svetovidov (1963). The group is essentially tropical (150 of 190 species; 37 of 50 genera there recognized) with an origin in the once nearly circumtropical Tethys and a long history related to the subdivision of that sea, the early development of cold-adapted and freshwater forms, and subsequent transgressions of boreal forms from the Atlantic into the Pacific and of temperate groups across the tropics. Cold-temperate herrings are not numerous in the South Pacific, and six of the eight genera (including freshwater and anadromous groups) are clearly related to essentially tropical genera (*Nematalosa, Ethmidium, Hyperolophus, Potamalosa, Spratelloides,* and *Etrumeus*). The remaining two, *Sprattus* and *Sardinops,* are bipolar in temperate latitudes and zonally distributed within the limits imposed by competition and reproductive needs. The pantemperate zonation of *Sprattus* is limited by reproductive requirements (Svetovidov, 1963), and *Sardinops,* by competition with *Sardina. Sardinops,* with an origin in the southern oceans from a tropical *Sardinella* during or prior to the Pliocene, reached and colonized the North Pacific later during the glacial periods, simultaneous establishment in the North Atlantic being blocked there by the older and already established European *Sardina.* Together these two form a zonal and bipolar distribution.

The general distributional picture is not dissimilar to that of other groups: tropical origin; tertiary evolution of cold-adapted forms that later recrossed the tropics; and a secondary spread of tropical forms into temperate latitudes with decreasing numbers of both genera and species poleward from the tropics, from west to east within the tropics, and in a similar direction, from Australia to Chile, in temperate latitudes. It does not, however, appear to be an "old" distribution in terms of the high level of taxonomic bipolarity and of adaptation to cold water better demonstrated by certain of the more advanced grades of fishes, such as the gadoids and scorpaenoids.

The second major group of clupeomorphs, the anchovies, are neritic, speciose in the tropics and particularly the American tropics (*ca.* 25 species along the Pacific American coast, Hildebrand, 1943, p. 6; 12 in the Philippines, Herre, 1953); more characteristic of continental than island situations (one species in Hawaii, Gosline and Brock, 1960, p. 96; none in the Galápagos, Rosenblatt and Walker, 1963); and represented in temperate waters of all oceans by the single genus *Engraulis.* Included here is *E. ringens,* the Peruvian anchovy. Although representatives of the genus have not been reported from the central American coast between Guatemala and Costa Rica (Hildebrand, 1943, p. 7), and the latitudinal distribution in the Indo-Pacific appears to be discontinuous, *Engraulis* is not wholly barred from warm waters, and the temperate species could have been relatively recently derived from tropical ancestors. It seems more probable, however, that *Engraulis,* like the temperate herring genera, colonized the temperate seas during preglacial times, recrossed the tropics during periods of cooling, and subsequently reinvaded them. Myers (1966) has also noted the abundance of anchovies in the Indo-Pacific and the American tropics, and has commented on the anomalous distribution of *Engraulis,* the only cool-water engraulid genus.

Protacanthopterygii

Most of the 50 families of this superorder are primary deep-sea groups that have few inshore relatives (e.g., the synodontid lizard fishes and allied Aulopidae; and the benthic *Argentina,* the partial distributional history of which was considered by Cohen, 1959). A few, such as the pikes and salmonids, are zonally distributed and relatively successful freshwater forms, but many others are surviving in their original freshwater environment in various parts of the world in a pattern suggestive of competitive weakness rather than strength: the Asiatic Salangidae (Fang, 1934), *Plecoglossus* in Japan and Formosa, *Phractolaemus, Kneria* and *Cromeria* in a few African rivers, the holarctic Umbridae, etc. (see Darlington, 1957, p. 106, for distributional data on the freshwater groups). *Chanos* is a euryhaline Indo-Pacific relict, and *Gornorhynchus* is a marine genus of the far southern hemisphere where its distribution, a zonal one, includes Africa, Australia, and New Zealand but not continental South America.* Of particular interest in the present connection are the northern true smelts, Osmeridae, and the southern marine and freshwater "smelts" in and allied to

---

*Dr. Richard H. Rosenblatt has recently informed me that *Gonorhynchus* was collected near the Juan Fernandez Islands off central Chile during Cruise XII, R.V. *Anton Bruun*

the family Galaxiidae. This southern group, which has recently been reviewed by McDowall (in press), is derived from an early osmerid stock close to the northern genus *Spirinchus* (Weitzman, 1967), and the two groups can be considered a pair bipolar at the subordinal level. *Spirinchus* is a Pacific genus, and the Pacific origin of *Galaxias* is further supported by parasitological data (Manter, 1955). Both are allied to the northern freshwater and anadromous salmonids. All are cold-adapted, and their occurrence in the shallow tropics during any period since the radiation that produced the southern families is unlikely (McDowall), as is transgression beneath tropic seas in deep water during any time during the Tertiary. Given this subordinal level of taxonomic bipolarity and the nature of the superorder as a whole, I would tend to look for a far earlier ancestry.

The distributional pattern of the entire superorder, 50 highly diverse families of fishes, few of which are speciose, leaves me with the impression that the primary radiation was an ancient one, as old as that of ancestral cods, earlier than that of Clupeomorphs, and one that occurred in or probably earlier than the Cretaceous and probably in fresh water; and that the present distributions are generally fugitive ones. Many groups were pushed into the deep sea by the superior competitive advantage of more recent forms, there to become either highly specialized or extinct; others became restricted to localized refugia, and a few became zonally distributed and cold-adapted. In an evolutionary sense, the entire superorder may be on the way out.

Paracanthopterygii

The paracanthopterygian order Gobiesociformes, the cling-fishes, has been reviewed by Briggs (1955 and subsequent papers), who found a close parallel between the phylogenetic history, which he established from morphology, and the distributional. The group is most speciose in the American Pacific but had its origin in the Indo-West-Pacific where there remain a few very advanced and specialized representatives (e.g., the slender *Diademichthys lineatus* that lives among the spines of the common black sea urchin) and a few not so specialized, the competitive success of which is unknown. The most primitive genus (*Creocele*) survives in Tasmania, and other relict genera live elsewhere in Australia and New Zealand. There are two small and restricted relict subfamilies, the Haplocylicinae in New Zealand and the Chorisochisminae in southernmost South Africa. Indo-Pacific genera are thought to be the diverse but not numerous descendants of the Indo-West-Pacific ancestral stock. A second radiation may have resulted from the immigration of an Indo-Pacific form to American waters and its radiation there, or by the separation of a once-widespread Tethyian ancestral stock with extinction in the Indo-Pacific. The American subfamily Gobiesocinae has a bipolar representa-

tive monotypic at the generic level off South Africa (*Eckloniaichthys*); the European Lepadogastrini genus *Diplecogaster* (*bimaculata*) has a bipolar species counterpart, *D. megalops* Briggs, in South Africa that was thought to have arrived via deep waters off Atlantic Africa. Zonal patterns at the generic level do not seem to have become established, although cold-adaptation by some, usually the more generalized genera and species, is evident. This may be a reflection of the level of generic splitting but, more probably, can be traced to the relatively low powers of dispersal of these small and highly adhesive fishes. While other more dispersible groups that have colonized New Zealand from Indo-Pacific tropical waters and Australia have found their way across the temperate South Atlantic to Chile, none of the several New Zealand cling-fishes have managed to do so, and the Chilean cling-fish fauna (some of commercial importance) is all of American origin.

The toadfishes (Batrachoidiformes, with its single family Batrachoididae) are worldwide but nowhere dominant. A few are deep-water but may, like the American midshipman, *Porichthys*, repair to the shallows to breed. They are notably scarce on coral-dominated islands (none in the Seychelles, Smith, 1963; or in Hawaii, Gosline and Brock, 1960; or in Fiji, Fowler, 1959; and but marginally represented in the Bahamas, Walters and Robins, 1961). Although common on continental shores, they do not extend poleward into truly cold-temperate seas. Of the three subfamilies, the largest, the Batrachoidinae, is worldwide but with few if any cosmopolitan genera.* (Generic confusion still reigns—Smith, 1965, p. 423, for example, both recognizing and synonymizing his own genus *Batrichthys* on the same page of print.) The other two subfamilies are of greater significance. These, the Porichthyinae and the Thalassophryninae, are both wholly American, are composed of genera that are Atlantic or Pacific but not both (*Thalassophryne* vs. *Daector; Porichthys* and *Aphos* vs. *Nautopaedium*), and both groups are clearly more advanced than the cosmopolitan Batrachoidinae. The Thalassophryninae have developed hollow dorsal spines and associated glands to form "the most highly developed venom apparatus of any fishes" (Collette, 1966, p. 846), while two of the three genera of Porichthyinae have evolved a system of photophores as magnificent as any found among the primary deep-sea fishes and superior to any luminescent system among the higher fishes. Regardless of the presence of these clearly derivative groups in American waters, I am disposed to look elsewhere, preferably to the Indo-Pacific, in the noncoralline tropics for the point of origin of the batrachoidids. The southwestern Pacific forms may easily have come from the Indo-Pacific, as may have the progenitors of the now typically American genera and subfamilies. Isolation of

---

*I have been unable to verify Smith's remarks (1965, p. 423) that *Austrobatrachus* occurs in South America as well as South Africa.

this progenitor from its Indo-Pacific parental stock is forti-
fied by the coral-island stepping stones, a habitat generally
eschewed by all members of the group. Transport eastward
from temperate Australia and New Zealand has apparently
not occurred, for the toadfishes of South America are of
tropical American origin.

Most of the Lophiiformes are deep-water and thus are
excluded here, but the one inshore group, the Antennariidae
or frogfishes, while it falls into my category of groups self-
successful but evolutionarily dead-ends, calls for a comment.
The family Brachionichthyidae, if valid, is endemic to
Australia, and the Antennariidae proper is circumtropical.
The group was reviewed by Schultz (1957). If one compares
the scanty distributional data provided with his phylogeny
(p. 53) it becomes apparent that the more primitive genera,
with the exception of the sargassum fish *Histrio*, are Australian, where the family is well represented. It is also apparent
that, like the toadfishes, the group is a tropical one but is
not in any major way associated with coral reefs.

The five suborders included within the archaic Gadiformes
by Greenwood *et al.* (1966) are diverse, and the relationships
among them are obscure. Most uncodlike are the zoarcids, a
deep-water and polar group well represented in the South
Pacific, especially in the Magellanic region, by both endemic
(e.g., *Maynea, Pogonolycus, Phucocoetes, Austrolycus*) and
bipolar (e.g., *Melanostigma, Lycodapus*) genera. The group
is poorly known but the origin is probably northern.
Andriashev (1965) derives them from the north via routes
along the eastern borders of the continents and at abyssal
or bathyal depths postulates their exclusion from more
shallow waters along the coasts of Africa and South America
through competition with an earlier established fauna there,
and would have them rise to lesser depths in the far south
where competition was less extreme (and preadaptation to
cold of greater value). I can only concur that "If our considerations are confirmed in the future by further investigations, the conclusion that the coastal fauna of the Patagonian Region (and partly of the Antarctic continental shelf)
were formed to a certain extent by deep-water elements will
be strengthened. This would represent an exceptional example amongst fishes of the formation of a littoral fauna
from a bathyal-abyssal one" (Andriashev, 1965, p. 545).

Although the origin of the suborder Ophidioidei, the
messmates, ass fishes and slippery dicks, is equally obscure
and is probably as ancient as that of the cods proper (Paleocene or before; Svetovidov, 1962, p. 36), a tropical origin is
not unlikely. Ancestral groups in an increasingly competitive
situation can do one of several things without perishing:
They can stay in the area but adopt an unoccupied and
highly specialized niche as indeed the messmates (Carapidae)
that live in the digestive tracts of sea cucumbers and mantle
cavities of oysters have done; they can move into more temperate waters, the deep-sea, or other more specialized habitats, a pattern that would go far to explain the present

distribution of the brotulid ophidioids; or they can move
into less-favored or open tropical areas, as perhaps the
ophidiid ophidioids have done through an eastward movement to the Americas where they now form a prominent
rather than a minor part of the fauna. (There is but one in
the Philippines according to Herre, 1953, p. 818; none in
New Guinea according to Munro, 1967, p. 465.) In contrast
to this apparent eastward movement through the tropics, the
ancestor of one genus, *Genypterus*, migrated south to
Australia and New Zealand and thence eastward, following
the route of *Congiopodus*, to South America. It is represented along the Chilean and cooler Peruvian coasts by two
species (*G. chilensis* and *G. congria*) of commercial importance and a third, *G. blacodes*, which lives in the Patagonian
Atlantic and is considered by Norman (1937, p. 112) and
Hildebrand (1943, p. 414) to be conspecific with the populations off Australia and New Zealand. This distribution suggests that the point of initial colonization was far south in
South America, followed by the migration of the colonizer
up the Atlantic coast, and radiation with speciation up the
Pacific. Both major branches of the ophidioids are now in
the hands of competent monographers, and hence we may
hope that authoritative evolutionary histories will soon be
forthcoming.

The diverse and speciose Macrouridae (rattails) are an
important part of the benthic and engybenthic fauna. The
group is nearly cosmopolitan, has formed endemic species
groups in at least one area, the Sulu Sea (Gilbert and Hubbs,
1920), has differentiated across species barriers far less pronounced (for example, the mid-Atlantic ridge, Marshall,
1963), but is excluded from further discussion here because
the macrourids are typically deep-sea rather than coastal
fishes. All the remaining gadids, however, are represented in
coastal or shallow waters of the South Pacific.

One family with its probably monotypic genus, *Breg-
maceros* (reviewed by d'Ancona and Cavinato, 1965), which
dates at least from the Eocene (Svetovidov, 1962, p. 28), is
unique among the gadids proper in being tropical and subtropical. It is also almost the only gadid that is oceanic and
estuarine, which may be the secret to its having survived in
otherwise codless tropical seas. At the other extreme are
the cold-restricted and zonally but not wholly sympatrically
distributed Antarctic cods, the Muraenolepidae. If derived
from their Arctic counterparts, these fishes took their leave
at an early date probably long before the emigration from
the Arctic polar basin into the North Atlantic of the ancestral gadid that gave rise there, during mid-Tertiary, to most
of the contemporary gadid genera (Svetovidov, 1962).

The true cods are essentially a Northern Hemisphere
Arctic and cold-temperate group, and the representatives
differ in the degree to which they have become cold-
stenothermal on the one hand and deep-water on the other.
These are the factors that have clearly determined which
have been able to reach the Southern Hemisphere and which

have not. The pan-temperate and tropical morids are well represented in the South Pacific by widespread genera. The Gadidae, fishes of shallower and more polar seas, are less so with true Arctic forms, such as *Boreogadus* and the cold-temperate cod (*G. morhua*) and haddock (*M. aeglefinus*) excluded from the south. *Gaidropsarus* and *Micromesistius* are bipolar, and the hake genus *Urophycis* and the whitings, *Merluccius*, which are commercially important off Chile, form species complexes that are very nearly continuous from the north-temperate to the south-temperate zones and beneath tropical seas (see Svetovidov, 1962). The apparent absence of *Merluccius* and *Gaidropsarus* from Australia but not New Zealand is enigmatic.

Atherinomorpha

The origin and relationships of the paracanthopterygian superorder Atherinomorpha and its single included order, Atheriniformes, have been reviewed by Rosen (1964) and revised in whole or in part by Jordan and Hubbs (1919) and Schultz (1948). The group had its origin somewhere in the ancestry of the Perciformes and includes as separate suborders the flying fishes (Exocoetoidei) and killifishes (Cyprinodontoidei) and the jack-smelts (Atherinoidei), although Gosline (1968) doubts the unity of the order. Rosen proposed a common origin for these in the fresh and brackish waters of Australasia, with subsequent spread around the world. When properly studied, the group will be of considerable zoogeographic significance. However, because of the large number of freshwater and estuarine forms among the atherinids, the group cannot be expected to reveal significant patterns in the present marine zoogeographic connection, although the coastal forms are both abundant and commercially important. Some extend poleward to southern Chile, New Zealand, South Australia, and Tasmania (Scott, 1962, p. 134; Mann, 1954; Graham, 1956), but none are bipolar. The cyprinodonts, being primarily (but not "primary") freshwater fishes, can be omitted here. The remaining group, the halfbeaks, flying fishes and their allies, are freshwater, estuarine, and oceanic. The oceanic genera are cosmopolitan in tropical latitudes but, again, representatives extend poleward to Tasmania and New Zealand and, in the southeastern Pacific, to the Juan Fernandez Islands, where large numbers of large individuals occur seasonally. There is even one sight record as far south as the Cape of Good Hope (Deacon, 1960, p. 446). Highly dispersible, cosmopolitan oceanic forms such as these are of limited instructional value here.

Acanthopterygii

Among the structurally more advanced groups, the mail-cheeked fishes of the order Scorpaeniformes are potentially a most significant group for historical analysis. Certain

groups that are essentially tropical and arose there (e.g., the nonsebastine Scorpaenidae, Synancejidae, Caracanthidae, the deeper-water and partially bipolar Triglidae, and the predominantly Indo-West-Pacific Platycephalidae) have radiated poleward into subtropical waters but have not generally become zonally distributed there. An origin other than a tropical one need not be postulated for the Hoplichthyidae, a small family that ranges, perhaps discontinuously, from Japan to temperate South Australia. The few members of the families Pataecidae and Aploactinidae, now endemic to Australia and New Zealand, may also be of tropical origin, although they are so highly modified that the relationships are obscure. A similar origin is probable for the related pig-fishes (Congiopodidae), with an origin in the tropics, subsequent establishment in the colder waters of Australia and New Zealand, and eventual spread across the cold-temperate South Pacific followed by the establishment of a congener, *Congiopodus peruvianus* (*Agriopus hispidus*) along the Pacific and thence to the Atlantic coasts of South America (Norman, 1937). The group has also spread south to Kerguelen (*Zanclorhynchus spinifer* Günther).

The cold-temperate North Pacific has clearly been the site of the radiation, if not the origin, of the cold-water scorpaenoids. There have been three primary radiations. All have resulted in the production of speciose groups in the North Pacific, and two, through tropical transgression, have established relatives in the cold-temperate South Pacific. The hexagrammoids (Hexagrammidae, Anoplopomatidae, and Zaniolepididae) are restricted to the temperate North Pacific, which is perhaps an indication of recent origin, for in both dispersal ability and warm-water tolerance they are apparently equivalent to representatives of other related groups that have managed to cross into the Southern Hemisphere. The distributional pattern of the hexagrammoids is more characteristic of the percoids than of other scorpaenoids. A second line, the sebastine scorpaenids, so notably speciose in the North Pacific, has been more successful in invading the Southern Hemisphere (*Sebastodes, Sebastapistes, Neosebastes,* and *Helicolenus;* see, e.g., Krefft, 1961), where they occur along all coasts, than they have in colonizing the North Atlantic. There the group (*Sebastes,* a genus probably not distinct from *Sebastodes*) is known from three species, or but two if the still-problematical *S. mentela* is excluded. These few Atlantic species, however, form as dominant a part of the North Atlantic fish fauna as do the many Pacific species in the fish fauna of that ocean. The history of the third lineage will, when fully understood, form a zoogeographical contribution of great importance. This is the suborder Cottoidei, with eight or nine included families. The lumpfishes (Cyclopteridae, incl. Liparidae) are of Arctic and cold-North Pacific origin and are there still most abundant but include some reef genera. They have transgressed the tropics to form bipolar pairs at the species level with southern cold-temperate forms (*Careproctus, Liparis*) and have become a part of the secondary deep-sea benthic fauna (data from Burke, 1930; see also Cohen, 1968). While the history of the successive origins of many scorpaeniformes is unknown, that of the Cottoidei is central to the problem and has been intensively investigated by Shmidt (1965, p. 284 *et seq.*). Shmidt places the origin of the group in the then-tropical or subtropical waters of an Oligocene Sea of Okhotsk, with successive waves of irreversibly cold-adapted cottoids emanating southward with subsequent cooling of the North Pacific. Five of the eight cottoid families have contributed, via the Arctic, to the relatively depauperate cold-temperate fauna of the North Atlantic, and four of the eight have crossed the tropics and have bipolar relatives at the generic level (Cottidae, Psychrolutidae, Agonidae, and Cyclopteridae)—a degree of bipolarity unsurpassed by any other coastal teleostean group. To add to an already complex zoogeographical situation, the cottoids include a holarctic freshwater genus (*Cottus*), another freshwater complex of Arctic origin in and adjacent to the North American Great Lakes (the populations of *Myoxocephalus* formerly included within the genus *Triglopsis*); and two endemic families (Comephoridae and Cottocomephoridae) with seven genera and 23 autochthonous endemic species in Lake Baikal, the two or three ancestors of which are thought to have reached the lake from the Pacific shores of Asia in Late Tertiary (Taliyev, cited by Kozhov, 1963, pp. 145, 291, the age of the lake given as 30 million years by Dmitriev as reported by Frey, 1968).

The South Pacific cold-temperate scorpaenoids are thus few but historically most interesting. Bolin (1952) derived his Tasman Sea bipolar cottid, *Antipodocottus galatheae*, from a relative of the western North American genus *Icelinus* through the Japanese *Stlengis*. Hence it becomes a derived form living external to the center of distribution of the Cottidae and presumed site of origin of that group. Within the same family, the Chilean *Normanichthys*, although recognized as a distinct family by Quast (1965) and Greenwood *et al.* (1966), was viewed as a form close to the ancestral cottid line by Norman (1938). If it is indeed a primitive cottid it is behaving in accord with the wishes of Matthew (1939) and his present disciples by living not just peripheral to the current range of the group as a whole or place of origin but off at the other end of the world.

There are 20 suborders and over 140 families in the immense order Perciformes, and both the evolutionary and the distributional history of this enormous group is a task for the future. A few matters of confusion, and others of interest, are, however, worthy of note.

First, I have viewed as fundamentally sensible and have here relied on the teleost classification provided by Greenwood *et al.* (1966) although no thoughtful ichthyologist, including its authors, would pretend to characterize it as definitive. I have also studied and admired the evolutionary study on Mesozoic acanthopterygians by Patterson (1964).

Both have been dismissed, in what would appear to be a rather cavalier way, in a recent paper by Gosline (1968) who has in most respects simply returned to prior classifications. The resultant uncertainties in classification, together with others of long standing, such as the relationships of the Antarctic notothenioids, naturally add further questions to the analysis attempted here.

The notothenioids include four endemic, speciose, and, I believe, autochthonous families that dominate Antarctic shores, have been there since the Paleocene (Andriashev, 1965), and have radiated back along the colder continental shores of the South Pacific. They do not have the general facies of a typical percoid and have been variously aligned (Gill, 1861; Boulenger, 1901; Regan, 1913a, 1913b; among others). Greenwood *et al.* (1966) refer them, with qualification, to the percoids. The question is of obvious importance here, for the group is cold-adapted and zonally distributed, in contrast to the pattern found in most other percoids. Hugh H. DeWitt (personal communication, 1968) has noted not only his intent to study the question of group relationship but his openness to gadiform relationship for these fishes, which, perhaps prophetically, have been commonly known as Antarctic cods. He refers as well to data inadvertently omitted from Freihofer's study (1963) of the *ramus lateralis accessorius* patterns in teleosts that suggest gadiform affinities. This evidence was also discussed by Gosline (1968), who considered the apparent concurrence in nerve pattern to be the result of convergence rather than close relationship. Similar arguments, with little convincing additional morphological data, were used to remove the Ophidioidei from the Gadiformes, the considered placement of Greenwood *et al.* (1966). I can agree neither with these transpositions of Gosline nor with those regarding the characterization of the Gobiesocidae and the Lophiiformes as percoid derivatives.

The order is not only the largest but the most widely dispersed of all primarily oceanic groups. Relatives are widespread in fresh waters, and many are oceanic (e.g., six genera of cheilodipterids; the trichiuroids, *Tetragonurus*, the bramids and pteraclids, etc.). (The pertinence of the highly specialized Chiasmodontidae, which in degree of modification as well as habitat appear to be primary deep-sea fishes, should be restudied before allocation to the percoids and to the secondary deep-water fauna.) There are also localized endemic families, in the Pacific as elsewhere: the North Pacific Embiotocidae, Trichodontidae, Bathymasteridae, Zaproridae, and several groups of eel-blennies, some of which have reached the Atlantic; the Australian-New Zealand Arripididae, Chironemidae, Odacidae, Gadopsidae, Cheimarichthyidae, Leptoscopidae, Notagraptidae; and the Aploactylidae, which also reaches the South American coast. Many of these are doubtless of relatively direct tropical ancestry. Others, a suspiciously large number, are trachinoids, which may be, according to Greenwood *et al.*

(1966, p. 389), like the notothenioids, of different and lower relationship. Gosline (1968), probably correctly, includes a part of these in his percoid suborder Blennioidei. The derivation of these blennioids and another percoid derivative, the flatfishes, from some ancient point on the percoid line prior to the invention of the more typical existing percoid lineages would appear to be sensible with respect to both evolution and distribution. Not all cold-adapted families can be so considered. The Embiotocidae, for example, are endemic North Pacific fishes and are of distributional interest because they are viviparous. Viviparity should aid rather than hinder dispersal, and a broad latitudinal distribution should favor passage from North Pacific to North Atlantic and South Pacific, but this has not occurred. Why?

Bipolarity among the percoids is limited and of a low taxonomic level. It appears to be largely limited to groups living in or near the subtropics (Labracoglossidae, Hemerocoetidae), and fishes of offshore distribution, such as certain carangids and the stromateoids *Centrolophus, Icichthys,* and *Schedophilus.*

It is also subjectively clear (1) that cold-temperate zonally distributed percoid groups are rare; (2) that the group dominates the tropical fauna, especially that of the coral reefs; and (3) that the number of species and of individuals decreases sharply poleward from the tropics both in number of species and in number of genera. The percentage of the total fish fauna contributed by the percoids declines poleward.

Although they are a structurally advanced grade of fishes in the sense of Patterson (1964), I cannot identify the adaptive feature, probably a subtle one, that led to the initial explosive radiation of the percoid fishes. Within the group, however, one important invention has occurred that may account for the success of a large segment of these fishes but one that has rarely, or at best indirectly, been considered in an evolutionary or zoogeographical context. This is the development of the ability to eat and to digest attached plant material. This ability is found among freshwater fishes, including some of the more primitive of the primary freshwater forms (e.g., *Brycon* among the Characidae) and occasionally among lower marine species such as the ballyhoo, *Hemiramphus brasiliensis,* that will eat floating sea-grass (Randall, 1965, p. 255; 1967, p. 826). Diatom feeders among the clupeiform fishes, such as the Peruvian anchovy, also occur but are uncommon.

Some of the northern eel blennies of the genera *Cebidichthys, Apodichthys, Epigeichthys, Phytichthys,* and *Xiphidion* are said to feed on the large attached brown algae among which they live (Donald W. Wilkie, personal communication; Jordan and Everman, 1898, pp. 2424-2426). Confirmation is required, for these fishes may be simply ingesting these fronds in order to digest the overburden of encrusting and free-living invertebrates. In any event, such examples among cold-temperate fishes are few.

Hiatt and Strasburg (1960) note the irregular occurrence of the herbivores in their outstanding study of tropical reef ecology (the families with herbivorous representatives: Mugilidae, Leiognathidae, Siganidae, Kyphosidae, Chaetodontidae, Acanthuridae, Pomacentridae, Labridae, Scaridae, Gobiidae, Blenniidae, and almost all the percoid derivatives among the Tetraodontoidei: the Balistidae, Monacanthidae, Ostracionidae, Tetraodontidae, and Canthigasteridae). The list of herbivorous elements in the West Indian fauna studied by Randall (1967) is a similar one. Randall (1965) has also shown the effect of browsing on strands of turtle grass. Stephenson and Searles (1960) and Randall (1961), by their experimental exclusion of these herbivores from reef areas, make it abundantly clear that the tropics and its attached marine flora must have been vastly different prior to the evolution of the browsing fish herbivores. The question deserves the attention of evolutionists as well as ecologists.

It is also noteworthy that it is just these advanced percoid forms that have not yet radiated in any substantial way into cold-temperate seas, and, among those that have, very few indeed eat plant materials. For example, Limbaugh (1955), in his analysis of the fish fauna associated with the temperate kelp beds off Southern California, found but one species, the half-moon *Medialuna californiensis* (Kyphosidae), that actually ate kelp as well as other algae. Quast (1968a, pp. 122, 125, 128) added the señiorita, *Oxyjulis californica;* the opaleye, *Girella nigricans;* and the zebra perch, *Hermosilla azurea.* He presented convincing if not conclusive evidence, both observational and experimental, elsewhere (1968b, p. 147) that kelp is incidentally ingested but not digested, by all these fishes and that they are physiologically carnivores. (There are, of course, herbivorous invertebrates, especially among echinoids, gastropods, and crabs; see, e.g., Leighton, 1966; Quast, 1968a.) Further speculation is tempting: The extent of the algal forest (but not its diversity) increases poleward. The abundance of the advanced percoid herbivores decreases poleward. Is there a causal correlation?

Other explanations do, of course, exist; and other factors, such as major taxonomic differences between tropical and temperate algae, must be considered.

Thus the reasonably advanced position among the teleosts that the percoids have long been accorded is reflected in the pattern of radiation, which is, I believe, relatively recent and certainly not ancient. The pattern: tropical evolution with probable displacement there of many earlier prepercoid groups; radiation poleward into temperate but usually not polar seas; little circumglobal cold-temperate zonality; and little time for tropical transgression and hence little evidence of it.

The percoid stem has produced two lineages worthy of ordinal rank, both of which, because of extreme specialization, may be called "successful and speciose evolutionary dead-ends." Most of the Tetraodontiformes have improved on the ability developed by their immediate ancestors to eat attached plant material as well as animal food. They have developed locomotory structures that enable them to maneuver with ease, if not grace, next to a food-covered substrate; dentition highly adapted for feeding on this substrate; and protective devices ranging from armature to inflatability that permit the survival of these slow and exposed animals that would otherwise provide an easy mark for any reef predator. Perhaps because of these specializations they have remained largely restricted to the tropics and more especially to that part of the tropics that provides a hard substrate. A few have gone into fresh water and others, chiefly the triacanthodids, into deep water. One group, the aracanine ostraciodontids, has radiated chiefly in the seas off southern Australia (perhaps one day to be followed by cold-adaptation and the trans-Pacific colonization of southern Chile by a representative or two?). Most genera, however, appear to be widely distributed in the coralline tropics.

In striking contrast are the flatfishes (Pleuronectiformes), derivatives of some conservative early percoid and masters of the smooth-bottom habitat. Although much work of a localized or fragmentary nature has been done since the publication of Norman's monograph (1934), his distributional analysis is instructive. Zoogeographically, the group does not appear to be a recent one if the occurrence of zonally distributed cold-adapted forms is used as a criterion. The pattern is to some extent intermediate between that of the scorpaenoids and the percoids, resembling the former by having both cold-adapted suprageneric groups and strong tropical representation, and the latter by the relative absence of bipolar pairs. The group closest to the ancestral line, the Psettodidae, contains but a single genus found in the Indo-Pacific and off West Africa. The primary radiation that formed the basis for the principal existing lineages was undoubtedly in the tropics where flatfishes, particularly the more generalized, are still most abundant. A comparison of Norman's phylogeny of families and subfamilies with his account of latitudinal range (1934, pp. 43, 49) reveals that most of the derived or advanced groups are cold-adapted and restricted to the higher latitudes; the Scophthalminae in the North Atlantic, the Pleuronectidae in Arctic and northern seas, and the Rhombosoleinae in the Southern Hemisphere. The Rhombosoleinae provide yet another example of a South Pacific cold-temperate distributional pattern found in other quite unrelated groups such as the scorpaenoids and the ophidioids: origin in the Indo-Pacific; colonization and radiation with adaptation to cold water in Australia, New Zealand, and the more southern isles; and subsequent eastward dispersal through cold-temperate seas by one representative, *Oncopterus*, and its establishment off Patagonia. The flatfishes are clearly of percoid origin, but I suspect that the ancestral flatfish took its leave from the percoid line a very long time ago.

DISCUSSION

I am concerned here primarily with a historical question: Do the distributional histories of selected but significant groups of South Pacific fishes show any generalized patterns? It is useful to approach the question by first considering problems of cold-adaptation and then by populating a hypothetical fishless South Pacific Ocean.

Age and the successive evolution of cold-adaptation   The polar environment is a severe one. Colonization of a severe environment requires more adaptation than does adaptation to a less extreme one. Adaptation usually takes time; thus adaptation to a severe environment should favor the presence there of representatives of older rather than younger groups. Darlington (1960, 1965) has shown a difference between terrestrial invertebrates and vertebrates in the occurrence of cold-adapted groups (Figure 1). He has also rightly argued that this proposition is often improperly or inadequately stated; that the proper question is: Why have certain groups become cold-adapted and others not? That question is fundamental, of course, but it is a physiological and not a distributional one. That group-related trends exist, however, is of zoogeographical interest, regardless of the underlying physiological reasons for the development of these trends and patterns.

A correlated question is that of the proper use of a classification. In the written presentation of a classification, the limitations of the printer's art require a linear sequence. The creation of such a sequence, such as the fish classification followed here, thus presents an unparalleled opportunity to assign a chronology to the development (radiation if not origin) of these groups unintended by the authors. This has often been done, the vociferous protests of the

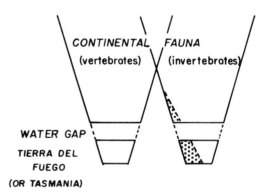

FIGURE 1   Diagram of the terrestrial fauna of Tierra del Fuego (or Tasmania) in relation to the fauna of the adjacent continent (South America or Australia). Convergence of lines southward indicates diminution of the total fauna. Stippling shows occurrence of independent southern cold-temperate groups. (Reprinted with permission from Darlington, 1960, p. 661.)

authors of such classifications notwithstanding. Among the teleosts it is clear that such linearity does not exist and that radiations by various groups have been at least in part chronologically parallel. This has been stated repeatedly and forcefully by systematists in part to protect their classifications from misuse by others less well informed on the facts underlying such classifications and on the lacunae.

Nevertheless, a subjective feeling for archaic versus non-archaic or advanced versus early "structural grades" (*sensu* Patterson, 1964) is emerging, and probably correctly so (see, for example, Myers's 1967 account of the "archaic and hence older" caracoid fishes with respect to Mesozoic geography and zoogeography). I provide these remarks not as an excuse for assigning relative chronology to major radiations but as evidence that I understand the dangers and qualifications of doing so.

A nonquantitative appraisal of poleward diversity and cold-adaptation has been attempted for four of the major conservative lineages of teleosts discussed above that clearly form an ascending series of structural grades. These are the Salmoniformes, Gadiformes (including the products, such as the ophidioids, of an early radiation, although the pattern would have been more pronounced without these elements), the Scorpaeniformes, and the Perciformes. Borrowing a diagrammatic technique from Darlington (1960, p. 661, Figure 1), I traced trends in poleward diversity and in cold-adaptation (Figure 2). Inspection is instructive. A relative increase in tropical importance through the series is clear. A decrease in the relative development of cold-adapted forms, i.e., the extent to which such groups contribute to the whole group, is also apparent. Also correlated with advance in structural grade are the extent to which these groups have penetrated into (or been pushed into) the deep sea (most of the salmoniform families; almost none of the perciform); a progressive decrease in the amount of cold-temperate zonality; and the progressive decrease in the amount of, and the taxonomic level of, bipolarity. Within this series a decrease in cold-adaptation and an increase in tropical representation are related to advance in structural grade and thus to age if, *and only if*, structural grade itself is related to age. I believe that within this series age-related cold-adaptation has occurred and is occurring. The scorpaenoids, for example, became important cold-adapted North Pacific (and thence South Pacific) fishes because they had both the time and the inherent ability to do so and were forced to do so by the rise and radiation in the tropics of superior competitive percoid types.

Degree of cold-adaptation is a trend correlated with structural advance in grade among marine bony fishes. To destroy this as a universal rule, however, one must only examine certain other fish groups. Why some groups become cold-adapted and others do not is a physiological and ecological question, and time is only a contributor to the process. Some animals can make the adjustment relatively

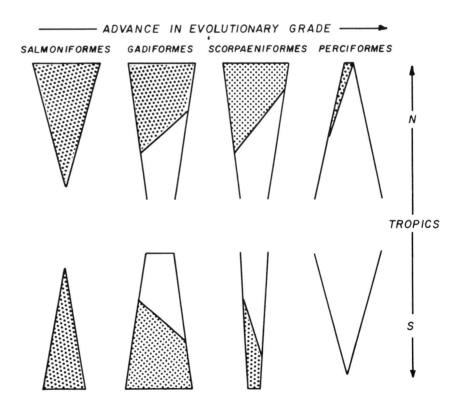

ADVANCE IN EVOLUTIONARY GRADE

SALMONIFORMES   GADIFORMES   SCORPAENIFORMES   PERCIFORMES

N

TROPICS

S

FIGURE 2 Diagram relating four groups of Pacific teleostean coastal fishes to their poleward diversity and the evolution of cold-adapted groups. No proportions are quantitative. Lines diverging from the equator indicate an increase in diversity poleward; convergent lines, a decrease. Stippled areas represent the relative importance poleward of nontropical cold-adapted groups.

fast; others never. Consider but two fish groups: the archaic eels, which have never adjusted to life in polar seas because, I believe, they have never been able to sever the reproductive ties to the central oceanic water masses. Adults become uncommon poleward, and even the mesopelagic eels are infrequently caught in the cold eastern boundary currents. Therefore, although they are old, their distributional pattern resembles that of the more modern groups such as the percoids. They have thrived by conquering the burrowing environment, but they have never become cold-adapted. Consider in contrast the flatfishes, a group derived from an early percoid, which abound in the tropics but have also evolved several cold-water lineages. Cold-adaptation is apparent and has permitted the evolution of the circumarctic halibut and the southern bothids that range as far south as the Falkland Islands and Antarctica (Norman, 1934; Andriashev, 1965, p. 513).

Time *favors* cold-adaptation or adaptation to any other rigorous environment for that matter, but time alone cannot *cause* it.

Radiation in the South Pacific   The Pacific south of the tropics appears to have been colonized by a series of radiations, each related to advance in evolutionary grade and to cold-adaptation.

First, the southern salmonoids and the Antarctic cods, the notothenioids and the muraenolepids, are probably of ancient origin. These are all cold-restricted and zonally distributed and have diverged from their northern ancestors to the extent that the bipolar relationship is at or above the family level. These fishes were probably resident in the South Pacific in the mid-Mesoic, whatever the configuration of that ocean might then have been.

Next are those groups, some of which have strong tropical representation and others not, that are less completely cold-adapted, less often or less completely zonally distributed, and with bipolar representation common at the generic or specific level—the gadids, clupeoids, and scorpaenoids. It is noteworthy that the southern representatives, with few exceptions, appear to have become cold-adapted in the Northern Hemisphere and to have then immigrated into the Southern Hemisphere. This is probably due to the relative scarcity of land and thus of a coastal habitat at subantarctic latitudes in the south as compared to that in equivalent northern latitudes.

Finally came the more recent population of the South Pacific by groups whose early evolution has been in the tropics with secondary radiation in temperate seas. These include supposedly old groups, such as the true eels and the clingfishes (Gobiesocidae), but primarily the more recent percoids and their derivatives, such as the tetraodontids. These dominate the tropics, become scarce poleward, are infrequently zoned in cold-temperate seas, and, save for a few subtropical and oceanic forms, are rarely bipolar.

The main source faunas for the temperate South Pacific fish fauna are four. The primary and oldest is that of the tropical Indo-Pacific and the secondary are as follows:

1. The American tropics, which is characterized in large measure by the Indo-Pacific elements that it lacks (Myers, 1941). It is composed of certain ancestral elements predating the division of the circumglobal Tethys Sea, and a few elements gradually and regularly contributed by eastward dispersal from the Indo-Pacific, and, to a considerable extent, by the secondary evolutionary radiation there of groups such as the batrachodids, gobiesocids, ophidiids, and certain blenniods that probably originated in the Indo-Pacific.

2. The cold-temperate North Pacific, with the evolution there of cold-adapted groups such as the true cods and the cottoids, and the eventual spread of representatives of these into cold waters of the South Pacific.

3. Cold-temperate Australia and New Zealand, with the development there of certain major taxa, such as some scorpaenoids, antennarioids, and flatfishes and the eastward dispersal of a few representatives to the Juan Fernandez Islands and coastal Chile.

The colonizing patterns of radiation from the Indo-Pacific, in order of importance, are these:

1. The spread of the Indo-Pacific tropical fauna eastward within the tropics, with progressive diminution along the route and with an apparent major break toward the east, the so-called East Pacific Barrier (Eckman, 1953; Briggs, 1961; Rosenblatt, 1967, among many others) that essentially separates the American fauna from that of Oceania. The nature of this faunal break has often been discussed and need not be considered here. According to Briggs (1966), it has been crossed successfully from the west by 62 species of shore fishes, while none of the American fauna has managed to cross in the opposite direction. The primary question, however, is one of habitat rather than dispersal ability. Eastward colonization from the Indo-Pacific has necessarily favored those species adapted to the stable, transparent waters of coral-rich Pacific isles, ecology standing in strong contrast to the seasonally fluctuating, cold, turbid, strongly tidal, and largely coral-free continental waters of, for example, Pacific Panama.

Although this so-called barrier is generally exemplified by the faunal break between Polynesia and the American mainland, it is apparent as well in the southeastern Pacific although the fauna there is much less well known. The fish fauna of the most eastern of the oceanic islands, Easter Island, which lies 2,000 miles from the American shore, has been studied by de Buen (1963) and his predecessors, and more recent collections are under study by Ian Efford and his associates (personal communication, 1967). About a quarter of that fauna is endemic. Relationship to the islands to the westward and the northwest is abundantly clear; that to mainland America, very slight. The faunal discontinuity between Oceania and the Americas is as real in the southeastern Pacific as it is farther north.

2. The second pattern is that extending from the Indo-Pacific to cold waters off Australia and New Zealand and, with a tremendous diminution of the fauna, across the cold-temperate South Pacific to the Juan Fernandez Islands and coastal Chile. Evidence for the cold-restricted nature of this fish fauna is shown by a comparison of the fishes of the Juan Fernandez, which lie adjacent to the northbound Humboldt Current, with those of Easter Island. The two island groups share few genera (*Pseudolabrus, Gymnothorax*) and fewer, if any, species. This eastward spread across the cold-temperate South Pacific is in the same direction as that of the prevailing surface-current system, while the spread across the tropics, in the same direction, is largely opposite to the predominant surface-water movements.

3. The third pattern is that from the Indo-Pacific into cold-temperate waters of the North Pacific, with subsequent recrossing of equatorial latitudes and the establishment of a few previously cold-adapted groups in the South Pacific.

4. The last is the simple spread, poleward and with diminution of the fauna, of the American continental coastal fauna.

Cold-temperate Chile clearly holds the answer to many distributional problems. It is the geographical end point of these radiations, together with a major but less significant one northward from Antarctica; and it is the place most distant from the several source faunas. This may explain why the fish fauna there is a relatively small one. Those fishes that are a part of it have had a varied and exciting zoogeographical history, and yet it is just that fauna that has not been, and is not being, seriously studied. Both the offshore and coastal fishes offer a scientific opportunity of potentially great significance.

ACKNOWLEDGMENTS

Several people have read one draft or another of this study and have provided valuable and provocative comments, for which I am most grateful. For their efforts I am particularly indebted to G. S. Myers of Stanford University; R. H. Backus, J. E. Craddock, and R. L. Haedrich of the Woods Hole Oceanographic Institution; P. J. Darlington, Jr., S. A. Earle, K. J. Boss, and R. M. McDowall of Harvard University; and I. Rubinoff of the Smithsonian Tropical Research Institute. All have contributed to this paper but none either agreed or disagreed with it in its entirety.

REFERENCES

d'Ancona, U., and G. Cavinato (1965) The fishes of the family Bregmacerotidae. *Dana Rep., 64*, 92 pp.

Andriashev, A. P. (1953) Ancient deep-water and secondary deep-water fishes and their importance in a zoogeographical analysis. Trans. by A. R. Gosline from: *Ocherki po obshchim voprocam ikhiologii, Aka. Nauk SSSR, Ikhiol. Kom.*, 58–64.

Andriashev, A. P. (1965) A general review of the Antarctic fish fauna. *Monographiae Biologicae, 15*, 491–550.

Aron, W. (1962) The distribution of animals in the eastern North Pacific and its relationship to physical and chemical conditions. *J. Fish. Res. Bd. Can., 19* (2), 271–314.

Bakus, G. J. (1964) The effects of fish-grazing on invertebrate evolution in shallow tropical waters. *Occ. Pap. Allan Hancock Fdn., 27*, 29 pp.

Bolin, R. L. (1952) Description of a new genus and species of cottid fish from the Tasman Sea, with a discussion of its derivation. *Vidensk. Medd. fra Dansk naturh. Foren., 114*, 431–441.

Boulenger, G. A. (1901) Notes on the classification of teleostean fishes. I. On the Trachinidae and their allies. *Ann. Mag. Nat. Hist.*, ser. 7, 8, 261–271.

Briggs, J. C. (1955) A monograph of the clingfishes (order Xenopterygii). *Stanford Ichthyol. Bull., 6*, 1–224.

Briggs, J. C. (1961) The East Pacific Barrier and the distribution of marine shore fishes. *Evolution, 15*, 545–554.

Briggs, J. C. (1966) Zoogeography and evolution. *Evolution, 20* (3), 282–289.

Briggs, J. C. (1967) Relationship of the tropical shelf regions. *Stud. Trop. Oceanogr., Miami, 5*, 569–578.

de Buen, F. (1963) Los peces de la Isla de Pascua. *Bol. Soc. Biol. Concepción, 35/36*, 3–80.

Burke, V. (1930) Revision of the fishes of the family Liparidae. *Bull. U. S. Nat. Mus., 150*, 204 pp.

Bussing, W. A. (1965) Studies of the midwater fishes of the Peru-Chile Trench. *Biol. Antarctic Seas, 2*, 185–227.

Cohen, D. M. (1959) The geographical history of the Argentininae. *Preprints, Int. Oceanogr. Congr., A.A.A.S.*, Washington, 259–261.

Cohen, D. M. (1968) The cyclopterid genus *Paraliparis*, a senior synonym of *Gymnolycodes* and *Eutelichthys*, with description of a new species from the Gulf of Mexico. *Copeia, 4*, 384–388.

Collete, B. B. (1966) A review of the venomous toadfishes, subfamily Thalassophryninae. *Copeia*, (4), 846–864.

Craddock, J. E., and G. W. Mead (in press) Midwater fishes from the eastern South Pacific Ocean. *Sci. Rep. Southeastern Pacific Exped. R/V "Anton Bruun."*

Darlington, P. J., Jr. (1957) *Zoogeography: The Geographical Distribution of Animals.* Wiley, N. Y., 675 pp.

Darlington, P. J., Jr. (1960) The zoogeography of the Southern cold temperate zone. *Proc. Roy. Soc. (B)*, No. 949, 152, 659–668.

Darlington, P. J., Jr. (1965) *Biogeography of the Southern End of the World-Distribution and History of Far-southern Life and Land, with an Assessment of Continental Drift.* Harvard Univ. Press, Cambridge, 236 pp.

Deacon, G. D. R. (1960) The southern cold temperate zone. *Proc. Roy. Soc. (B)*, No. 949, 152, 441–447.

Ebeling, A. E. (1962) Melamphaeidae I. Systematics and zoogeography of the species in the bathypelagic genus *Melamphaes* Günther. *Dana Rep., 58*, 164 pp.

Ebeling, A. W. (1967) Zoogeography of tropical deep-sea animals. *Stud. Trop. Oceanogr., Miami, 5*, 595–613.

Ekman, S. (1953) *Zoogeography of the Sea.* Sidgwick and Jackson, London, 417 pp.

Fang, P. W. (1934) Study on the fishes referring to Salangidae of China. *Sinensia, 4*, 231–268.

Fowler, H. W. (1945) Fishes of Chile – systematic catalog. *Revista Chilena de Hist. Nat.*, años 45–47, 36 & 171 pp.

Fowler, H. W. (1959) *Fishes of Fiji.* Gov't of Fiji, Suva. 670 pp.

Freihofer, W. C. (1963) Patterns of the ramus lateralis accessorius and their systematic significance in teleostean fishes. *Stanford Ichthyol. Bull., 8* (2), 80–189.

Frey, D. G. (1968) Paleolimnology. *Science, 159,* 1262-1264.

Gilbert, C. H., and C. L. Hubbs (1920) The macrourid fishes of the Philippine Islands and the East Indies. *Bull. U. S. Nat. Mus.,* no. 100, *1* (7), 369–587.

Gill, T. N. (1862) Synopsis of the notothenioids. *Proc. Acad. Nat. Sci. Philadelphia, 1861,* 512–522.

Gosline, W. A. (1960) Mode of life, functional morphology, and the classification of modern teleostean fishes. *Syst. Zool., 8,* 160–164.

Gosline, W. A. (1968) The suborders of perciform fishes. *Proc. U. S. Nat. Mus., 124,* 1–78.

Gosline, W. A., and V. E. Brock (1960) *Handbook of Hawaiian Fishes.* Univ. Hawaii Press, Honolulu, 372 pp.

Graham, D. H. (1956) *A Treasury of New Zealand Fishes,* ed. 2, Reed, Wellington, 424 pp.

Greenfield, D. W. (1968) The zoogeography of *Myriprists* (Pisces: Holocentridae). *Syst. Zool., 17* (1), 76–87.

Greenwood, P. H., D. E. Rosen, S. H. Weitzman, and G. S. Myers (1966) Phyletic studies of teleostean fishes, with a provisional classification of living forms. *Bull. Amer. Mus. Nat. Hist., 131* (4), 341–455.

Greenwood, P. H., G. S. Myers, D. E. Rosen, and S. H. Weitzman (1967) Named main divisions of teleostean fishes. *Proc. Biol. Soc. Wash., 80,* 227–228.

Gressitt, J. L. (editor; 1963) *Pacific Basin Biogeography,* based on a symposium convened during the 10th Pacific Science Congress, Honolulu, 1961, Bishop Museum Press, Honolulu, 563 pp.

Haedrich, R. L. (1967) The stromateoid fishes: systematics and a classification. *Bull. Mus. Comp. Zool., Harvard Univ., 135* (2), 31–129.

Herre, A. W. C. T. (1953) Check list of Philippine fishes. *Res. Rep. U. S. Fish Wildlife Serv., 20,* 977 pp.

Hiatt, R. W., and D. W. Strasburg (1960) Ecological relationships of the fish fauna on coral reefs of the Marshall Islands. *Ecol. Monogr., 30,* 65–127.

Hildebrand, S. F. (1943) A review of the American anchovies (family Engraulidae). *Bull. Bingham Oceanog. Coll., 8* (2), 1–165.

Hildebrand, S. F. (1946) A descriptive catalog of the shore fishes of Peru. *Bull. U. S. Nat. Mus., 189,* 530 pp.

Jordan, D. S., and B. W. Everman (1898) The Fishes of North and Middle America, *Bull. U.S. Nat. Mus., 47* (3):2183–3136.

Jordan, D. S., and C. L. Hubbs (1919) Studies in ichthyology – a monographic review of the family of Atherinidae or silversides. *Stanford Univ. Pubs., Univ. Ser.,* 87 pp., 12 pls.

Knox, G. A. (1960) Littoral ecology and biogeography of the southern ocean. *Proc. Roy. Soc. (B),* No. 949, 152, 577–624.

Kozhov, M. (1963) Lake Baikal and its life. *Monographiae Biologicae, 11,* 344 pp.

Krefft, G. (1961) A contribution to the reproductive biology of *Helicolenus dactylopterus* (de la Roche, 1809) with remarks on the evolution of the Sebastinae. Cons. Internat. Explor. de la Mer, *Rapp. P.-v.Reun. Cons. Perm. Int. Explor. Mer, 150,* 243-244.

Lavenberg, R. J., and A. W. Ebeling (1967) Distribution of midwater fishes among deep water basins of the southern California shelf. *Proc. Symp. Biol. Calif. Ids.,* pp. 185–201, Santa Barbara Botanic Garden.

Leighton, D. L. (1966) Studies of food preference in algivorous invertebrates of Southern California kelp beds. *Pacific Science, 20* (1), 104–113.

Limbaugh, C. (1955) Fish life in the kelp beds and the effects of kelp harvesting. Univ. Calif., Inst. Mar. Res., *IMR Ref.,* 55–9:1-158.

MacArthur, R. H., and E. O. Wilson (1967) *The Theory of Island Biogeography.* Monogr. in Popul. Biol., No. 1, Princeton Univ. Press, 203 pp.

Mann, G. (1954) *La Vida de los Peces en Aguas Chilenas.* Ministerio de Agricultura, Instituto Investigaciones Veterinarias, Santiago, 342 pp.

Manter, H. W. (1955) The zoogeography of trematodes of marine fishes. *Exper. Parisitol., 4,* 62–86.

Marshall, N. B. (1963) Diversity, distribution and speciation of deep-sea fishes. In *Speciation in the Sea, Publ. Systematics Assn., 5,* 181–195.

Matthew, W. D. (1939) Climate and Evolution, *Special Pub. New York Acad. Sci., 1,* 1–223.

McCulloch, A. R. (1929-30) A check-list of the fishes recorded from Australia. *Mem. Australian Mus., 5* (1–4), 534 pp.

McDowall, R. M. (1964) The affinities and derivation of the New Zealand fresh-water fish fauna. *Tuatara, 12* (2), 59–67.

McDowall, R. M. (in press) The galaxiid fishes of New Zealand. *Bull. Mus. Comp. Zool., Harvard University.*

Mead, G. W. (1966) Cruise report, R/V *Anton Bruun,* Cruise 13, *Spec. Rep.* No. 3, Mar. Lab., Texas A & M Univ., 50 pp.

Mead, G. W., E. Bertelsen, and D. M. Cohen (1964) Reproduction among deep-sea fishes. *Deep-Sea Res., 11,* 569–596.

Munro, I. S. R. (1967) *The Fishes of New Guinea.* Dept. of Agr., Stock and Fisheries, Port Moresby, New Guinea, 650 pp., 78 pls.

Myers, G. S. (1941) The fish fauna of the Pacific Ocean, with special reference to zoogeographical regions and distribution as they affect the international aspects of the fisheries. *Proc. Sixth Pac. Sci. Cong., 3,* 201–210.

Myers, G. S. (1960) The endemic fish fauna of Lake Lanao, and the evolution of higher taxonomic categories. *Evolution, 14* (3), 323–333.

Myers, G. S. (1966) The zoogeography of anchovies. Amer. Soc. Ichthyol. Herpetol.; Abstracts of papers, meeting of 1966, 35–36.

Myers, G. S. (1967) Zoogeographical evidence of the age of the South Atlantic Ocean. *Stud. Trop. Oceanogr., Miami, 5,* 614–621.

Norman, J. R. (1934) *A Systematic Monograph of the Flatfishes (Heterosomata),* Vol. 1 – Psettodidae, Bothidae, Pleuronectidae. British Museum, London, 459 pp.

Norman, J. R. (1937) Coast fishes – Part II. The Patagonian Region. *Discovery Rep., 16,* 1–150, pls. 1-5.

Norman, J. R. (1938) On the affinities of the Chilean fish, *Normanichthys crockeri* Clark. *Copeia,* (1), 29–32.

Pantin, C. F. A. (convenor) (1960) A discussion on the biology of the southern cold temperate zone. *Proc. Roy. Soc. (B),* No. 949, 152, 429–677.

Patterson, C. (1964) A review of Mesozoic acanthopterygian fishes, with special reference to those of the English chalk. *Phil. Trans. Roy. Soc. (B),* No. 739, 247, 213–482.

Quast, J. C. (1965) Osteological characteristics and affinities of the hexagrammid fishes, with a synopsis. *Proc. Calif. Acad. Sci.,* ser. 4, 31 (21), 563–600.

Quast, J. C. (1968a) Observations on the food of the kelp-bed fishes; pp. 109–142 *in* W. J. North and C. L. Hubbs, eds., Utilization of kelp-bed resources in Southern California. Calif. Dept. Fish and Game, Fish. Bull. 139, 264 pp.

Quast, J. C. (1968b) The effects of kelp harvesting on the fishes of the kelp beds; pp. 143-149 *in* W. J. North and C. L. Hubbs, eds., Utilization of kelp-bed resources in Southern California. Calif. Dept. Fish and Game, *Fish. Bull.* 139, 264 pp.

Randall, J. E. (1961) Overgrazing of algae by herbivorous marine fishes. *Ecology, 42,* 812.

Randall, J. E. (1965) Grazing effect on sea grasses by herbivorous reef fishes in the West Indies. *Ecology, 46,* 255–260.

Randall, J. E. (1967) Food habits of reef fishes of the West Indies. *Stud. Trop. Oceanogr., Miami, 5*, 665–847.

Regan, C. T. (1913a) The Antarctic fishes of the Scottish National Antarctic Expedition. *Trans. Roy. Soc. Edin., 49* (2), 229–292, 11 pls.

Regan, C. T. (1913b) The classification of the percoid fishes. *Ann. Mag. Nat. Hist.*, ser. 8, *12*, 111–145.

Romer, A. S. (1966) *Vertebrate Paleontology.* Univ. Chicago, 468 pp.

Rosen, D. E. (1964) The relationships and taxonomic position of the half-beaks, killifishes, silversides, and their relatives. *Bull Amer. Mus. Nat. Hist., 127* (5), 219–267.

Rosenblatt, R. H. (1967) The zoogeographic relationships of the marine shore fishes of tropical America. *Stud. Trop. Oceanogr., Miami, 5*, 579–592.

Rosenblatt, R. H., and B. W. Walker (1963) The marine shore fishes of the Galápagos Islands. *Occ. Pap. Calif. Acad. Sci., 44*, 97–106.

Schultz, L. P. (1948) A revision of six subfamilies of atherine fishes, with descriptions of new genera and species. *Proc. U. S. Nat. Mus., 98*, 1–48, 2 pls.

Schultz, L. P. (1957) The frogfishes of the family Antennariidae. *Proc. U. S. Nat. Mus., 107*, 47–105.

Scott, T. D. (1962) *The Marine and Fresh Water Fishes of South Australia.* Gov't Printer, Adelaide, 338 pp.

Shmidt, P. Iu. (1965) Fishes of the Sea of Okhotsk. 392 pp., Nat. Sci. Found., Washington; trans. from: Akad. Nauk SSSR, *Trudy Tikho-okeanskogo Komiteta*, vol. 6, 1950.

Simpson, G. G. (1944) *Tempo and Mode in Evolution.* Columbia Univ. Press, 237 pp.

Smith, J. L. B. (1963) *The Fishes of Seychelles.* Rhodes Univ., Grahamstown, 215 pp., 98 pls.

Smith, J. L. B. (1965) *The Sea Fishes of Southern Africa*, ed. 5. Central News Agency, S. Africa, 580 pp., 107 pls.

Stephenson, W., and R. B. Searles (1960) Experimental studies on the ecology of intertidal environments at Heron Island. *Aust. J. Mar. Freshwater Res., 2* (2), 241–267, 3 pls.

Svetovidov, A. N. (1962) Gadiformes, 304 pp., Nat. Sci. Found., Washington; trans. from: Zool. Inst. Akad. Nauk SSSR, *Fauna SSSR*, N. S. No. 34, 1948.

Svetovidov, A. N. (1963) Clupeidae, 428 pp., Nat. Sci. Found., Washington; trans. from: Zool. Inst. Akad. Nauk SSSR, *Fauna SSSR*, N. S. No. 48, 1952.

Walters, V., and C. R. Robins (1961) A new toadfish (Batrachoididae) considered to be a glacial relict in the West Indies. *Amer. Mus. Novit.*, No. 2047, 24 pp.

Weitzman, S. (1967) The origin of the stomiatoid fishes with comments on the classification of salmoniform fishes. *Copeia*, (3), 507–540.

Whitley, G. P. (1965) A survey of Australian ichthyology. *Proc. Linn. Soc. N. S. W.*, 89, 11–127.

Hiroshi Kasahara

UNITED NATIONS DEVELOPMENT PROGRAMME,
NEW YORK

COMMERCIAL FISHERIES

INTRODUCTION

The region under review includes waters bordered by the equator, the west coast of South America, the Antarctic Convergence, and the coasts of eastern New Guinea and Australia, excluding the Arafura Sea and the Great Australian Bight. This region may not coincide exactly with the definition of the South Pacific given in other papers. The equator is not a natural boundary for most of the species of major commercial importance found in the tropics of the Pacific Ocean and has to be ignored from time to time. The oceanographic characteristics of the South Pacific are discussed fully in other papers in this volume; hence, they are not covered in this paper. Kesteven (1967) discusses the natural regions of the South Pacific. Whales and other marine mammals are excluded from the present review.

A review of the commercial fisheries of any area should deal with two aspects, namely, the current status of exploitation and the potential resources. For this particular region, it is difficult to discuss the latter with reliability. Except for a few species, the resources of the South Pacific are exploited only lightly or not at all. Data from resource surveys and exploratory fishing are very scanty.

FISHERIES ALONG THE COAST OF SOUTH AMERICA

The coast of South America constitutes the complete eastern boundary of the South Pacific and borders waters of tropical, temperate, and subarctic characteristics. The narrow strip of water along this coast is the most productive part of the South Pacific and actually produces 95 percent of the estimated total catch in the South Pacific. This does not mean that the natural productivity of all other parts of the South Pacific is extremely low. The lack of large catches in other parts of this ocean region is due mainly to the absence of intensive commercial fishing.

Fishing is extremely selective, even within the coastal waters of South America, as reflected in the following statistical summary for the year 1966 (in thousands of metric tons; FAO 1967):

	Ecuador	Peru	Chile	Total
Total Marine Catch	48	8,709	1,384	10,141
Anchovy	—	8,530	1,091	9,621
Tunas*	12	83	13	108
Hake	—	—	94	94
All other species	36	96	186	318

Thus, 95 percent of the combined total marine catch in the three countries was of one species of fish, anchovy (*Engraulis ringens*). This, again, is not an indication of the absence of substantial stocks of other species in this area. Although the stock of *E. ringens* is undoubtedly the largest of a single species of fish not only in this area but perhaps in the entire World Ocean, data from existing fisheries and limited exploratory fishing indicate large resources of many other species in waters off the west coast of South America. Most of these, however, are either untapped or underutilized.

An enormous population of *E. ringens* has long been known to exist, supporting the populations of guano birds. The Peruvian fishmeal industry based on this species, however, has developed largely during the last 10 years. The total catch of anchovies by Peru increased from 119,000 MT in 1956 to 8,863,000 MT in 1964, the record high. Catches in 1965 and 1966 were 7,242,000 MT and 8,530,000 MT, respectively. These annual catches of one species of fish from a narrow coastal strip exceed *in volume* the total annual catch of the entire Japanese fishing industry, which consists of a great variety of species caught not only in the North Pacific and adjacent seas but also in many other parts of the world. Among the natural factors contributing to this

*Not including the catches landed outside the three countries.

enormous yield of the Peruvian anchovy are the high productivity of the waters of the Peru Current and the trophic level of the species, which is among the lowest for fish.

The Chilean fishmeal industry has developed during the same period, also largely based on the anchovy. The Chilean anchovy catches in 1964-1966 were 934,000 MT, 439,000 MT, and 1,091,000 MT, respectively. Thus, the total amount of *E. ringens* caught by Peru and Chile reached 9.8 million MT in the year 1964, dropped significantly in 1965, but recovered to 9.6 million MT in 1966. A little less than half of the world's total supply of fishmeal comes from these two countries.

The biology and population dynamics of *E. ringens* are discussed in various publications, based on data collected by the Peruvian Sea Institute (Instituto del Mar del Peru) and the Fishery Development Institute (Instituto de Fomento Pesquero, Chile), e.g., Boerema *et al.* (1967), Brandhorst and Rojas (1967), Jordan and Vildoso (1965), Schaefer (1967), etc. The species occurs in coastal waters from northern Peru to central Chile. Spawning takes place during almost the entire year, with the main peak in the southern spring. Natural mortality appears to be fairly high; so does the current fishing mortality. Fish enter the fishery at age 0, and few fish in the present catch are older than 24 months. Year-class fluctuations are substantial, but compared with some of the other species studied, the recruitment may be considered relatively stable. Changes in abundance are more noticeable in the Chilean waters, partly because they are nearer the southern boundary of the geographic distribution of the species. Such environmental factors as the strength of coastal upwelling or the extent of intrusion of warm offshore water seem to have significant effects on the distribution and apparent abundance of anchovies.

There are some differences of opinion among authors as to the level of maximum sustainable yield. It might be assumed, however, that the present level of total catch, 8-10 million MT a year, roughly represents the sustainable maximum catch.

Except for a substantial catch of bonito (*Sarda*), in the neighborhood of 70,000 MT a year, the catches of other marine species in Peru are rather insignificant, their annual total (excluding anchovy and bonito) being only about 100,000 MT. This low level of exploitation of other resources is due to social and economic factors. The anchovy also makes up an overwhelmingly large portion of the landings in Chile, but the industry there is developing in a more diversified way than in Peru. Among the important species are hake, Chilean sardines (*Clupea* and *Sardinops*), jack mackerel (*Trachurus*), bonito, sierras (Gempylidae), shrimp (*Heterocarpus*), langostinos (*Cervimunida* and others), mussels, and clams. The coast of Chile stretches from about 18°30′ to 56°S, and its fishing areas include subtropical, temperate, and subantarctic waters. Primary productivity appears to be high not only in coastal waters but also in some offshore areas. Since the stocks other than the anchovy are generally underexploited, the potential of the Chilean fishing industry is great. The lack of large domestic markets and the great distances to major foreign markets are among the factors hampering development.

The fishing industry of Ecuador consists of the following sectors: a shrimp fishery that annually catches and processes about 5,000 MT of shrimp, mainly penaids; a live-bait tuna fishery, largely for skipjack, landing about 10,000 to 17,000 MT a year; and traditional inshore fisheries catching roughly 30,000 MT of fishes and invertebrates. The potential resources of both pelagic and demersal species appear substantial.

FISHERIES OF EASTERN AUSTRALIA AND NEW ZEALAND

Except for the above-mentioned fisheries along the South American coast and the tuna fisheries in offshore waters, which will be discussed briefly in the next section, the level of fishing activity is extremely low in the South Pacific. The productivity of the waters around New Zealand and the Tasman Sea is relatively high, though not comparable to that of the South American coastal waters or the subarctic areas of the North Pacific. But the intensity of fishing in this part of the South Pacific is generally very low. The total marine catch by New Zealand was 48,000 MT in 1965; the total marine catch from the entire coast of Australia was about 89,000 MT in 1966, of which some 50,000 to 60,000 MT might have been taken from the waters of eastern Australia. All fisheries are coastal for both countries.

Among the relatively important species caught in eastern Australian waters are flathead (*Neoplatycephalus*), morwong (*Nemadactylus*), mullets, tunas, barracouta (*Leionura*), shrimps, rock lobsters, oysters, and scallops. In New Zealand waters, a variety of groundfishes (of which the porgy *Chrysophrys* is the most important), rock lobsters, and oysters constitute the bulk of the catch. Substantial stocks of coastal pelagic species (sardine *Sardinops*, anchovy *Engraulis*, jack mackerel *Trachurus*, etc.) are known to exist, but they are practically untapped in the seas around both Australia and New Zealand. There has been a Japanese trawl fishery on the continental shelf of New Zealand, which in 1965 produced approximately 4,000 MT of groundfish, the most important species being the porgy (*Chrysophrys*).

OFFSHORE TUNA FISHERIES

There are three basic types of tuna fishing: longlining for large, deep-swimming tunas, namely larger specimens of albacore (*Thunnus alalunga*), yellowfin (*T. albacares*), bigeye (*T. obesus*), and bluefin (*T. thynnus*), as well as such other species as marlins, swordfish, and sharks; purse-seining for surface schools of tuna, mainly yellowfin and skipjack (*Euthynnus pelamis*) and partly bluefin; and pole-and-line fishing with live bait for skipjack, yellowfin, and albacore. Bonitos (*Sarda*) are also caught in considerable quantities by purse-seining or live-bait fishing in coastal waters. Little tunas (*Euthynnus* other than *E. pelamis*) and frigate mackerel (*Auxis thazard*) are also caught in some areas, though in smaller quantities. Tuna longlining is carried out in tropical and temperate waters throughout the world. Tuna purse-seining is used most actively in the eastern tropical Pacific; it is also used in waters off Japan and along the west coast of Africa. Pole-and-line fishing with live bait is done mainly in waters relatively close to the continental coasts or to oceanic islands in many parts of the world. All three basic types of tuna fishing are carried out in the South Pacific.

Longline tuna fishing has an old history in Japan, but its extension to other parts of the world began only about 1950. By 1960, the Japanese tuna fleet was covering the main longline grounds in the North and South Pacific, as well as in the Indian Ocean. Tuna longline fishing in the Atlantic started in 1957, and the major Atlantic grounds were being fished by 1962. Longlining grounds continued to expand after 1962, though more slowly, and now include most ocean waters between roughly 40°N and 40°S. Japanese longline fishing reached its highest level in 1962; it has since been on the decline. Two other countries, Formosa and South Korea, started long-distance tuna fishing later than Japan but have expanded their fleets very rapidly. In addition to these three major tuna longlining nations, Ryukyu has a sizeable fleet of longliners operating in the Pacific and Indian oceans; a fleet of Cuban longliners fishes mainly in the Atlantic; and the USSR operates factory mother ships with deck-loaded longline catchers.

As of January 1967, an estimated total of roughly 500 tuna longliners, including Japanese, Formosan, and Korean boats, were operating in the tropical North Pacific and the South Pacific. Aside from vessels running from their home countries, particularly Japan, a large number of boats have been using local bases in the South Pacific, namely, American Samoa, Espiritu Santo (New Hebrides), Fiji, and Nouméa (New Caledonia). In 1967 American Samoa, which has two large canneries, received about 40,000 MT of fish caught by the longliners of the three countries.

Fishing grounds of the Japanese longline fishery in the Pacific Ocean are described in a number of publications. A paper by Rothschild *et al.* (MS) gives a good summary for albacore, yellowfin, and bigeye. Japanese longline fishing is now conducted across the entire Pacific and, in the western Pacific, from 45°N to about 40°S. The best grounds for yellowfin are found in the western tropical Pacific; for bigeye, in the central and eastern tropical Pacific, as well as in some temperate areas; and for albacore, generally in more

northerly and southerly areas (south of 5°S in the South Pacific). The longline catches of southern bluefin tuna are mainly from waters south of 30°S and west of 165°W, including the Tasman Sea. Because this species brings a high price on the Japanese market, fishing for bluefin has been intensive in recent years.

The total quantity of tunas caught by longliners in the South Pacific cannot be estimated, for catches by area are not shown in published statisitcs. It may be assumed that a total of 150,000 to 200,000 MT a year, including the above four species, are landed by the vessels of the three nations from the Pacific waters south of 15°–20°N, including the entire South Pacific. This figure may be incorrect, however.

Among the other commercially important species caught by tuna longliners are marlins (*Makaira*), swordfish (*Xiphias*), sailfish (*Istiophorus*), and various species of large sharks. The proportion of the combined catch of these and other nontuna species in the entire catch of the Japanese longline fishery (from the World Ocean) was about 29 percent in 1965.

One of the common characteristics of tuna longlining is that the catch per unit of effort decreases rapidly in a new fishing area. Longlining is obviously an efficient method of catching deep-swimming tuna, and the size of the stock in a new fishing area is reduced quickly as fishing effort mounts. This does not mean that tuna longlining causes over-fishing wherever it is done. In a new area, it is only natural for a stock of fish of relatively high age groups to go down as intensive fishing starts. There has not been convincing evidence that the total sustainable yield of any tuna species from any region has decreased as a result of over-fishing. However, there is no question but that the productivity of tuna longlining decreases rapidly as the amount of fishing increases. It is also unlikely that the total longline catches of yellowfin and albacore from the World Ocean, and perhaps of bigeye and bluefin as well, will show substantial increases as fishing is further intensified, although the proportions taken by the fisheries of different nations will change.

Another important development in tuna fishing has taken place in the eastern tropical Pacific region, initially through expansion of pole-and-line fishing (with live-bait) by U.S. tuna clippers for yellowfin and skipjack. From 1958 through 1960 increasing tuna imports from Japan affected prices paid to U.S. fishermen so badly that the fishery almost collapsed. The introduction of nylon webbing and the power block in purse-seining, however, started a rapid technological change, and most of the U.S. tuna vessels were converted into purse-seiners within 2 years; many new seiners have since been constructed. After this metamorphosis, the U.S. tuna fishery has been in a position to supply tuna cheaper than imports. As shown in Table 1, U.S. flag vessels still take the bulk of the catch of yellowfin and skipjack from this region,* the remainder being shared by vessels from other countries, including, Ecuador, Canada, Mexico, and Japan. Fishing for yellowfin is done largely in waters north of the equator, while the best skipjack grounds are found mainly in areas south of 5°N. Since 1966, the Inter-American Tropical Tuna Commission† has been regulating fishing for yellowfin by setting a total catch limit. In 1967, yellowfin fishing (except for incidental catch) was closed to vessels leaving port after June 24, and the catch limit, approximately 77,000 MT was exceeded by about 6 percent. The quota was increased for 1968.

A short summary on the skipjack tuna is perhaps appropriate. The skipjack is no doubt the most abundant species of tuna worldwide as well as in the Pacific Ocean. Unlike the other four major species (albacore, bluefin, yellowfin, and bigeye), its stocks still appear to be underexploited in many areas of the world. A good review of the situation in the mid-Pacific is found in Rothschild et al. (MS).

*The eastern Pacific tuna grounds, as defined by the Inter-American Tropical Tuna Commission, include all waters east of 125°W between 40°N and 20°N, east of 120°W between 20°N and 5°N, east of 110°W between 5°N and 10°S, and east of 90°W between 10°S and 30°S.

†Members: United States, Canada, Costa Rica, Ecuador, Mexico, and Panama. Ecuador withdrew as of August 1968.

The most important skipjack fishing grounds in the Pacific are the waters off Japan and the eastern tropical Pacific region mentioned above. The former area supports large Japanese fisheries (mainly live-bait fishing and partly purse-seining). The latter is fished largely by vessels from the United States (mainly purse-seining and partly live-bait fishing) and to a much lesser degree by tuna boats from Ecuador (live-bait fishing) as well as other Latin American countries and Canada (Table 1). Unlike fishing for yellowfin, skipjack fishing in the eastern tropical Pacific is unrestricted. It is generally believed that fishing has had no noticeable effects on the skipjack stock in this area and that a further increase in the total catch is attainable. There have been substantial fluctuations in the catch, the total from the area varying from 47,000 to 120,000 MT during 1960–1967. The high availability of fish and the early closure of fishing for yellowfin resulted in a record catch of skipjack in 1967.

Before World War II, active live-bait fishing for skipjack was carried out by the Japanese fishermen operating from some of the western Pacific islands of the present U.S. Trust Territory, such as Saipan, Yap, Palau, Truk, Ponape, and Jaluit. The same area is now fished by skipjack boats from Japan. In 1964, a U.S. company started a small skipjack fishery based in Palau, using Ryukyu live-bait fishing ves-

TABLE 1 Annual Catch of Yellowfin and Skipjack from the Eastern Tropical Pacific Ocean by Flag of Vessel (in metric tons)

Year	Flag of Vessel										Total
	Canada	Chile	Colombia	Costa Rica	Ecuador	Japan	Mexico	Panama	Peru	U.S.A.	
YELLOWFIN											
1960	–	–	–	–	–	–	–	–	–	–	110,828
1961	0	88	91	190	722	2,303	1,301	0	685	99,054	104,434
1962	0	372	72	0	849	5,287	1,770	2,339	4,944	63,339	78,972
1963	0	390	72	2	916	3,710	1,808	1,572	2,463	54,997	65,930
1964	0	176	45	0	962	3,174	1,912	794	3,306	82,111	92,480
1965	418	78	174	43	652	3,100	1,680	80	1,547	73,915	81,687
1966	1,466	500	120	0	1,220	2,357	1,991	120	119	74,795	82,688
1967[a]	1,194	188	110	0	2,627	1,852	1,906	191	31	73,186	81,285
SKIPJACK											
1960	–	–	–	–	–	–	–	–	–	–	46,706
1961	0	60	29	17	11,651	0	855	0	4,646	50,647	67,905
1962	9	70	161	0	10,348	0	1,040	957	2,742	55,429	70,756
1963	0	191	249	0	12,874	0	1,494	2,856	4,526	74,010	96,200
1964	0	6	206	0	9,018	0	1,444	374	1,528	46,676	59,252
1965	1,077	146	227	10	14,865	0	1,921	0	932	58,951	78,129
1966	2,191	97	303	0	10,292	0	1,337	175	43	45,924	60,362
1967[a]	4,002	225	428	0	17,279	0	3,714	400	140	94,071	120,259

[a]Preliminary.

Source: Inter-American Tropical Tuna Commission.

sels. An Indonesian shipjack fishery operates from northern Sulawesi. There is also a long established live-bait fishery in Hawaii, which lands several thousand tons of shipjack a year.

Relationships between skipjack populations found in different areas are discussed by Kawasaki (1965) for the western North Pacific and by Rothschild *et al.* (MS) and other authors for the central and eastern Pacific.

Except for the eastern tropical Pacific fishery, all of the skipjack fisheries mentioned above are in the Pacific waters north of the equator. In the western and central South Pacific, there are only an Australian tuna fishery and small local fisheries around some of the tropical islands, for example, Tahiti. Skipjack are observed, sometimes in large schools, around most of the tropical oceanic islands and off the larger land masses of the western South Pacific, but commercial fishing for skipjack in these waters is generally insignificant.*

A practical way of catching skipjack in tropical oceanic areas is by pole-and-line with live bait. (Smaller quantities of yellowfin are also caught by this method.) Trolling is effective, but not for large catches. One of the serious limitations for live-bait fishing is the generally low availability of bait fish around small islands. In Tahiti, pole-and-line fishing is carried out with lures instead of live bait. To secure enough live bait, however, should not be a difficult problem for fishing in waters near larger land masses, in this case many areas of the western South Pacific. Purse-seining has been tried by the U.S. Bureau of Commercial Fisheries around some of the mid-Pacific islands, with generally poor results. The Japanese have recently started commercial purse-seine operations in an area north of New Guinea, with good initial results.

OTHER FISHERIES

Excluding landings in New Zealand and Australia, and long-line tuna catches delivered to such bases as American Samoa, Fiji, and Espiritu Santo for canning or trans-shipment, only a few thousand tons of fish are recorded in FAO's statistics as local catches in the entire central and western South Pacific. Although statistics are incomplete or nonexistent for most of the subsistence fisheries, this indicates a very low level of fishing activity by local fishermen in this general area.

---

*An example of intensive fishing for skipjack by the inhabitants of oceanic islands is found in the Maldive Islands of the Indian Ocean. An estimated total of some 18,000–20,000 MT of tuna, mainly skipjack but also including small yellowfin, little tuna, and frigate mackerel, is landed all by unpowered local craft (United Nations Development Programme, 1966)

POTENTIAL RESOURCES

As the South Pacific in general represents one of the most underexploited regions of the World Ocean, it is appropriate to give a short summary of the resources available for future development, although information is very meager.

With the exception of hake (*Merluccius*) in Chilean waters and a few crustacean and molluscan forms (shrimps, lobsters, Chilean langostinos, Chilean mussels, etc.), the demersal stocks on the long, narrow continental shelf along the west coast of South America are only lightly exploited or not used at all. Even the potential of the hake stock is not known. Little use is made of the apparently large stock of hake in northern Peru. Not much information is available in published papers regarding the resources of such other demersal forms as sciaenids, sparids, and flatfishes. Data on coastal pelagic fishes are even scantier, and there is no way of estimating the potential yields of such common forms as clupeids, engraulids (other than *E. ringens*), carangids, and mackerels. General surveys of demersal and coastal pelagic species along the entire coast are needed.

Skipjack resources have been reviewed above. Even for the larger tuna species of commercial importance, not much is known about distribution and abundance in areas outside the present range of operation of tuna fishing vessels, such as the southeastern corner of the South Pacific (off Chile south of about 25°S). Surface schools of albacore are found around the offshore islands of Chile. There might also be a substantial bluefin stock in this vicinity. The magnitude of the saury (*Cololabis*) stock in the South Pacific is entirely unknown; they are found in large quantities in the waters off Chile. Enormous stocks of squids are also present in the waters off Chile. If markets can be developed for them, they will become valuable fishery resources.

An even more extreme situation is found on the western side of the South Pacific. As mentioned previously, the combined total catch in New Zealand and along the east coast of Australia amounts to roughly 100,000 MT, with most of the resources remaining undeveloped. The potential harvest from waters around New Zealand and the Tasman Sea must be many times the present catch, although it is perhaps not comparable to those of the most productive parts of the Pacific, such as the northwest Pacific, the northeast Pacific, and the waters off Peru and Chile.

No statistics are available for the local fisheries of New Guinea. Laskaridis (1967) describes the fisheries of West Irian and estimates the total yearly production of marine fish at about 5,000 MT (including landings from the south coast). It is obvious that most of the resources around New Guinea are virtually untapped. Statistics are also unavailable, or are very incomplete, for the larger islands in the northwestern part of the South Pacific. But in general, the level of exploitation of fishery resources appears quite low.

No one has attempted to estimate the potential yields

from the stocks of reef and lagoon fishes commonly found around the tropical oceanic islands, such as carangids, serranids, lutjanids, pomadasyids, and mullids. In the offshore waters of the tropical Pacific, flying fishes (Exocoetidae) represent some of the most abundant resources; these remain unexploited in most areas. The distribution of dolphin (*Coryphaena*) and some other species is discussed by Rothschild *et al.* (MS).

Eventually some of the enormous stocks of bathypelagic forms (gonostomatids, myctophids, etc.) may become harvestable. Findings from midwater trawling in the tropical Pacific are summarized by King and Iversen (1962). Idyll (1968) reviews available information about squids, which also constitute great potential resources of the ocean on a worldwide basis.

Oceanographic data indicate relatively high primary productivity and standing crops of zooplankton in the broad midoceanic zone of subantarctic water between the Subtropical Convergence and the Antarctic Convergence, but no reliable information is available on the dominant forms of fish and invertebrates at higher trophic levels.

REFERENCES

Boerema, L. K., G. Saetersdal, I. Tsukiyama, J. E. Valdivia, and B. Alegre (1967) Informe sobre los efectos de la pesca en el recurso peruano de anchoveta. *Boln. Inst. Mar Peru, 1*(4), 133–186.

Brandhorst, W., and O. Rojas (1967) Distribucion geografica de la pesca de la anchoveta en el norte de Chile y su composicion del tamaño, de marzo de 1961 a julio de 1963. *Publnes Inst. Fom. Pesq., 24,* 69 pp.

FAO (1967) Yearbook of Fishery Statistics, 1966.

Gromov, A., H. Kasahara, B. R. Devarajan, and P. Kung (1966) Report of a mission to the Maldive Islands. United Nations Development Programme, New York, 125 pp.

Idyll, C. P. (1968) Food resources of the sea beyond the continental shelf excluding fish. United Nations, document E/4449/Add. 2. 145 pp.

Jordan, R., and A. Chirinos de Vildose (1965) La anchoveta (*Engraulis ringens J.*) Conocimiento actual sobre su biologia, ecologia y pesqueria. *Inf. Inst. Mar Peru, 6,* 52 pp.

Kawasaki, T. (1965) Ecology and dynamics of the skipjack population. *Nihon Suisan Shigen Hogo Kyokai, Suisan Kenkyu Sosho, 8*(1–2) 108 pp.

Kesteven, G. L. (1967) Fisheries of the South Pacific. *Proc. Seventh Int. Cong. Nutr., 4,* 1032–1038.

King, J. E., and R. T. B. Iversen (1962) Midwater trawling for forage organisms in the central Pacific 1951–1956. *Fish. Bull., U.S., 62,* 271–321.

Laskaridis, K. (1967) Sea fisheries in West Irian. United Nations Development Programme, N.Y., 31 pp.

Rothschild, B. J., R. N. Uchida, and H. O. Yoshida (MS) The pelagic fishery resources of the Pacific Ocean. Paper presented at the Conference on the Future of the U.S. Fishing Industry, March 1968, Seattle, Wash.

Schaefer, M. B. (1967) Dynamics of the fishery for the anchoveta, *Engraulis ringens*, off Peru. *Boln. Inst. Mar Peru, 1*(5), 191–303.